Reducing greenhouse gas emissions from livestock production

It is widely recognised that agriculture is a significant contributor to global warming and climate change. Agriculture needs to reduce its environmental impact and adapt to current climate change whilst still feeding a growing population, i.e. become more 'climate-smart'. Burleigh Dodds Science Publishing is playing its part in achieving this by bringing together key research on making the production of the world's most important crops and livestock products more sustainable.

Based on extensive research, our publications specifically target the challenge of climate-smart agriculture. In this way we are using 'smart publishing' to help achieve climate-smart agriculture.

Burleigh Dodds Science Publishing is an independent and innovative publisher delivering high quality customer-focused agricultural science content in both print and online formats for the academic and research communities. Our aim is to build a foundation of knowledge on which researchers can build to meet the challenge of climate-smart agriculture.

For more information about Burleigh Dodds Science Publishing simply call us on +44 (0) 1223 839365, email info@bdspublishing.com or alternatively please visit our website at www.bdspublishing.com.

Related titles:

Climate change and agriculture
Print (ISBN 978-1-78676-320-4); Online (ISBN 978-1-78676-322-8, 978-1-78676-323-5)

Improving rumen function
Print (ISBN 978-1-78676-296-2); Online (ISBN 978-1-78676-334-1, 978-1-78676-335-8)

Assessing the environmental impact of agriculture
Print (ISBN 978-1-78676-228-3); Online (ISBN 978-1-78676-228-3, 978-1-78676-231-3)

Advances in breeding of dairy cattle
Print (ISBN 978-1-78676-296-2); Online (ISBN 978-1-78676-298-6, 978-1-78676-299-3)

Achieving sustainable production of milk Volume 1
Print (ISBN 978-1-78676-044-9); Online (ISBN 978-1-78676-046-3, 978-1-78676-047-0)

Achieving sustainable production of milk Volume 3
Print (ISBN 978-1-78676-052-4); Online (ISBN 978-1-78676-054-8, 978-1-78676-055-5)

Chapters are available individually from our online bookshop:
https://shop.bdspublishing.com

climate•SMART•publishing
IN AGRICULTURAL SCIENCE

BURLEIGH DODDS SERIES IN AGRICULTURAL SCIENCE
NUMBER 95

Reducing greenhouse gas emissions from livestock production

Edited by Dr Richard Baines, Royal Agricultural University, UK

burleigh dodds
SCIENCE PUBLISHING

Published by Burleigh Dodds Science Publishing Limited
82 High Street, Sawston, Cambridge CB22 3HJ, UK
www.bdspublishing.com

Burleigh Dodds Science Publishing, 1518 Walnut Street, Suite 900, Philadelphia, PA 19102-3406, USA

First published 2021 by Burleigh Dodds Science Publishing Limited

Library of Congress Control Number: 2020945585

British Library Cataloguing in Publication Data
A catalogue record for this book is available from the British Library

ISBN 978-1-78676-439-3 (Print)
ISBN 978-1-78676-442-3 (PDF)
ISBN 978-1-78676-441-6 (ePub)
ISSN 2059-6936 (print)
ISSN 2059-6944 (online)

DOI 10.19103/AS.2020.0077

Typeset by Deanta Global Publishing Services, Dublin, Ireland

Contents

Part 2 Breeding, animal husbandry and manure management

Series list

Title	Series number
Achieving sustainable cultivation of maize - Vol 1 From improved varieties to local applications Edited by: Dr Dave Watson, CGIAR Maize Research Program Manager, CIMMYT, Mexico	001
Achieving sustainable cultivation of maize - Vol 2 Cultivation techniques, pest and disease control Edited by: Dr Dave Watson, CGIAR Maize Research Program Manager, CIMMYT, Mexico	002
Achieving sustainable cultivation of rice - Vol 1 Breeding for higher yield and quality Edited by: Prof. Takuji Sasaki, Tokyo University of Agriculture, Japan	003
Achieving sustainable cultivation of rice - Vol 2 Cultivation, pest and disease management Edited by: Prof. Takuji Sasaki, Tokyo University of Agriculture, Japan	004
Achieving sustainable cultivation of wheat - Vol 1 Breeding, quality traits, pests and diseases Edited by: Prof. Peter Langridge, The University of Adelaide, Australia	005
Achieving sustainable cultivation of wheat - Vol 2 Cultivation techniques Edited by: Prof. Peter Langridge, The University of Adelaide, Australia	006
Achieving sustainable cultivation of tomatoes Edited by: Dr Autar Mattoo, USDA-ARS, USA; and Prof. Avtar Handa, Purdue University, USA	007
Achieving sustainable production of milk - Vol 1 Milk composition, genetics and breeding Edited by: Dr Nico van Belzen, International Dairy Federation (IDF), Belgium	008
Achieving sustainable production of milk - Vol 2 Safety, quality and sustainability Edited by: Dr Nico van Belzen, International Dairy Federation (IDF), Belgium	009
Achieving sustainable production of milk - Vol 3 Dairy herd management and welfare Edited by: Prof. John Webster, University of Bristol, UK	010
Ensuring safety and quality in the production of beef - Vol 1 Safety Edited by: Prof. Gary Acuff, Texas A&M University, USA; and Prof. James Dickson, Iowa State University, USA	011
Ensuring safety and quality in the production of beef - Vol 2 Quality Edited by: Prof. Michael Dikeman, Kansas State University, USA	012
Achieving sustainable production of poultry meat - Vol 1 Safety, quality and sustainability Edited by: Prof. Steven C. Ricke, University of Arkansas, USA	013
Achieving sustainable production of poultry meat - Vol 2 Breeding and nutrition Edited by: Prof. Todd Applegate, University of Georgia, USA	014

Achieving sustainable production of poultry meat - Vol 3 015
Health and welfare
Edited by: Prof. Todd Applegate, University of Georgia, USA

Achieving sustainable production of eggs - Vol 1 016
Safety and quality
Edited by: Prof. Julie Roberts, University of New England, Australia

Achieving sustainable production of eggs - Vol 2 017
Animal welfare and sustainability
Edited by: Prof. Julie Roberts, University of New England, Australia

Achieving sustainable cultivation of apples 018
Edited by: Dr Kate Evans, Washington State University, USA

Integrated disease management of wheat and barley 019
Edited by: Prof. Richard Oliver, Curtin University, Australia

Achieving sustainable cultivation of cassava - Vol 1 020
Cultivation techniques
Edited by: Dr Clair Hershey, formerly International Center for Tropical Agriculture (CIAT), Colombia

Achieving sustainable cultivation of cassava - Vol 2 021
Genetics, breeding, pests and diseases
Edited by: Dr Clair Hershey, formerly International Center for Tropical Agriculture (CIAT), Colombia

Achieving sustainable production of sheep 022
Edited by: Prof. Johan Greyling, University of the Free State, South Africa

Achieving sustainable production of pig meat - Vol 1 023
Safety, quality and sustainability
Edited by: Prof. Alan Mathew, Purdue University, USA

Achieving sustainable production of pig meat - Vol 2 024
Animal breeding and nutrition
Edited by: Prof. Julian Wiseman, University of Nottingham, UK

Achieving sustainable production of pig meat - Vol 3 025
Animal health and welfare
Edited by: Prof. Julian Wiseman, University of Nottingham, UK

Achieving sustainable cultivation of potatoes - Vol 1 026
Breeding improved varieties
Edited by: Prof. Gefu Wang-Pruski, Dalhousie University, Canada

Achieving sustainable cultivation of oil palm - Vol 1 027
Introduction, breeding and cultivation techniques
Edited by: Prof. Alain Rival, Center for International Cooperation in Agricultural Research for Development (CIRAD), France

Achieving sustainable cultivation of oil palm - Vol 2 028
Diseases, pests, quality and sustainability
Edited by: Prof. Alain Rival, Center for International Cooperation in Agricultural Research for Development (CIRAD), France

Achieving sustainable cultivation of soybeans - Vol 1 029
Breeding and cultivation techniques
Edited by: Prof. Henry T. Nguyen, University of Missouri, USA

Achieving sustainable cultivation of soybeans - Vol 2 030
Diseases, pests, food and non-food uses
Edited by: Prof. Henry T. Nguyen, University of Missouri, USA

Acknowledgements

We wish to acknowledge the following for their help in reviewing particular chapters:

- Chapter 1: Dr Pekka Huhtanen, Natural Resources Institute Finland, Finland
- Chapter 2: Dr Nick Hutchings, Aarhus University, Denmark; and Dr Andre Bannink, Wageningen University, The Netherlands
- Chapter 3: Professor Mike Coffey, Scotland's Rural College (SRUC), UK; and Dr Nicolas Friggens, INRA, France
- Chapter 4: Professor Jeff Lakritz, The Ohio State University, USA
- Chapter 5: Dr April Leytem, USDA-ARS, USA
- Chapter 6: Dr Sofia Mai and Dr Barampouti Elli, National Technical University of Athens, Greece
- Chapter 7: Professor Richard Dewhurst, Scotland's Rural College (SRUC), UK
- Chapter 8: Dr Nancy McLean, Dalhousie University, Canada
- Chapter 9: Professor Marcello Mele, University of Pisa, Italy
- Chapter 11: Dr Emilio M. Ungerfeld, Científico Nutrición de Rumiantes - INIA, Chile; Dr Karen Beauchemin, Agriculture and Agri-Food Canada, Canada; and Dr Ros Gilbert, Queensland Alliance for Agriculture and Food Innovation - University of Queensland, Australia

Introduction

Recent IPCC reports have highlighted the environmental impact of livestock production as a major source of non-CO_2 emissions: methane (CH_4), nitrous oxide (N_2O) and ammonia (NH_3). The livestock sector must react to these reports and develop or implement methods that can reduce greenhouse (GHG) emissions from livestock production.

Part 1 of this volume focuses on the analysis of greenhouse gas emissions from livestock, specifically drawing attention to the range of methods that can be used to reduce these emissions. Chapters in Part 2 discuss breeding, animal husbandry and animal management and how improving these elements can help to reduce the environmental impact of livestock production. Part 3 concentrates on nutritional approaches such as improving feed efficiency, forage quality and using plant bioactive compounds to reduce GHG emissions. Chapters also review the use of feed supplements and how modifying the rumen environment can also help to reduce GHG emissions.

Part 1 Analysis

Chapter 1 looks at the key techniques used for measurement of CH_4 and other gas emissions from livestock production, ranging from individual animal measurements to herd scale measurements for grazing animals and whole farm emissions such as feedlots. Individual animal measurement techniques discussed include whole-animal respiration chambers and head capture measurement. Herd scale measurements include micrometeorological methods and the eddy covariance technique.

Expanding on topics previously covered in Chapter 1, Chapter 2 discusses greenhouse gas emissions in livestock production, focusing specifically on modelling methods, methane emission factors and mitigation strategies. The chapter begins by reviewing systems analysis and how it can be used to quantify GHG emissions from livestock. It then looks at the various stages of life cycle assessment and how it can be used to analysis the environmental effects of livestock production. Modelling applications and the importance national greenhouse gas inventory submissions are also considered in the chapter.

Part 2 Breeding, animal husbandry and manure management

The first chapter of Part 2 analyses the contribution of animal breeding to reducing the environmental impact of livestock production. Chapter 3 begins by addressing the impact of livestock production on the environment. It then goes on to discuss the environmental impact of broilers, layer hens, pigs and dairy cattle and how improving breeding techniques for all of these species can help to reduce the emissions they produce. The chapter also highlights future research directions and provides resources for further information on the subject.

Chapter 4 focuses on the environmental impact consequences of endemic livestock health challenges that lead to deterioration in animal health, and on the potential impacts arising from their mitigations. The first part of the chapter concentrates on the potential of animal health to affect the environmental impact of livestock systems. It also reviews the literature to date which has quantified the impact of health challenges for the environmental impacts of livestock systems. The potential of successful health interventions to mitigate negative environmental impacts represents a point of synergy between concerns around environmental sustainability and animal welfare, both of which represent 'hot topics' in the discourse surrounding the livestock industry and its sustainability. The chapter concludes by highlighting the challenges associated with modelling health interventions and their potential to mitigate environmental impacts.

The subject of Chapter 5 is sustainable nitrogen management for housed livestock, manure storage and manure processing. The chapter begins by discussing the various forms nitrogen can take, focusing specifically on ammonia, nitrous oxide and di-nitrogen. It then goes on to review livestock feeding and housing for dairy and beef cattle, pigs and poultry. The chapter also examines manure storage, treatment and processing by discussing the principles of emissions produced from these processes as well as mitigation measures that can be used. It also addresses the best practices and priority measures for livestock feeding, housing and manure storage, treatment and processing.

Chapter 6 discusses developments in anaerobic digestion (AD) to optimize use of livestock manure, particularly the use of livestock manure in the production of biogas. The chapter begins by reviewing the quantities and risks of livestock manure, which is then followed by a discussion of the biogas potential of livestock manure. The chapter also examines mono- and co-digestion and the various factors that can affect the efficiency of anaerobic digestion. It also discusses the use of biogas slurry and residues. The chapter shows how AD can play an important role in promoting circular agriculture. A case study on the use of AD in practice in Henan Province in China is also included.

Part 3 Nutrition

Part 3 opens with a chapter that examines the impact of improving feed efficiency on the environmental impact of livestock production. Chapter 7 starts by discussing the relation between greenhouse gases and dairy production, highlighting how important it is to the dairy sector to find ways of decreasing greenhouse gas output. The chapter then moves on to discuss the origins of methane and reactive nitrogen excretions in ruminants. A section on improving feed conversion efficiency is also included, which is then followed by a review of the nutritional practices that can be used to enhance feed conversion efficiency and decrease methane excretion. The chapter also examines the nutritional practices that can be used to increase milk protein efficiency and nitrous oxide excretion as well. Discussions on genetics and feed conversion efficiency and postabsorptive metabolism and feed conversion efficiency are also provided.

Chapter 8 reviews grazing management strategies that can contribute to reducing livestock greenhouse gas emissions. Strategies discussed include grazing season length and timing as well as sward structure and quality, including dry matter and clover content. The chapter also discusses the use of condensed tannin legumes such as chicory and plantain, as well as measurement issues including life cycle assessment.

Chapter 9 focuses on the opportunity to use plant bioactive compounds in ruminant diets for their potential to mitigate greenhouse gas emissions, particularly enteric methane. Nitrous oxide emissions related to urinary nitrogen waste are addressed when information is available. The main families considered are plant lipids and plant secondary compounds (tannins, saponins, halogenated compounds and essential oils). The effects of these compounds in vivo, their mechanisms of action, and their potential adoption on farms are discussed, and future trends in this research area are highlighted.

The next chapter looks at the use of feed supplements to reduce livestock greenhouse gas emissions, specifically focusing on direct-fed microbials. Chapter 10 outlines the strategy of using feed supplements for the reduction of greenhouse gas emissions in ruminants, including methane (CH_4), carbon dioxide and nitrous oxide, given that feed intake is an important variable in predicting these emissions. The chapter focuses on direct-fed microbials, a term reserved for live microbes which can be supplemented to feed to elicit a beneficial response. The viability of such methods is also analysed for their use in large scale on-farm operations.

Chapter 11 focuses on modifying the rumen environment to reduce greenhouse gas emissions. Ruminants were among the first domesticated animals and have been providing food, leather, wool, draft and by-products to humanity for at least 10000 years. However, rumen methanogens reduce CO_2 to CH_4 in association with other rumen microbes that generate substrates for methanogenesis. Consequently, other rumen microbiota can directly and indirectly impact the abundance and activity of methanogens. Enteric methanogenesis from ruminants accounts for approximately 6% of total anthropogenic greenhouse gases emissions and can represent from 2% to 12% of the host's gross energy intake. A myriad of strategies to mitigate CH_4

emissions have been investigated, but few have been adopted by industry. This chapter reviews rumen- and feed-associated factors affecting CH_4 production and outlines the challenges associated with achieving a reduction in enteric CH_4 emissions. The pros and cons of these strategies are discussed in an attempt to define the best approaches to mitigate CH_4 emissions from ruminant production systems.

Part 1

Analysis

Chapter 1

Measuring methane emissions from livestock

Trevor Coates, Agriculture and Agri-Food Canada, Canada; and Deli Chen and Mei Bai, University of Melbourne, Australia

1 Introduction

Methane (CH_4) gas was first isolated by the Italian physicist Alessandro Volta in 1776 and described as the 'inflammable air native of marshes'. Although recognized as a local gas associated with decaying biological matter, CH_4 was thought to be a relatively static and minor component of the atmosphere. It would be another 200 years before advances in gas chromatography (GC) allowed Rasmussen and Khalil (1981) to show that CH_4 concentration in the atmosphere was not static but was increasing by an estimated 2% per year. With continued atmospheric monitoring and a lengthening historic record derived from ice-core data, the nature of rising CH_4 concentration and its relevance to global warming became increasingly apparent.

The establishment of the Intergovernmental Panel on Climate Change (IPCC) in 1988 and a growing awareness of the need to curb emissions sparked a flurry of research related to cattle CH_4 emissions beginning in the 1990s. Mitigation of enteric CH_4 emissions and the development of measurement techniques to validate the effectiveness of mitigation practices continue to be ongoing areas of research. Strategies that alter the rumen environment and the digestion process (Hristov et al., 2013) through the use of feed additives

http://dx.doi.org/10.19103/AS.2020.0077.03
Published by Burleigh Dodds Science Publishing Limited, 2021.

(Grainger et al., 2008) can improve feed efficiency and decrease enteric CH_4 emissions per kg of meat produced. Farm management and grazing strategies can also influence CH_4 emissions and the genetic selection of more efficient cattle (kg CH_4 per kg live weight) is also an ongoing promising area of research (Basarab et al., 2013).

Mitigation of CH_4 emissions requires emission measurements that are sensitive enough to measure the difference between standard practices and proposed mitigation strategies. Emission measurement methods also need to be operational at a range of spatial scales from the individual animal to *in situ* measurements under typical animal management conditions. This chapter reports on key techniques used for measuring CH_4 emission in agriculture at a variety of spatial scales. Advances in these techniques and promising new approaches to measure CH_4 emissions are also discussed.

2 Individual animal measurement techniques: whole-animal respiration chambers and head capture measurement

2.1 Whole-animal respiration chambers

The first estimates of CH_4 emissions from ruminants came from chamber studies, long before CH_4 had ever been measured in the atmosphere. The animal nutrition laboratory of the Pennsylvania State College constructed a respiration calorimeter in 1902 as a key component to better understanding animal physiology and ruminant nutrition (Armsby and Fries, 1903). Animal CH_4 production was recognized as a loss in feed energy and the Armsby respiration chamber, as it came to be known, was instrumental in generating feed ration and nutrition guidelines for America's expanding cattle industry. Measurement of CH_4 emissions was accomplished by routing a known volume of chamber air to a combustion furnace (Fries, 1910) and recording the change in combustion end products. After many years of experiments, Bratzler and Forbes (1940) developed a simple model relating animal CH_4 production with carbohydrate intake.

Modern chambers are more sophisticated, offering fine control of temperature, humidity and airflow, and advances in sensor technology have allowed for analysis of more components. Construction of chambers represents a considerable expense, but the importance of CH_4 as a source of agricultural greenhouse gas (GHG) emission has hastened their development. Whole-animal respiration chambers are now found at animal research facilities throughout the world. Chamber results have contributed to the current understanding of animal energetics and are considered a 'gold standard' measurement technique. Chambers offer a direct measure of emissions with few assumptions and a methodology that can be easily validated through gas release and recovery tests (McLean and Tobin, 1988). While whole-animal respiration chambers have been extremely valuable for mitigation work and quantifying treatment effects, the chamber represents a constrained environment, and it is less certain to what extent results can be extrapolated to actual cattle production systems (Johnson et al., 1994).

2.2 Head capture measurement

Measurement techniques capable of operating within real agricultural production environments are necessary for validating methane mitigation measures under typical animal management conditions. As an alternative to large animal chambers, a variety of systems have been designed to measure emissions, principally through focused airflow and concentration measurements of the area around the animal's head. These methods include:

- sniffer methods, where a sampling unit is incorporated into feed troughs;
- ventilated hood or headbox systems, which provide a more controlled environment but allow the animal access to food and water; and
- mask systems, which are fitted to the animal's nose and mouth.

The latter two techniques are also known as flux methods since they involve greater control of the airflow to capture emitted gases and measure CH_4 fluxes. A different approach is the use of handheld laser methane detectors (LMD) which are pointed by an operator at an animal's nostrils to measure methane column density along the length of the laser beam.

These techniques can be used within existing barn facilities and, depending upon the design, can be used to measure emissions continuously over a 24-h period or through spot measurements over the course of the day (Hammond et al., 2016a; Kebreab, 2015). Similar to respiration chamber measurement, these techniques can be affected by decreased feed intake, and intensive training is required for animals to become familiar with the hood apparatus, making it impractical to measure large numbers of animals (NASEM, 2018).

Sniffer methods are based on continuous breath analysis of exhaled air from animals using feed troughs in environments such as automated milking systems. A sampling unit is placed in the feed trough, and the air around the animal's muzzle is continuously monitored during feeding. Sensor systems detect the animal and activate breath analyzers located in the troughs, including Fourier transform infrared (FTIR) and non-dispersive infrared (NDIR) techniques. Measurements can then be used to develop an index of CH_4 emissions during milking as a product of peak frequency and mean peak area of CH_4 concentration (Garnswothy et al., 2012), or using the ratio of CO_2 to CH_4 (Lassen et al., 2012; Lassen and Løvendahl, 2016; Bell et al., 2014). Sniffer methods may be more affected by variable air-mixing conditions due to factors such as the geometry of the feed trough, muzzle position and movement, suggesting that flux techniques are more reliable (Huhtanen et al., 2015). However, recent research suggests that results from sniffer and flux methods are both comparable with each other and with respiration chambers, suggesting a growing degree of accuracy (Sorg et al., 2018; Difford et al., 2018).

The GreenFeed system (GF) (C-Lock Inc, Rapid City SD, USA) incorporates elements of the ventilated hood chamber into an automated feeder that dispenses a programmed amount of pelletized feed as bait to encourage visits to the GF. The GF is a robust system and can be incorporated into the production environment with one GF unit capable of measuring many animals consecutively. A proximity sensor in the head chamber identifies the visiting animal through its ear tag and initiates a gas sampling routine during which bait pellets are dispensed to keep the animals head in the feeder for 3–7 min during which time an emission rate is calculated. The procedure for deriving emission rates is reported by McGinn et al. (2021).

The GF unit can be programmed to limit the number of permitted visits per day but this measurement method is dependent on the animal's desire for the bait in the feeder, and the actual number of visits each animal makes per day will vary as will the number of animals that visit the device. For this reason, measurements are typically accumulated over several weeks to establish a daily emissions pattern for each animal that regularly uses the GF (Hammond et al., 2016b; Hristov et al., 2015; Huhtanen et al., 2019). When using spot measurements to determine daily emissions, care must be taken to prevent sampling bias by ensuring sampling times are appropriate for the daily feeding cycle of the animals using the device (Hammond et al., 2016a). As with ventilated hood chambers and head masks, the GF system is unable to capture the small emission eructed through the rectum, which has been reported to be between 4% and 8% of the total emission from cattle nose, mouth and rectum (Grainger et al., 2007b; McGinn et al., 2006a; Ulyatt et al., 1999). Muñoz et al. (2012) assumed 3% ± 1.5% emission was from the rectum.

3 Individual animal measurement techniques: tracer techniques

Tracer techniques rely on the co-location of a tracer gas source (with a known release rate) and the source to be measured, based on the assumption that both gases will be transported in the atmosphere in the same manner. Concentration measurements of the tracer gas and the source gas are made at some distance downwind. The ratio of gas concentrations is used with the known release rate of the tracer to determine the emission rate. Tracer techniques offer the advantage of a strictly ratiometric measure independent of meteorological conditions.

3.1 Sulfur hexafluoride tracer technique

The sulfur hexafluoride (SF_6) tracer technique (Johnson et al., 1994) was developed to overcome the limitations of chamber measurements, providing individual animal emission estimates without constraints on the animal's typical behavior in the production environment. The technique requires placing a small permeation tube in the animal's rumen which emits SF_6 at a pre-calibrated rate. This tracer gas is expelled through eructation along with the CH_4 produced in the rumen, while a collection device attached to the animal slowly draws air, usually for 24 h, from the nose and mouth region through an intake affixed to a halter. This technique provided the first measurements of animal emissions from grazing systems (Lassey et al., 1997; McCaughey et al., 1997). It also provided a means to monitor the dynamics of pasture conditions by conducting short measurement programs over a year (Pavao-Zuckerman et al., 1999; Ulyatt et al., 2002).

The technique gained in popularity as many agricultural research facilities already had the capacity for gas chromatographic analysis of the air samples collected with the SF_6 technique. With a modest outlay to construct the sampling apparatus and prepare the permeation tubes, a technique was now available to obtain measurements of animal CH_4 emissions from their typical production environment. With an increasing number of users, the technique was improved through a better understanding of animal emission variability compared with chamber measurements (Grainger et al., 2007a), the release characteristics of the permeation tubes with time (Lassey et al., 2001), the effect of permeation rate on emissions estimates (Vlaming et al., 2007), the effect of background measurements (Williams et al., 2011) and the importance of the sample collection rate (Deighton et al., 2014).

The SF_6 tracer technique has proven valuable for mitigation work and is well suited to measurements on small groups of animals, particularly within dairy systems where animals are accustomed to daily handling and sampler changes can be coordinated with daily milking. Implementation in grazing systems is more problematic as cattle require extensive training to become accustomed to handling and the fitting of yokes (DeRamus et al., 2003). The technique is also limited by higher between-cow variability in measurement accuracy (Pinares-Patiño et al., 2011). In addition to the permeation tube approach, SF_6 has also been used as a tracer gas to estimate CH_4 emissions from whole farms by releasing gas along barn vents and pen railings (McGinn et al., 2006b) and collecting downwind air samples for GC analysis of SF_6 and CH_4.

3.2 Other tracer techniques: nitrous oxide-tracer Fourier transform infrared spectroscopy

Tracer studies have also been carried out using open-path FTIR and nitrous oxide (N_2O) as tracer gas (Griffith et al., 2008). The open-path FTIR has proven to be a robust instrument for trace gas studies with stable performance and high precision of concentration measurement. Concentrations of CH_4 and N_2O can be measured by one FTIR concentration sensor at a short interval (e.g. 3-min interval). The CH_4 emission rate (Q_{CH4}) is calculated following Eq. 1:

$$Q_{CH_4} = Q_{N_2O} * (\Delta CH_4 / \Delta N_2O) \quad (1)$$

where Q_{N2O} is the known N_2O release rate and ΔCH_4 and ΔN_2O are the enhanced mixing ratios of CH_4 and N_2O above the local background level.

Griffith et al. (2008) monitored grazing dairy animals by affixing release points along a fence line of a grazing paddock. Animals were confined within a narrow paddock with respect to the prevailing wind so as to minimize the tracer to animal distance and reduce errors arising from the separation of the emission source and the tracer. This technique has also been utilized with gas release canisters affixed to halters on individual animals (e.g. dairy cattle and sheep) (Bai, 2010; Jones et al., 2011) to better collocate the tracer with the emission source. The tracer canisters emit tracer gas N_2O at a known rate similar to SF_6 permeation tubes, and the concentrations of CH_4 and N_2O are measured simultaneously downwind of the animals with an open-path FTIR instrument. This provides a herd-emission rate rather than individual animal emissions. Similar to the SF_6 technique, the method requires daily animal handling for canister replacement. The N_2O tracer-FTIR approach and its need for animal confinement and daily animal handling make this technique more suited to short-term intensive field campaigns. Care must also be taken to ensure the measurement paddock is not a significant source of N_2O as might arise with an irrigated/fertilized grazing paddock.

Another application of the ratiometric technique was reported by Laubach et al. (2014) and McGinn et al. (2019). The ratio of the above background concentration of a target gas measured downwind from two different sources (a control and a treatment) is used to infer the emission reduction of the treatment (numerator) as a fraction of the control (denominator). The assumption used was that the wind flow and physical dimensions of the two sources were identical, thus satisfying the requirement of a single gas transfer coefficient for the calculation of both emissions.

4 Herd-scale measurement techniques: micrometeorological methods

Micrometeorological methods (MM) are by nature non-interference techniques and the most applicable tool for herd-scale emission measurement. In principle, they can be used to study animals at the farm scale without the requirement for animal handling. MM require measurements of atmospheric CH_4 concentration

Published by Burleigh Dodds Science Publishing Limited, 2021.

and meteorological variables above or downwind of the animals to infer emissions.

McGinn (2006), in a review of MM techniques, suggested that the mass difference (MD) technique, the integrated horizontal flux (IHF) technique and inverse dispersion (ID) methods were all appropriate choices for measuring emissions of free-ranging animals. The MD and IHF techniques are both mass balance (MB) methods where differences in concentration profiles upwind and downwind of the source are used with a wind speed profile to infer emissions. The MD technique was first proposed by Denmead et al. (1998) and subsequently used for estimating CH_4 emissions from sheep (Leuning et al., 1999) and from grazing animals (Harper et al., 1999). Instrumentation and set-up demands were significant in each of the studies, requiring pumps and switches to manage multiple sampling lines on each face of the pen running to a central CH_4 analyzer. The IHF technique similarly requires logistically complex profile measurements, and as with MD, it is spatially constrained by the downwind profile height, and the source area must be kept small. Application on a grazing landscape requires the animals to be confined, and this effectively restricts both the number of animals and the length of time for measurement.

ID methods rely on atmospheric dispersion models to compute the theoretical relationship between concentration and emission rate from a defined source. Flesch et al. (1995) described a backward-time Lagrangian stochastic model and its application for deriving emissions requiring only a single-concentration measurement of the plume, a known background concentration and wind statistics recorded with a three-axis sonic anemometer. This model has since been incorporated into a commercial software platform WindTrax® (http://www.thunderbeachscientific.com/) providing a convenient tool for mapping source and sensor locations and visualizing model runs. A model run consists of thousands of 'particles' (fluid elements) released from the sensor and followed backward in time with trajectories that are consistent with the averaged wind and turbulence statistics of the surface. Particles that 'touch down' within the defined source area are accumulated and their velocity at touchdown is used in the final calculation of the ratio of concentration to flux. The model can also operate in a forward mode where particles are 'released' from the source and any trajectories that intercept with the sensor volume are accumulated. This technique has since become the most popular MM technique for estimating animal methane emissions (Table 1).

Open-path sensors such as tunable diode lasers (TDL) and FTIR spectrometers have been used because of their ability to spatially integrate concentration measurements and are ideally paired with the WindTrax software. Similar to the MB methods, ID modeling requires a known background measurement and sensors with sufficient sensitivity to accurately sense the increase above background due to the source. Confinement of animals may be

Table 1 Micrometeorological techniques used for monitoring cattle methane emissions

Author	Technique[†]	Sensor	# of animals	Confinement	Stocking density*	Duration
Judd et al. (1999)	FG	GC	55‡	Paddock (3 ha)	18.6	5 d
Leuning et al. (1999)	MD	FTIR (closed path)	14‡	Pen (500m^2)	289	5 d
Harper et al. (1999)	MD	Gas analyser/tubing	4	Pen (500m^2)	80	3-4 d
Laubach and Kelliher (2005)	ID/IHF	Open-path laser	269–287	Paddock (1.7 ha)	164	2 wks
McGinn et al. (2006b)	ID	Open-path laser	321	Dairy farm	186	1 mon
Griffith et al. (2008)	Tracer/IHF	FTIR (closed path)	23–50	Paddock (210 m^2)	1095–2381	2 wks
McGinn et al. (2009)	ID	Open-path laser	60	Pens (2 × 252 m^2)	1190	3 wks × 3
Bjorneberg et al. (2009)	ID	Open-path FTIR	780	Dairy farm	167	1 wk × 4
Laubach (2010)	ID	Open-path laser	58	Paddock (0.35 ha)	166	4 d × 4
Dengel et al. (2011)	EC	Open-path analyser	10–120‡	Paddock (5.4 ha)	1.8–22	60–277 d
Gao et al. (2011)	ID	Open-path laser	700	Feedlot	316–346	4 wks × 2
Tomkins et al. (2011)	ID	Open-path laser	18	Paddock (1ha)	18	1wk × 5
Tallec et al. (2012)	EC	Closed-path analyser	5	Pen (400 m^2)	125	8 d
Laubach et al. (2013)	MB/ID ID	Closed-path analyser, Open-path FTIR Open-path laser	61	Paddock (3772 m^2)	162	1 wks × 3
Laubach et al. (2014)	ID	Closed-path analyser	30	Paddock (2000 m^2)	150	16 d
McGinn et al. (2014)	ID	Open-path laser	40	Paddock (1 ha)	40	1 wk × 4
VanderZaag et al. (2014)	ID	Open-path laser	149–245	Dairy farms	-	3 wks × 4
McGinn et al. (2015)	ID	Open-path laser	60–40	Paddock (6 × 1 ha)	30–10	5 d × 3
Tomkins and Charmley (2015)	ID	Open-path laser	60	Pen (600 m^2)	1029	2 wks

Bai et al. (2015)	ID	Open-path FTIR	17500	Feedlot	738	2 wks
Felber et al. (2015)	EC	Closed-path analyser	17–20	Paddock (0.6 ha)	28–33	7 mon
Bai et al. (2016)	ID	Open-path FTIR	28	Pen (400 m^2)	700	6 wks
Dumortier et al. (2017)	EC	Closed-path analyser	~30	Paddock (4.2 ha) Confined 1.7 ha	~7 >17	19 mon
Taylor et al. (2017)	EC	Closed-path analyser Open-path analyser	264	Paddock (6–30 ha)	3–18	12 mon
Prajapati and Santos (2017)	EC	Closed-path analyser	58000	Feedlot (59 ha)	983	10 mon
Prajapati and Santos (2018)	EC/ID	Closed-path analyser	24116	Feedlot (59 ha)	455–529	10 mon
Coates et al. (2018) McGinn and Flesch (2018)	EC ID	Open-path analyser Open-path laser	20 5190	Paddock (15 ha) Feedlot A	1.3 12.5–16.2	7 wks 90 d
			8110	Feedlot B	10.0–13.6	20 d
McGinn et al. (2019)	ID	Open-path laser	666-678	Pens (2 × 1200 m^2)	560	23 wks
Prajapati and Santos (2019)	EC	Open-path FTIR Closed-path analyser	24116	Feedlot (59 ha)	455–529	7 d 10 mon
Todd et al. (2019)	EC/Tracer ID	Open-path analyser Open-path laser	12 50–12	Paddocks	1.89 1.95 75	8 9 7
Dumortier et al. (2019)	EC/Tracer	Open-path analyser	9	Paddock (4.2 ha)	2.2	27 d

* Stocking density (cattle/ha) refers to stocking density of measurement area (i.e. pen or paddock) † EC = eddy covariance, MD = mass difference, MB = mass balance, FG = flux gradient, IHF = integrated horizontal flux, ID = inverse dispersion ‡ denotes sheep.

required on a grazing landscape to achieve a sufficient rise in concentration and to ensure an uncontaminated background (prevention of wandering animals). McGinn et al. (2015) demonstrated that ID was effective with confined grazing animals at a stocking density of ten animals/ha but measurement of grazing cattle emissions remains a challenge due to the spatial limitations of sensor footprints and the need to manage cattle according to the changing pasture conditions over the grazing season (Todd et al., 2019).

Overall, reviews of MM suggest that results are comparable with other emission monitoring methods and verification of MM can be achieved through controlled gas release/recovery experiments (McGinn, 2013). The use of MM to distinguish treatment effects is more challenging due to the number of variables that must be considered but with care this can be achieved (Laubach et al. 2013; McGinn et al., 2019).

5 Herd-scale measurement techniques: the EC technique

The theory behind the EC technique is based on fundamentals of turbulent transport based on the early work of Reynolds (1894) and Taylor (1915) (Taylor, 1938) that set the theoretical framework for EC. However, it would be many decades before sensors were developed that could confirm the theoretical functions and explore potential applications. The flux of a given scalar quantity (F_s) can be described simply by Eq. 2:

$$F_s = w\rho_s \tag{2}$$

where w is the vertical wind velocity and ρ_s is the molar density of the trace gas of interest. Applying Reynolds decomposition, the above can be separated into a mean component (product of the mean vertical wind and the mean gas density over an averaging period) and a fluctuating component (the product of the deviations from the mean for the same variables $w' \rho_s'$) to give Eq. 3:

$$F_s = \overline{w\rho_s} + \overline{w'\rho_s'} \tag{3}$$

where $\overline{w\rho_s}$ represents the product of the mean vertical wind and the mean molar gas density, and $\overline{w'\rho_s'}$ represents the mean product of the instantaneous deviations from the mean for the same components.

One of the assumptions made to simplify the above equation is that, over a set averaging period, the mean vertical wind speed should be zero and that, as part of processing, the turbulence components are processed through a coordinate rotation such that the mean vertical wind velocity for the measurement period is zero. With the first component removed, the basic equation for EC becomes (Eq. 4):

$$F_s = \overline{w'\rho_s'} \tag{4}$$

where the flux of the gas component is calculated as the mean covariance between the instantaneous fluctuations of the vertical wind velocity and the molar density of the trace gas.

5.1 The use of the technique in agriculture

Although the theoretical underpinnings of EC were in place early in the 20th century, the instrumentation to measure the fluctuating components of wind and concentration with a time response sufficiently fast to capture the range of eddy sizes would not be available for several decades. The first EC measurements associated with agriculture explored fluxes of heat, momentum and water (Swinbank, 1951). Early attempts were made to measure CO_2 fluxes over an agricultural crop using a propeller-type anemometer to record vertical wind fluctuations (Desjardins, 1974) but a loss of high-frequency data associated with the slow response times of the instrumentation led to an underestimate of fluxes.

EC techniques increased in popularity throughout the 1980s as the development of the microprocessor facilitated the manufacture of three-axis sonic anemometers. This coupled with sensor improvements and digital data acquisition systems led to further use of EC to generate flux data over crops (Anderson et al., 1984; Desjardins et al., 1984). As the availability of robust commercially available sonic anemometers and stable fast-response sensors increased, so too did the number of monitoring sites. The accumulation of data from carbon balance studies and the need for spatial integration across various ecosystems led to the development of regional and global flux networks throughout the '90s (Baldocchi, 2003).

With the rapid expansion of flux networks and the accumulation of long-term flux data, there came a need for a more standardized approach to data analysis to allow meaningful comparison between sites. Quality control testing and assessment of data quality became an increasing concern (Foken and Wichura, 1996; Vickers and Mahrt, 1997). This emphasis on the documentation of analysis procedures led to the convergence of analysis protocols and to the development of a number of software packages designed to incorporate analytical processing options, quality-control tests and sensor-dependent corrections (Table 2). The use of a standardized analysis platform allows for a more consistent approach to data processing. Data can be shared in a manner that clearly documents the analysis steps taken. A programmed approach also aids in the reanalysis of datasets when software is updated to incorporate improvements in processing steps or additional corrections.

5.2 The use of EC techniques to measure livestock emissions

The vast flux networks that began in the '90s were focused on characterizing ecosystem productivity through measurement of the carbon, water and

Table 2 A selection of freely available software packages for eddy covariance analysis

Software Package	Maintained by	Website
AltEddy	Wageningen University, The Netherlands	http://www.climatexchange.nl/projects/alteddy/
EdiRe	University of Edinburgh, Scotland	http://www.geos.ed.ac.uk/homes/rclement/micromet/EdiRe/
TK3	University of Bayreuth, Germany	https://zenodo.org/record/20349#.WKfwAG995hE
EddyUH	University of Helsinki, Finland	https://www.atm.helsinki.fi/Eddy_Covariance/EddyUHsoftware.php
ECO_2S	University of Tuscia, Italy	http://gaia.agraria.unitus.it/eco2s
EddyPro	LI-COR Biosciences, Lincoln Nebraska, USA	https://www.licor.com/env/products/eddy_covariance/compute.html#eddypro

energy exchanges between the homogenous extensive surfaces (grasslands, forests, etc.) and the atmosphere. This decade, however, also saw the first measurements of CH_4 fluxes using EC with closed-path CH_4 analyzers (Denmead, 1991; Verma et al., 1992). These first-generation analyzers were lab-based instruments requiring temperature control, mains power and pumps to draw air through the analyzer, and field measurements were limited to short-term intensive testing. The development of a low-power open-path CH_4 analyzer (McDermitt et al., 2011) has since enabled the deployment of EC systems to monitor CH_4 fluxes from a variety of landscapes including wetlands (Matthes et al., 2014), rice paddies (Alberto et al., 2014) and tundra (Raz-Yaseef et al., 2016).

Although EC has grown to be a well-tested micrometeorological technique for measuring exchanges between the surface and the ground, this is typically done in association with an extensive homogenous source, and fluxes are calculated on an area basis. In fact, this calculated flux will represent a spatially weighted average of surface fluxes from a defined portion of the underlying surface. Areas of the surface that contribute to the calculated flux (and their relative contributions) constitute the flux footprint (Schuepp et al., 1990). Typically, the footprint encompasses the immediate vicinity of the sensor and extends upwind. The shape of the footprint and its upwind extent vary with sensor height, aerodynamic roughness of the surface and atmospheric conditions but are generally in the range of a few hundred meters. Atmospheric stability plays a large role, however, with the footprint contracting substantially in unstable conditions and increasing potential to hundreds of meters during stable conditions (Göckede et al., 2004). When EC is used over a homogenous extensive surface, the calculated vertical flux is representative of the surface-emission rate, expressed in terms of g m^{-2} s^{-1}. When EC is used to infer emissions from point sources (or spatially limited area sources), the relationship between emission and flux is more complex as the contribution of these distinct sources to the EC flux will be dependent on the source location within the footprint.

Grazing systems represent a challenging and complex environment for the application of EC. The sensitivity of EC to the presence of grazing animals on the landscape has created difficulties for those who are interested in ecosystem fluxes. Herbst et al. (2011) found CH_4 peaks in their flux data from wetland emissions and utilized a camera system to show that these anomalies were correlated with grazing animals upwind of their measurement tower. Baldocchi et al. (2012) found that the presence of cows on the landscape was easily detected in their flux measurements and became a challenge in monitoring the emissions from a peat landscape as the influence of nearby cows resulted in fluxes many times higher than the typical landscape fluxes they sought to measure. A tower-mounted camera was installed so that periods with animals in the vicinity could be eliminated from the dataset, although the authors suggested they could, at a future date, use this data to estimate animal emissions.

Intentional measurement of ruminant emissions using EC was first performed by Dengel et al. (2011) who used an EC flux tower to capture CH_4 fluxes of a sheep-grazed paddock. The movement of sheep in and out of the measurement footprint led to high variability in the individual analysis periods, and no means of recording animal locations were available. However, by integrating the fluxes over a 7-month measurement period and dividing by the stocking density, a rough estimate of emissions was obtained. Grazing animals introduce extreme temporal and spatial heterogeneity on the landscape, and CH4 fluxes measured by EC fluxes will fluctuate widely depending on the number of animals in the footprint and their locations relative to the sensor (Fig. 1).

The application of EC to estimate emissions in the grazing environment will generally require an analysis of the footprint and its overlap with animal positions. Tallec et al. (2012) accounted for animal position by confining groups of animals in small pens around a central tower. Felber et al. (2015) used EC with free-ranging cattle and tracked animal position with global positioning system (GPS) collars. Both Tallec et al. (2012) and Felber et al. (2015) used the footprint weighting tool of Neftel et al. (2008), based on the 2D analytical footprint model of Kormann and Meixner (2001), as the basis of interpretation of EC fluxes to generate emission estimates. While the authors concluded that EC was sufficiently accurate for animal emission studies, a systematic underestimation of emissions when animals are far from the tower was noted. Felber et al. (2015) proposed that the use of a more sophisticated Lagrangian footprint model (as used in the present work) could yield more accurate estimates. Recent studies highlight the value of the technique in

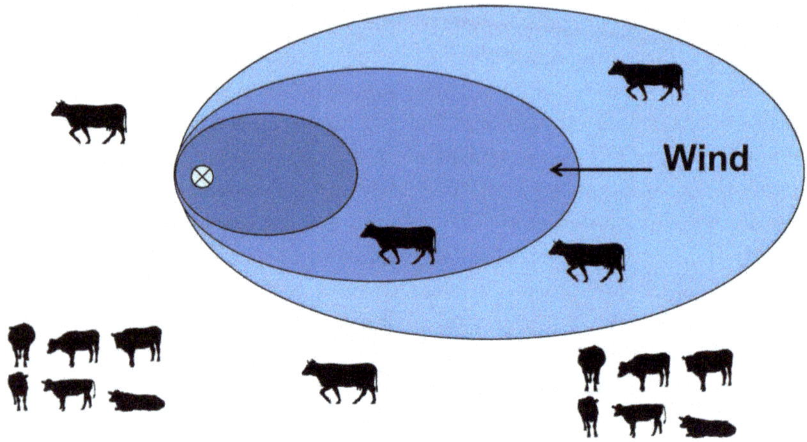

Figure 1 Representation of the flux footprint on a grazing landscape. In this representation only three animals contributed to the flux. Animals closer to the measurement point (marked by the x) contributed more than those further away.

assessing emissions in grazing environments but the challenges in accounting for variables such as wind and other atmospheric variables as well as animal distribution and movement continued (Prajapati and Santos, 2017, 2018; Coates et al., 2018).

6 Conclusion and future trends

Measuring methane emissions from livestock remains challenging yet there is a compelling urgency to continue to refine existing techniques and explore new opportunities as technologies evolve. The ability to reliably and repeatedly make valid measurements under a range of animal production scenarios is vital to support ongoing GHG mitigation efforts from the livestock sector. The expanding options for measurement are encouraging and the research community will continue to explore techniques best suited to their unique access to available equipment and expertise. The precision of respiration chambers will continue to play an important role in demonstrating effectiveness of new diet strategies for methane mitigation. Other techniques more applicable to actual farm conditions and a higher throughput of measurements have a role to play by accounting for animal variability. Recent studies comparing respiration chambers with the LMD, sniffer, SF_6 and GF techniques have found that results from the latter techniques correlated well overall with those from respiration chambers, but there was a lower correlation between techniques and degree of repeatability within individual techniques (with head enclosure techniques such as GF and SF_6 performing best) (Garnsworthy et al., 2019; Zhao et al., 2020). This suggests that a way forward is to combine techniques with appropriate weightings and target them for specific applications, whether emissions monitoring (using techniques such as respiration chambers and head enclosure methods) or using applications such as genetic evaluation in breeding cattle with lower emissions (using techniques such as GF, sniffer or LMD methods). MM methods will continue to be important for herd scale emissions and in understanding the relationship between animals and the rangeland ecosystem. Results from analytical methods can also be used both to validate and inform the range of increasingly sophisticated models predicting GHG emissions from livestock, allowing rapid analysis of a range of scenarios and parameters to help farmers improve their operations (Jose et al., 2016).

7 Where to look for further information

The following books provide helpful overviews on respiration chamber measurements, hood measurements and the SF_6 tracer technique:

- Animal and Human Calorimetry (McLean, J. A. and Tobin, G. 1988. Cambridge University Press, Cambridge).
- Measuring methane production from ruminants (Makkar, H. P. and Vercoe, P. E. 2007. Springer).

The following references provide valuable insights into meteorological approaches for flux measurement:

- Harper, L. A., Denmead, O. T. and Flesch, T. K. 2011. Micrometeorological techniques for measurement of enteric greenhouse gas emissions. *Animal Feed Science and Technology*, 166–67: 227–239.
- Rochette, P. and McGinn, S. M. 2005. Methods for measuring soil-surface gas fluxes. In Alvarez-Benedí, J. and Munoz-Carpena, R. (eds), *Soil-water-solute process characterization: an integrated approach*. CRC Press, Boca Raton: Florida, USA.

For up to date information on research initiatives and international conferences and workshops related to greenhouse gas emissions from agriculture:

- https://globalresearchalliance.org/.

8 References

Alberto, M. C. R., Wassmann, R., Buresh, R. J., Quilty, J. R., Correa, T. Q., Sandro, J. M. and Centeno, C. A. R. 2014. Measuring methane flux from irrigated rice fields by eddy covariance method using open-path gas analyzer. *Field Crops Research* 160: 12-21.

Anderson, D. E., Verma, S. B. and Rosenberg, N. J. 1984. Eddy correlation measurements of CO_2, latent heat, and sensible heat fluxes over a crop surface. *Boundary-Layer Meteorology* 29(3): 263-272.

Armsby, H. P. and Fries, J. A. 1903. *The Available Energy Of Timothy Hay*. US Government Printing Office.

Bai, M. 2010. Methane emissions from livestock measured by novel spectroscopic techniques. PhD Thesis. University of Wollongong, 303 pp.

Bai, M., Flesch, T. K., McGinn, S. M. and Chen, D. 2015. A snapshot of greenhouse gas emissions from a cattle feedlot. *Journal of Environmental Quality* 44(6): 1974-1978.

Bai, M., Sun, J., Dassanayake, K. B., Benvenutti, M. A., Hill, J., Denmead, O. T., Flesch, T. and Chen, D. 2016. Non-interference measurement of CH_4, N_2O and NH_3 emissions from cattle. *Animal Production Science* 56(9): 1496-1503.

Baldocchi, D., Detto, M., Sonnentag, O., Verfaillie, J., Teh, Y. A., Silver, W. and Kelly, N. M. 2012. The challenges of measuring methane fluxes and concentrations over a peatland pasture. *Agricultural and Forest Meteorology* 153: 177-187.

Baldocchi, D. D. 2003. Assessing the eddy covariance technique for evaluating carbon dioxide exchange rates of ecosystems: past, present and future. *Global Change Biology* 9(4): 479-492.

Basarab, J. A., Beauchemin, K. A., Baron, V. S., Ominski, K. H., Guan, L. L., Miller, S. P. and Crowley, J. J. 2013. Reducing GHG emissions through genetic improvement for feed efficiency: effects on economically important traits and enteric methane production. *Animal*7(Suppl. 2): 303-315.

Bell, M. J., et al., 2014. Methane emissions among individual dairy cows during milking quantified by eructation peaks or ratio with carbon dioxide. *Journal of Dairy Science* 97(10): 66536-6546.

Bjorneberg, D. L., Leytem, A. B., Westermann, D. T., Griffiths, P. R., Shao, L. and Pollard, M. J. 2009. Measurement of atmospheric ammonia, methane, and nitrous oxide at a concentrated dairy production facility in southern Idaho using open-path FTIR spectrometry. *Transactions of the ASABE* 52(5): 1749-1756.

Bratzler, J. W. and Forbes, E. B. 1940. The estimation of methane production by cattle. *The Journal of Nutrition* 19(6): 611-613.

Coates, T. W., Benvenutti, M. A., Flesch, T. K., Charmley, E., McGinn, S. M. and Chen, D. 2018. Applicability of eddy covariance to estimate methane emissions from grazing cattle. *Journal of Environmental Quality* 47(1): 54-61.

Deighton, M. H., Williams, S. R. O., Hannah, M. C., Eckard, R. J., Boland, T. M., Wales, W. J. and Moate, P. J. 2014. A modified sulphur hexafluoride tracer technique enables accurate determination of enteric methane emissions from ruminants. *Animal Feed Science and Technology* 197(0): 47-63.

Dengel, S., Levy, P. E., Grace, J., Jones, S. K. and Skiba, U. M. 2011. Methane emissions from sheep pasture, measured with an open-path eddy covariance system. *Global Change Biology* 17(12): 3524-3533.

Denmead, O., Harper, L. A., Freney, J. R., Griffith, D. W. T., Leuning, R. and Sharpe, R. R. 1998. A mass balance method for non-intrusive measurements of surface-air trace gas exchange. *Atmospheric Environment* 32(21): 3679-3688.

Denmead, O. T. 1991. Sources and sinks of greenhouse gases in the soil-plant environment. *Vegetatio* 91(1-2): 73-86.

DeRamus, H. A., Clement, T. C., Giampola, D. D. and Dickison, P. C. 2003. Methane emissions of beef cattle on forages: efficiency of grazing management systems. *Journal of Environmental Quality* 32(1): 269-277.

Desjardins, R. L. 1974. A technique to measure CO_2 exchange under field conditions. *International Journal of Biometeorology* 18(1): 76-83.

Desjardins, R. L., Buckley, D. J. and Amour, G. S. 1984. Eddy flux measurements of CO_2 above corn using a microcomputer system. *Agricultural and Forest Meteorology* 32(3-4): 257-265.

Difford, G. F., Olijhoek, D. W., Hellwing, A. L. F., Lund, P., Bjerring, M. A., de Haas, Y., Lassen, J. and Løvendahl, P. 2018. Ranking cows' methane emissions under commercial conditions with sniffers versus respiration chambers. *Acta Agriculturae Scandinavica, Section A–Animal Science* 68(1): 25-32.

Dumortier, P., Aubinet, M., Beckers, Y., Chopin, H., Debacq, A., Gourlez de la Motte, L., Jérôme, E., Wilmus, F. and Heinesch, B. 2017. Methane balance of an intensively grazed pasture and estimation of the enteric methane emissions from cattle. *Agricultural and Forest Meteorology* 232: 527-535.

Dumortier, P., Aubinet, M., Lebeau, F., Naiken, A. and Heinesch, B. 2019. Point source emission estimation using eddy covariance: validation using an artificial source experiment. *Agricultural and Forest Meteorology* 266-267: 148-156.

Felber, R., Münger, A., Neftel, A. and Ammann, C. 2015. Eddy covariance methane flux measurements over a grazed pasture: effect of cows as moving point sources. *Biogeosciences* 12(12): 3925-3940.

Flesch, T. K., Wilson, J. D. and Yee, E. 1995. Backward-time Lagrangian stochastic dispersion models and their application to estimate gaseous emissions. *Journal of Applied Meteorology* 34(6): 1320-1332.

Foken, T. and Wichura, B. 1996. Tools for quality assessment of surface-based flux measurements. *Agricultural and Forest Meteorology* 78(1-2): 83-105.

Fries, J. A. 1910. Electric combustion furnace for methane determination. *Journal of the American Chemical Society* 32(8): 949-953.

Gao, Z., Yuan, H., Ma, W., Liu, X. and Desjardins, R. L. 2011. Methane emissions from a dairy feedlot during the fall and winter seasons in Northern China. *Environmental Pollution* 159(5): 1183-1189.

Garnsworthy, P., Craigon, J., Hernandez-Medrano, J. and Saunders, N. J. J. o. d. s. 2012. On-farm methane measurements during milking correlate with total methane production by individual dairy cows. *Journal of Dairy Science* 95(6): 3166-3180.

Garnsworthy, P. C., Difford, G. F., Bell, M. J., Bayat, A. R., Huhtanen, P., Kuhla, B., Lassen, J., Peiren, N., Pszczola, M., Sorg, D., Visker, M. H. P. W. and Yan, T. 2019. Comparison of methods to measure methane for use in genetic evaluation of dairy cattle. *Animals (Basel)* 9(10): 837.

Göckede, M., Rebmann, C. and Foken, T. 2004. A combination of quality assessment tools for eddy covariance measurements with footprint modelling for the characterisation of complex sites. *Agricultural and Forest Meteorology* 127(3-4): 175-188.

Grainger, C., Clarke, T., Beauchemin, K. A., McGinn, S. M. and Eckard, R. J. 2008. Supplementation with whole cottonseed reduces methane emissions and can profitably increase milk production of dairy cows offered a forage and cereal grain diet. *Australian Journal of Experimental Agriculture* 48(2): 73-76.

Grainger, C., Clarke, T., McGinn, S. M., Auldist, M. J., Beauchemin, K. A., Hannah, M. C., Waghorn, G. C., Clark, H. and Eckard, R. J. 2007. Methane emissions from dairy cows measured using the sulfur hexafluoride (SF6) tracer and chamber techniques. *Journal of Dairy Science* 90(6): 2755-2766.

Griffith, D. W. T., Bryant, G. R., Hsu, D. and Reisinger, A. R. 2008. Methane emissions from free-ranging cattle: comparison of tracer and integrated horizontal flux techniques. *Journal of Environmental Quality* 37(2): 582-591.

Hammond, K. J., Crompton, L. A., Bannink, A., Dijkstra, J., Yáñez-Ruiz, D. R., O'Kiely, P., Kebreab, E., Eugène, M. A., Yu, Z., Shingfield, K. J., Schwarm, A., Hristov, A. N. and Reynolds, C. K. 2016a. Review of current in vivo measurement techniques for quantifying enteric methane emission from ruminants. *Animal Feed Science and Technology* 219: 13-30.

Hammond, K. J., Waghorn, G. C. and Hegarty, R. S. 2016b. The GreenFeed system for measurement of enteric methane emission from cattle. *Animal Production Science* 56(3): 181-189.

Harper, L. A., Denmead, O. T., Freney, J. R. and Byers, F. M. 1999. Direct measurements of methane emissions from grazing and feedlot cattle. *Journal of Animal Science* 77(6): 1392-1401.

Herbst, M., Friborg, T., Ringgaard, R. and Soegaard, H. 2011. Interpreting the variations in atmospheric methane fluxes observed above a restored wetland. *Agricultural and Forest Meteorology* 151(7): 841-853.

Hristov, A. N., Oh, J., Firkins, J. L., Dijkstra, J., Kebreab, E., Waghorn, G., Makkar, H. P., Adesogan, A. T., Yang, W., Lee, C., Gerber, P. J., Henderson, B. and Tricarico, J. M. 2013. Special topics - mitigation of methane and nitrous oxide emissions from animal operations: I. A review of enteric methane mitigation options. *Journal of Animal Science* 91(11): 5045-5069.

Hristov, A. N., Oh, J., Giallongo, F., Frederick, T., Weeks, H., Zimmerman, P. R., Harper, M. T., Hristova, R. A., Zimmerman, R. S. and Branco, A. F. 2015. The use of an automated system (GreenFeed) to monitor enteric methane and carbon dioxide emissions from ruminant animals. *Journal of Visualized Experiments* 103: e52904.

Huhtanen, P., Cabezas-Garcia, E. H., Utsumi, S. and Zimmerman, S. 2015. Comparison of methods to determine methane emissions from dairy cows in farm conditions. *Journal of Dairy Science* 98(5): 3394-3409.

Huhtanen, P., Ramin, M. and Hristov, A. N. 2019. Enteric methane emission can be reliably measured by the GreenFeed monitoring unit. *Livestock Science* 222: 31-40.

Johnson, K., Huyler, M., Westberg, H., Lamb, B. and Zimmerman, P. 1994. Measurement of methane emissions from ruminant livestock using a sulfur hexafluoride tracer technique. *Environmental Science and Technology* 28(2): 359-362.

Jones, F. M., Phillips, F. A., Naylor, T. and Mercer, N. B. 2011. Methane emissions from grazing Angus beef cows selected for divergent residual feed intake. *Animal Feed Science and Technology* 166-167: 302-307.

Jose, V. S., Sejian, V., Bagath, M., Ratnakaran, A. P., Lees, A. M., Al-Hosni, Y. A. S., Sullivan, M., Bhatta, R. and Gaughan, J. B. 2016. Modeling of greenhouse gas emission from livestock. *Frontiers in Environmental Science* 4(27). doi: 10.3389/fenvs.2016.00027.

Judd, M. J., Kellier, F. M., Ulyatt, M. J., Lassey, K. R., Tate, K. R., Shelton, I. D., Harvey, M. J. and Walker, C. F. 1999. Net methane emissions from grazing sheep. *Global Change Biology* 5(6): 647-657.

Kebreab, E. 2015. Calorimeters for enteric methane measurements. In: Gerrites, W. and Labussière, E. (Eds), *Indirect Calorimetery: Techniques, Computations and Applications*. Wagenigen Academic, Wageningen, Netherlands, pp. 294.

Kormann, R. and Meixner, F. X. 2001. An analytical footprint model for non-neutral stratification. *Boundary-Layer Meteorology* 99(2): 207-224.

Lassen, J. and Løvendahl, P., 2016. Heritability estimates for enteric methane emissions from Holstein cattle measured using noninvasive methods. *Journal of Dairy Science*, 99(3): 1959-1967.

Lassen, J. and Løvendahl, P. J. A. A. B. 2013. Heritability for enteric methane emission from Danish Holstein cows using a non-invasive FTIR method. 4: 280.

Lassey, K. R., Ulyatt, M. J., Martin, R. J., Walker, C. F. and Shelton, I. D. 1997. Methane emissions measured directly from grazing livestock in New Zealand. *Atmospheric Environment* 31(18): 2905-2914.

Lassey, K. R., Walker, C. F., McMillan, A. M. S. and Ulyatt, M. J. 2001. On the performance of SF6 permeation tubes used in determining methane emission from grazing livestock. *Chemosphere - Global Change Science* 3(4): 367-376.

Laubach, J. 2010. Testing of a Lagrangian model of dispersion in the surface layer with cattle methane emissions. *Agricultural and Forest Meteorology* 150(11): 1428-1442.

Laubach, J. and Kelliher, F. M. 2005. Methane emissions from dairy cows: comparing open-path laser measurements to profile-based techniques. *Agricultural and Forest Meteorology* 135(1-4): 340-345.

Laubach, J., Bai, M., Pinares-Patiño, C. S., Phillips, F. A., Naylor, T. A., Molano, G., Cárdenas Rocha, E. A. and Griffith, D. W. T. 2013. Accuracy of micrometeorological techniques for detecting a change in methane emissions from a herd of cattle. *Agricultural and Forest Meteorology* 176: 50-63.

Laubach, J., Grover, S. P. P., Pinares-Patiño, C. S. and Molano, G. 2014. A micrometeorological technique for detecting small differences in methane emissions from two groups of cattle. *Atmospheric Environment* 98: 599-606.

Leuning, R., Baker, S. K., Jamie, I. M., Hsu, C. H., Klein, L., Denmead, O. T. and Griffith, D. W. T. 1999. Methane emission from free-ranging sheep: a comparison of two measurement methods. *Atmospheric Environment* 33(9): 1357-1365.

Matthes, J. H., Sturtevant, C., Verfaillie, J., Knox, S. and Baldocchi, D. 2014. Parsing the variability in CH_4 flux at a spatially heterogeneous wetland: integrating multiple eddy covariance towers with high-resolution flux footprint analysis. *Journal of Geophysical Research: Biogeosciences* 119(7): 1322-1339.

McCaughey, W. P., Wittenberg, K. and Corrigan, D. 1997. Methane production by steers on pasture. *Canadian Journal of Animal Science* 77(3): 519-524.

McDermitt, D., Burba, G., Xu, L., Anderson, T., Komissarov, A., Riensche, B., Schedlbauer, J., Starr, G., Zona, D., Oechel, W., Oberbauer, S. and Hastings, S. 2011. A new low-power, open-path instrument for measuring methane flux by eddy covariance. *Applied Physics B* 102(2): 391-405.

McGinn, S. M. 2006. Measuring greenhouse gas emissions from point sources in agriculture. *Canadian Journal of Soil Science* 86(3): 355-371.

McGinn, S. M. 2013. Developments in micrometeorological methods for methane measurements. *Animal* 7(Suppl. 2): 386-393.

McGinn, S. M., Beauchemin, K. A., Coates, T. and McGeough, E. J. 2014. Cattle methane emission and pasture carbon dioxide balance of a grazed grassland. *Journal of Environmental Quality* 43(3): 820-828.

McGinn, S. M., Beauchemin, K. A., Flesch, T. K. and Coates, T. 2009. Performance of a dispersion model to estimate methane loss from cattle in pens. *Journal of Environmental Quality* 38(5): 1796–1802.

McGinn, S. M., Beauchemin, K. A., Iwaasa, A. D. and McAllister, T. A. 2006a. Assessment of the sulfur hexafluoride (SF_6) tracer technique for measuring enteric methane emissions from cattle. *Journal of Environmental Quality* 35(5): 1686–1691.

McGinn, S. M., Flesch, T. K., Harper, L. A. and Beauchemin, K. A. 2006b. An approach for measuring methane emissions from whole farms. *Journal of Environmental Quality* 35(1): 14–20.

McGinn, S. M., Coulombe, J. F. and Beauchemin, K. A. 2021. Technical note: validation of the GreenFeed system for measuring enteric gas emissions from cattle. *Journal of Animal Science* 99(3): 1–6.

McGinn, S. M. and Flesch, T. K. 2018. Ammonia and greenhouse gas emissions at beef cattle feedlots in Alberta, Canada. *Agricultural and Forest Meteorology* 258: 43–49.

McGinn, S. M., Flesch, T. K., Beauchemin, K. A., Shreck, A. and Kindermann, M. 2019. Micrometeorological methods for measuring methane emission reduction at beef cattle feedlots: evaluation of 3-Nitrooxypropanol feed additive. *Journal of Environmental Quality* 48(5): 1454–1461.

McGinn, S. M., Flesch, T. K., Coates, T. W., Charmley, E., Chen, D., Bai, M. and Bishop-Hurley, G. 2015. Evaluating dispersion modeling options to estimate methane emissions from grazing beef cattle. *Journal of Environmental Quality* 44(1): 97–102.

McLean, J. A. and Tobin, G. 1988. *Animal and Human Calorimetry*. Cambridge University Press, Cambridge.

Muñoz, C., Yan, T., Wills, D., Murray, S. and Gordon, A. 2012. Comparison of the sulfur hexafluoride tracer and respiration chamber techniques for estimating methane emissions and correction for rectum methane output from dairy cows. *Journal of Dairy Science* 95(6): 3139–3148.

NASEM. 2018. *Improving Characterization of Antrhopogenic Methane Emissions in the United States*. The National Academies Press, Washington, DC, 250 pp.

Neftel, A., Spirig, C. and Ammann, C. 2008. Application and test of a simple tool for operational footprint evaluations. *Environmental Pollution* 152(3): 644–652.

Pavao-Zuckerman, M. A., Waller, J. C., Ingle, T. and Fribourg, H. A. 1999. Methane emissions of beef cattle grazing tall fescue pastures at three levels of endophyte infestation. *Journal of Environmental Quality* 28(6): 1963–1969.

Pinares-Patiño, C. S., Lassey, K. R., Martin, R. J., Molano, G., Fernandez, M., MacLean, S., Sandoval, E., Luo, D. and Clark, H. 2011. Assessment of the sulphur hexafluoride (SF6) tracer technique using respiration chambers for estimation of methane emissions from sheep. *Animal Feed Science and Technology* 166–167: 201–209.

Prajapati, P. and Santos, E. A. 2017. Measurements of methane emissions from a beef cattle feedlot using the eddy covariance technique. *Agricultural and Forest Meteorology* 232: 349–358.

Prajapati, P. and Santos, E. A. 2018. Comparing methane emissions estimated using a backward-Lagrangian stochastic model and the eddy covariance technique in a beef cattle feedlot. *Agricultural and Forest Meteorology* 256–257: 482–491.

Prajapati, P. and Santos, E. A. 2019. Estimating herd-scale methane emissions from cattle in a feedlot using eddy covariance measurements and the carbon dioxide tracer method. *Journal of Environmental Quality* 48(5): 1427–1434.

Rasmussen, R. A. and Khalil, M. A. K. 1981. Increase in the concentration of atmospheric methane. *Atmospheric Environment (1967)* 15(5): 883-886.

Raz-Yaseef, N., Torn, M. S., Wu, Y., Billesbach, D. P., Liljedahl, A. K., Kneafsey, T. J., Romanovsky, V. E., Cook, D. R. and Wullschleger, S. D. 2016. Large CO_2 and CH_4 emissions from polygonal tundra during spring thaw in northern Alaska. *Geophysical Research Letters*: n/a-n/a.

Reynolds, O. 1894. II. On the dynamical theory of incompressible viscous fluids and the determination of the criterion. *Proceedings of the Royal Society of London* 56(336-339): 40-45.

Schuepp, P. H., Leclerc, M. Y., MacPherson, J. I. and Desjardins, R. L. 1990. Footprint prediction of scalar fluxes from analytical solutions of the diffusion equation. *Boundary-Layer Meteorology* 50(1 4): 355-373.

Sorg, D., Difford, G. F., Mühlbach, S., Kuhla, B., Swalve, H. H., Lassen, J., Strabel, T., Pszczola, M. J. C. and Agriculture, E. I. 2018. Comparison of a laser methane detector with the GreenFeed and two breath analysers for on-farm measurements of methane emissions from dairy cows. *Computers and Electronics in Agriculture* 153: 285-294.

Swinbank, W. C. 1951. The measurement of vertical transfer of heat and water vapor by eddies in the lower atmosphere. *Journal of Meteorology* 8(3): 135-145.

Tallec, T., Klumpp, K., Hensen, A., Rochette, Y. and Soussana, J. F. 2012. Methane emission measurements in a cattle grazed pasture: a comparison of four methods. *Biogeosciences Discussions* 9(10): 14407-14436.

Taylor, A. M., Amiro, B. D., Tenuta, M. and Gervais, M. 2017. Direct whole-farm greenhouse gas flux measurements for a beef cattle operation. *Agriculture, Ecosystems and Environment* 239: 65-79.

Taylor, G. I. 1915. Eddy motion in the atmosphere. 1. *Monthly Weather Review* 43(7): 315-316.

Taylor, G. I. 1938. The spectrum of turbulence. *Proceedings of the Royal Society A: Mathematical, Physical and Engineering Sciences*. The Royal Society 164(919): 476-490.

Todd, R. W., Moffet, C., Neel, J. P. S., Turner, K. E., Steiner, J. L. and Cole, N. A. 2019. Enteric methane emissions of beef cows grazing tallgrass prairie pasture on the southern Great Plains. *Transactions of the ASABE* 62(6): 1455-1465.

Tomkins, N. W. and Charmley, E. 2015. Herd-scale measurements of methane emissions from cattle grazing extensive sub-tropical grasslands using the open-path laser technique. *Animal: An International Journal of Animal Bioscience* 9(12): 2029-2038.

Tomkins, N. W., McGinn, S. M., Turner, D. A. and Charmley, E. 2011. Comparison of open-circuit respiration chambers with a micrometeorological method for determining methane emissions from beef cattle grazing a tropical pasture. *Animal Feed Science and Technology* 166-167: 240-247.

Ulyatt, M. J., Lassey, K. R., Shelton, I. D. and Walker, C. F. 2002. Seasonal variation in methane emission from dairy cows and breeding ewes grazing ryegrass/white clover pasture in New Zealand. *New Zealand Journal of Agricultural Research* 45(4): 217-226.

Ulyatt, M. J., McCrabb, G. J., Baker, S. K. and Lassey, K. R. 1999. Accuracy of SF_6 tracer technology and alternatives for field measurements. *Australian Journal of Agricultural Research* 50(8): 1329-1334.

VanderZaag, A. C., Flesch, T. K., Desjardins, R. L., Baldé, H. and Wright, T. 2014. Measuring methane emissions from two dairy farms: seasonal and manure-management effects. *Agricultural and Forest Meteorology* 194: 259–267.

Verma, S. B., Ullman, F. G., Billesbach, D., Clement, R. J., Kim, J. and Verry, E. S. 1992. Eddy correlation measurements of methane flux in a northern peatland ecosystem. *Boundary-Layer Meteorology* 58(3): 289–304.

Vickers, D. and Mahrt, L. 1997. Quality control and flux sampling problems for tower and aircraft data. *Journal of Atmospheric and Oceanic Technology* 14(3): 512–526.

Vlaming, J. B., Brookes, I. M., Hoskin, S. O., Pinares-Patiño, C. S. and Clark, H. 2007. The possible influence of intra-ruminal sulphur hexafluoride release rates on calculated methane emissions from cattle. *Canadian Journal of Animal Science* 87(2): 269–275.

Williams, S. R. O., Moate, P. J., Hannah, M. C., Ribaux, B. E., Wales, W. J. and Eckard, R. J. 2011. Background matters with the SF6 tracer method for estimating enteric methane emissions from dairy cows: a critical evaluation of the SF6 procedure. *Animal Feed Science and Technology* 170(3–4): 265–276.

Zhao, Y., Nan, X., Yang, L., Zheng, S., Jiang, L. and Xiong, B. 2020. A review of enteric methane emission measurement techniques in ruminants. *Animals (Basel)* 10(6): 1004.

Chapter 2

Greenhouse gas emissions from livestock production: modelling methods, methane emission factors and mitigation strategies

Donal O'Brien, Environment, Soils and Land Use Department, Teagasc, Ireland; and Laurence Shalloo, Animal and Grassland Research and Innovation Department, Teagasc, Ireland

1 Introduction

An integral component for quantifying greenhouse gas (GHG) emissions from livestock production systems is whole-farm modelling. In contrast to the International Panel on Climate Change (IPCC) method, whole-farm modelling does not specify the estimation of GHG emissions by sector, but by the definition of system boundaries (Schils et al., 2005). This allows a complete analysis of methane and GHG emissions that is not possible within the framework of the IPCC method, as on-farm GHG emissions emanating from livestock farming systems are reported in three different sectors (Soussana et al., 2010). These sectors are agriculture, land-use change and forestry, and energy. Furthermore, the IPCC method is limited to national GHG emissions. Thus, even if GHG emissions from national sectors are combined to quantify total emissions, any emissions generated outside of the national boundaries are not included (Cerri et al., 2009). For example, a large proportion of concentrate feeds used within winter milk production systems are sometimes produced outside a nation's borders, but the associated GHG emissions from

http://dx.doi.org/10.19103/AS.2020.0077.04

cultivation and harvesting are included in the inventories of the nation(s) that produce the concentrate feeds instead of the inventory where they are consumed.

An alternative to the production-focussed IPCC method is consumption-based GHG accounting. Peters and Hertwich (2008) outlined the methodology in detail and argued that emissions are driven by consumption, not by production. Estonia has demonstrated it is possible to use the consumption approach at a national scale to quantify GHG emissions (Gavrilova and Vilu, 2012). Quantifying national GHG emissions using both methods, that is, production and consumption could potentially identify any major transfers of GHG emissions from one nation to another (carbon leakage). This is an issue as some nations have less focus on reducing GHG emissions than others. The consumption approach, however, requires the development of harmonized farm models capable of quantifying GHG emissions associated with the life cycle of goods and services, that is, carbon footprint and methane emission intensity (Peters, 2008). Whole-farm models often use LCA methodologies to quantify emissions, and LCA methodologies often use farm models to complete the LCAs. Both of these modelling methods use a systems approach to quantify GHG emissions. When these modelling approaches are applied using the same set of assumptions (e.g. boundaries, unit of expression, methane emission factors), the results should be similar. This chapter will discuss the systems analysis and life cycle assessment modelling approaches and will go on from there to look at a range of model applications. These applications include use within the national inventories of various countries across species, including discussions around the use of different emission factors. The chapter concludes with applications to quantify emissions at the farm level and a discussion around some of the mitigation strategies that have been modelled.

2 Systems analysis

Systems analysis has been widely used by researchers to quantify GHG emissions from the ruminant system and assess methane mitigation strategies (Beukes et al., 2010; O'Brien et al., 2011; Clarke et al., 2013). The main stages of system analysis focus on the definition of a conceptual framework, with the precise aim and boundary of the model and its application.

2.1 Conceptual framework

In general, the first step of systems analysis is to formulate a conceptual model of the farming system of interest. The delimitation of the system boundaries of conceptual models is determined by the objective of the study and the precise aim of the model use. The objective of the analysis required should be used when decided on an appropriate model to complete an analysis. Generally, whole farm models that have quantified GHG emissions from livestock production systems estimate emissions from on- and off-farm sources related to the livestock product(s) up to a point it is exported from the farm (Olesen et al., 2006). Off-farm emissions from the production of external farm inputs such as concentrate feed and fertilizers are included with methane and other GHG emissions.

2.2 Model development

The second step of systems analysis entails creating a mathematical model of the production system defined. During this stage a series of algorithms are assembled and connected to mathematically model the farming system. In most cases, these equations are based on empirical relationships from representative field studies (Shalloo et al., 2004). Occasionally, experimental data from the farming system under study is used to estimate methane and other GHG emissions (Schils et al., 2005; del Prado et al., 2013). Within livestock system simulation models, emission factors in conjunction with livestock activity data are used to compute farm emissions. Sometimes these emission factors are obtained from the IPCC (2006) guidelines for on-farm emission sources. However, the IPCC emission factors are designed to enumerate national-level emissions. Thus, they often lack the refinement, model functions and emission factors, necessary to quantify the effect that changes to the production systems have on GHG emissions from individual farms (Schils et al., 2006). As a result, direct emission measurements, published in the scientific literature, are sometimes used to generate emission factors for subsequent inclusion in models in the pursuit of assessing on-farm emissions. In the case of off-farm emissions, almost all emission factors are obtained from literature sources or databases, for example, Ecoinvent (2010). Normally, data collected on-farm or representative information such as regional statistics are used as input data to operate whole-farm GHG models. For bovine production systems, several models have been developed to quantify GHG emissions from conventional pasture-based or confinement systems and organic systems (Beukes et al., 2010; Rotz et al., 2010; O'Brien et al., 2011). Results from previous livestock GHG models have been expressed per farm, per hectare of farmland and per unit of product, for example, per kg of fat and protein corrected milk (FPCM) or kg carcass weight (Thomassen et al., 2008; Clarke et al., 2013).

3 Life cycle assessment

Life cycle assessment considers the environmental effects of a product or service system (ISO, 2006a). The method also adopts a systems approach, but in contrast to systems analysis, the general step and principles of the LCA methodology are internationally standardized (ISO, 2006a,b). The International Organisation for Standardisation (ISO) originally developed standards for LCA in 1997, which were subsequently revised in 2006 (ISO 14040-14044). The main phases of LCA accordingly are goal and scope definition, life cycle inventory analysis, life cycle impact assessment and life cycle interpretation.

3.1 Goal and scope definition

This stage requires clearly stating the aims and objectives of an LCA project and the intended audience (ISO, 2006a). The goal definition determines what level of detail and what accuracy is needed for prediction/analysis. The scope of an LCA study should clearly describe the system under study and define the boundaries of the studied system (ISO, 2006a). Typically, the system boundaries of livestock LCA studies are defined to assess GHG emissions from all processes up until the point the primary product is sold from the farm (Beauchemin et al., 2011). This is commonly referred to as a 'cradle to farm gate' LCA. Some studies have also analysed further production stages, for instance, the processing stage and distribution to the retailer (Berlin, 2002; Hessle et al., 2017). The main environmental impacts evaluated in previous LCA studies of livestock are GHG emissions (global warming potential), acidification potential, eutrophication potential, land use and energy use (de Vries and de Boer, 2010; de Vries et al., 2015).

3.2 Life cycle inventory analysis

The second phase of LCA involves the compilation of inputs, outputs and emissions for a given product system throughout its lifecycle (ISO, 2006b). The aim of this stage of LCA modelling is to develop a model that quantifies the different resources used and the amount of waste and emissions generated per functional unit (Rebitzer et al., 2004). Resources used on-farm are normally collected directly or computed using relevant data sources. Emissions from on-farm processes are mainly estimated by applying emission factors from the literature or the IPCC (IPCC, 1997, 2006). For most LCA studies of livestock systems, international databases, for example, Ecoinvent (2010) or literature sources are used to estimate the resources used and emissions generated from processes that are indirectly related to the production system of interest, for example, data on fuel and fertilizer production.

3.3 Life cycle impact assessment

The inventory analysis phase lists the various substances used and pollutants emitted from a livestock production system (Thomassen et al., 2008). These results are generally difficult to interpret. Thus, a further stage known as life cycle impact assessment is needed to complete LCA of livestock systems (ISO, 2006a,b). This phase aggregates resources and emissions from the inventory analysis phase and computes (characterization) various potential environmental effects (Guinee et al., 2002). Environmental impacts are computed by converting the results of the inventory analysis stage using relevant characterization or equivalency factors. For instance, the global warming potential metric in CO_2-equivalents is applied during this stage to assess the climate impact of methane and other GHG emissions (Basset-Mens et al., 2009). The life cycle impact assessment stage allows the environmental effects of a livestock system to be assessed in a more interpretable way.

4 Modelling applications

Recent LCA and whole-farm GHG models from cool or temperate livestock regions were assessed. Table 1 provides a description of the modelling methods and emission factors used for analysis in 11 dairy studies, 9 beef studies, 3 sheep studies and 2 pig studies. In many situations, models calculated GHG emissions according to the approaches reported in national GHG inventories and the IPCC guidelines. A few studies used alternative equations for these sources such as Capper et al. (2009) or measured emissions directly as part of an on-farm research trial, for example, Doreau et al. (2011). Whole farm or LCA models were used for the following purposes:

- Quantify the environmental performance of farm systems and the effect of mitigation strategies on GHG emissions;
- Estimate the environmental sustainability of commercial farmers, for example, Bord Bia sustainable livestock assurance schemes; and
- Investigate the effect modelling decisions have on GHG emissions through sensitivity analysis.

4.1 Farm systems considered

Studies modelling GHG emissions from livestock farms have compared a variety of different systems including organic and conventional production systems, extensive and intensive systems, and confinement and grazing systems. These comparisons were generally representative of farms for a particular region. Many modelling studies that compare contrasting production systems aimed to assess the effect of intensification, defined as increased use of inputs per ha (e.g. fertilizer) have on GHG emissions. The results of these studies highlight that the effect of intensification on GHG emissions varies depending on the unit of expression. For instance, when dairy-farm GHG emissions are quantified per

Table 1 Summary of studies modelling greenhouse gas (GHG) emissions from livestock systems since 2009

Livestock species	Study	Farm description	Study goal(s) and methodology	GHG emission factors
Beef	Beauchemin et al. (2011)	Average farm in Western Canada	Life cycle assessment - multi-year model that assessed strategies to mitigate GHG emissions intensity from suckler beef	IPCC (2006)
Beef	Capper (2011)	Average US farm	Whole farm model - compared 1977 beef system GHG emissions to modern conventional systems	IPCC (2006) and literature sources
Beef	Clarke et al. (2013)	Irish research farm	Life cycle assessment - combined LCA model with a suckler beef bio-economic model and examined emissions from farms differing in stocking rate and finishing system	Ireland national GHG inventory method
Beef	Doreau et al. (2011)	French research farm	Life cycle assessment - completed an LCA of finishing or feedlot beef bull systems and estimated the effect different diets have on their GHG emission	Enteric methane emissions were measured. IPCC (2006) was used for manure
Beef	Hessle et al. (2017)	Average Swedish farm in Vastra Gotaland region	Life cycle assessment - assessed the effect four environmental improvement scenarios have on GHG emissions, nutrient loss and energy use from beef and milk	Swedish national GHG inventory method and IPCC (2006)
Beef	Murphy et al. (2017)	Irish research farm	Whole farm model - coupled a beef GHG model with a dairy beef bio-economic model to quantify effect of diet and slaughter age on GHG emissions	Ireland national GHG inventory method and literature sources
Beef	Ridoutt et al. (2013)	Commercial Australian farms	Life cycle assessment - assessed carbon, water and land-use footprints of contrasting cow-calf beef systems	Australian national GHG inventory method
Beef	Stackhouse et al. (2012)	Industry-simulated Californian farm	Life cycle assessment - quantified effect a growth promoter has on GHG emission from suckler beef	

Beef	Veysset et al. (2014)	Commercial French farms	Life cycle assessment - conducted a technical-economic survey on 53 suckler beef farms and quantified GHG emissions and fossil fuel use	French literature sources
Dairy	Bell et al. (2011)	UK research farm	Life cycle assessment - a Markov chain was used with LCA to compare GHG emissions from cows offered different levels of forage	IPCC (1997) and IPCC (2006)
Dairy	Beukes et al. (2010)	Average New Zealand farm in Waikato region	Whole farm model - integrated three models to investigate the effect different management scenarios have on dairy farming systems GHG emissions.	New Zealand national GHG inventory method
Dairy	Capper et al. (2009)	Average US farm	Whole farm model - compared GHG emissions from mid-1940 dairy system and modern conventional system	IPCC (2006) and literature sources
Dairy	Chobtang et al. (2016)	Commercial New Zealand farms in Waikato region	Life cycle assessment - evaluated the cradle-farm gate environmental performance of 53 commercial farms using dairy-base dataset.	New Zealand national GHG inventory method
Dairy	Del Prado et al. (2013)	Commercial farms in northern Spain	Life cycle assessment - coupled LCA and nutrient cycling models to estimate 17 dairy farms GHG emissions and nutrient-use efficiency	IPCC (2006) and literature sources
Dairy	Dolle et al. (2009)	Average farms for four French regions	Life cycle assessment - evaluated the effect different allocation methods have on dairy farms GHG emission intensity	Literature sources
Dairy	Mc Geough et al. (2012)	Average farm in Eastern Canada	Life cycle assessment - multi-year model that assessed the effect different allocation methods have on the GHG emissions intensity of raw milk	IPCC (2006)
Dairy	O'Brien et al. (2011)	Irish research farm	Whole farm model - combined a bio-economic dairy model with a GHG model to examine the effect genetic potential and grazing system has on dairy GHG emissions	Ireland national GHG inventory method

(Continued)

Table 1 (*Continued*)

Livestock species	Study	Farm description	Study goal(s) and methodology	GHG emission factors
Dairy	O'Brien et al. (2015)	Commercial Irish farms	Life cycle assessment - related the GHG emission intensity of 221 dairy farms to economic performance using the Teagasc national farm survey	Ireland national GHG inventory method
Dairy	van Middelaar et al. (2013)	Average Dutch farm	Life cycle assessment - coupled LCA with an economic model to assess the effect different feed strategies have on GHG emissions	Dutch national GHG inventory method
Dairy	Zehetmeier et al. (2012)	Average German farm	Whole farm model - assessed the effect increasing dairy cow milk yield has on GHG emissions from the dairy and beef industries	German national GHG inventory method
Pig	McAuliffe et al. (2017)	Commercial Irish farms	Life cycle assessment - compared GHG emissions and environmental metrics of average and top-performing pig farms	Ireland national GHG inventory method
Pig	Reckmann et al. (2013)	Average farm in northern Germany	Life cycle assessment - quantified GHG emissions and environmental metrics for a typical pig farm	IPCC (2006) and literature sources
Sheep	Jones et al. (2014)	Commercial sheep farms	Whole farm model- analysed GHG emissions from 64 sheep farms and identified areas for improvement	IPCC (2006) and literature sources
Sheep	O'Brien et al. (2016)	Average and commercial sheep farm	Life cycle assessment - evaluated the effect of intensification on GHG emissions and resource use of sheep farms	IPCC (2006) and literature sources
Sheep	Wiedemann et al. (2015)	Average Australian, UK, and New Zealand sheep farms	Life cycle assessment - evaluated the effect allocation methods have on GHG emissions from typical sheep farms	Literature sources

hectare of land, whole-farm models usually show that reducing the intensity of dairy systems reduces GHG emissions (Beukes et al., 2010; O'Brien et al., 2011). However, when GHG emissions are assessed on per unit of product basis (GHG emission intensity), intensification usually reduces GHG emissions (Capper et al., 2009; O'Brien et al., 2011). Given the rising demand for livestock products such as milk, this implies that GHG emissions should not be assessed in isolation from milk or meat production. The impact of livestock systems on emissions should also be contextualized with other production systems and countries.

A potential option for reducing GHG emissions from livestock, according to (Capper et al. 2009; Capper, 2011), is sustainable intensification. Capper et al. (2009) modelled GHG emissions from US beef and dairy systems and demonstrated increasing livestock output reduces GHG emissions through improved productive efficiency, defined as 'units of milk or meat produced per unit resource inputs'. Improving productive efficiency facilitates the dilution of maintenance effect, whereby the total resource cost per unit of product is reduced (Bauman et al., 1985). The land the strategy spares can potentially further reduce emissions by sequestering carbon in soil or woody biomass. The effect of sustainable intensification and livestock productivity on GHG emissions is not accounted for in national inventories that use the basic IPCC tier 1 approach but can be partially captured by inventories that use higher tiers. Many inventories could better reflect the influence livestock efficiency has on GHG emission intensity by updating important nutritional parameters. Further improvements may be possible to make once appropriate emission factors are established to determine the benefits of methane-reducing feed additives. Including additives or supplements within the inventory calculations as well as having the activity data captured could cut methane emissions from enteric fermentation of feed by 15–30% (Hristov et al., 2013).

Intensification may have undesirable effects on emissions as well and can lead to declines in soil organic matter or carbon levels (Zehetmeier et al., 2012; van Middelaar et al., 2013). For instance, van Middelaar et al. (2013) reported that converting grassland to arable land to support higher stocking rates improved dairy-farm production levels, but also dramatically increased soil carbon losses. Consequently, intensification nearly doubled GHG emissions per unit of product. A production system in a steady state, in terms of on-farm land use, is less likely to cause a land-use change effect, unless the origin of feed ingredients contained in compound concentrate feeds brought into the system change. O'Brien et al. (2016) noted that GHG emissions from livestock systems were sensitive to changes in these ingredients and showed including carbon sequestration in grassland resulted in extensive hill sheep farms emitting less GHG emissions per unit of output than intensive lowland farms. The inconsistent effect of intensification on livestock farms GHG emissions indicates that emissions and carbon sequestration should be assessed together. Within this context, the short-term effect of, for example, CO_2 versus the long-term effects captured through the GWP coefficients need to be taken into account in any computations.

Beef modelling studies report that GHG emission intensity can be mitigated by rearing beef from the dairy herd instead of the suckler herd. Switching to a dairy calf to beef production system eliminates suckler cows from the beef herd, which significantly reduces the GHG emission intensity of live weight or carcass weight as 80–90% of dairy cows' emissions are allocated to milk (Zehetmeier et al., 2012). The reduction is lower in terms of live weight because the terminal traits of surplus dairy cattle are inferior to suckler cattle. Some farmers generally select sires that are easy calving and have shorter gestation, which has a negative influence on carcass weight production. This may be possible to change by using new animal breeding technologies, for example, genomic selection.

5 National greenhouse gas inventory

Livestock methane emission factors reported in national GHG inventory submissions to the United Framework Convention on Climate Change (UNFCCC) were reviewed for annex 1 (developed) countries and 5 non-annex 1 countries, that is, Brazil, China, India, South Africa and Uruguay. The review was carried out on 2017 annex 1 national inventory submissions, which estimated GHG emissions for the period 1990–2015 (UNFCCC, 2017a). Unlike annex 1 countries, non-annex 1 nations do not report emissions annually. The latest reports available for these nations were 2–3 years older and estimated GHG emissions generated prior to 2013 (UNFCCC, 2017b). The appraisal of annex 1 and non-annex 1 GHG inventories considered the methods that nations use to estimate emission factors for enteric fermentation in livestock and livestock manure. The findings of this international methodological evaluation were summarized for livestock categories using the IPCC tier(s) (tier 1, tier 2 or 3) that each nation uses for emission calculations. The categories of livestock evaluated were: dairy and non-dairy cattle (bovine), sheep, pig, poultry and other livestock (e.g. rabbits, horses, mules). The results of the review for each category were compared to pertinent Irish livestock methane emission factors.

5.1 Bovine

Differences in dairy cow methane emission factors were related to the calculation and reporting methodology used, as well as cow productivity. For example, Ireland's tier 2 methane emission factors for bovines were generally lower than nations with heavier and higher-yielding cows such as the United States and higher than nations with lighter and lower-yielding cows (Table 2).

Most countries used tier 2 enteric methane emission factors for dairy and non-dairy cattle (Table 3). In official inventory terminology, anything more detailed than a tier 2 methodology is termed tier 3. However, it is not uncommon for people to use the term tier 2+ to describe situations where the tier 2 equations are used, but country-specific parameters are used. The tier 3 model can be an empirical or mechanistic model describing the fermentative and digestive processes in the gastrointestinal tract of dairy cattle. Tier 3 models operate with greater levels of activity data allowing regional, system and animal-based simulations. Tier 1 emission factors were not applied to compute dairy cow's enteric methane emission and were only used by Cyprus and the United Kingdom (UK) for mature beef cattle and younger stock (together known as non-dairy). Tier 3 emission factors were used by Ukraine and France (not implemented yet) to quantify enteric methane emissions from dairy cows and non-dairy cattle. Tier 3 emission factors were not used for estimating methane emission from dairy cow manure (exception being the Netherlands having a model capable of simulating methane, ammonia and manure methane at a tier 3 level) and rarely used for this source for non-dairy cattle.

Generally, nations that reported a higher tier method to estimate bovine enteric methane emissions used more data-intensive and detailed emission algorithms. Tier 3 equations were normally derived from published national research projects that measured and/or modelled methane. For example, a French project derived tier 2 and 3 methane emission factors for several categories of dairy cows and other cattle, considered as representative of the nation's breeding situations. Each category was associated with a breed, an average mass, a milk yield if necessary, as well as energy needs.

The Swiss tier 2 approach estimates cow feed and gross energy requirements using recommended national feeding standards that are widely used by Swiss farmers because they are a basis for their basic support payment. The Swiss Ym, methane conversion rate expressed as a fraction (i.e. the fractional loss of GEI as combustible CH_4), depends on the diet. The Ym value comes from national projects that measured methane from dairy cows in open and closed calorimeter chambers. The Swiss and French methods to estimate enteric methane emission factors for dairy cows aligned better with inventories like the Irish one rather than the complex tier 3 approaches applied by the Netherlands and Germany. Briefly, the Netherlands used Bannink et al. (2011) mechanistic, dynamic model of the rumen fermentation process to estimate methane from enteric fermentation in dairy cows. The inputs required to operate the Dutch model were feed intakes, the chemical composition of feed and degradation characteristic of the constituents of feed (e.g. crude protein). Cow feed intakes were estimated according to national feeding standards, and nutritional data was provided by a widely used Dutch agricultural laboratory.

Table 2 Live weight, milk yield, energy requirement and methane emission factors from selected annex 1 parties to the United Nations Framework Convention on Climate Change reported for dairy cows in production year 2015 (UNFCCC, 2017a)

	Ireland	United Kingdom	USA	New Zealand
Average live weight, kg	535	608	680	448
Milk yield, kg/cow per year	5458	7705	10268	4362[a]
Gross energy required, MJ/d	261	300	IE[b]	IE
Methane from enteric fermentation, kg/year	117.2	130.0	146.0	84.3[c]
Methane from manure, kg/year	11.2	17.4	74.0	5.8[c]

[a] Obtained from dairy NZ (2016) https://www.dairynz.co.nz/media/5788611/quickstats_new_zealand_web_2017.pdf.
[b] Not reported.
[c] Includes dairy heifers.

Detailed nutrition data was used in the German approach to estimate dairy cow enteric methane as well. Their method, described by Rosemann et al. (2017), accounted for the effects of feed composition and feed properties using a German model developed by Kirchgessner et al. (1994). This model, like the Dutch model, was data-intensive and required information that is unlikely to be feasible to collect in the short term at a national level in some other countries that practice extensive grazing and/or where detailed quality information is not available most of the time.

Bovine feed intakes and OM digestibility estimates were required to estimate methane emissions from manure in most national inventories reviewed. The tier 3 method was only used to estimate methane from the manure of Australian beef cattle fed on feedlots. The calculation used measured methane conversion factors from Redding et al. (2015) for Australian manure storage systems. Australia was not the only nation that used country-specific methane conversion factors for manure management. Other examples included New Zealand, Denmark and Austria. A few national inventories such as the UK reported the proportion of cattle slurry systems that form a natural crust cover. In reality, it can be rather arbitrary whether a country declares themselves as operating at a tier 2 or tier 3 level within their inventories.

Several countries recognize that methane emissions from bovine and livestock manure are linked to other GHG emissions from this source, for example, nitrous oxide. For consistency, some countries use a comprehensive model that simultaneously quantified GHG and ammonia emissions from livestock. This approach was recommended in the EU 2017 submission to the UNFCCC for its member states but is only applied by a few nations, for example, the German GAS-EM model, Denmark's IDA model and Ireland.

Table 3 Method[a] applied to quantify bovine methane emission factors for annex 1 and non-annex 1 nations

			Dairy		Non-dairy	
Annex	Continent	Nation	Enteric	Manure	Enteric	manure
Annex 1	North America	Canada	T2	T2	T2	T2
		USA	T2	T2	T2	T2
Annex 1	Asia	Japan	T2	T2	T2	T2
		Kazakhstan	T2	T2	T2	T2
		Turkey	T2	T1	T2	T1
Annex 1	Asia/Europe	Russia	T2	T2	T2	T2
Annex 1	Europe	Austria	T2	T2	T2	T2
		Belarus	T2	T2	T2	T2
		Belgium	T2	T2	T2	T2
		Bulgaria	T2	T2	T2	T2
		Croatia	T2	T2	T2	T2
		Cyprus	T2	T2	T1	T2
		Czech Republic	T2	T2	T2	T2
		Denmark	T2	T2	T2	T2
		Estonia	T2	T2	T2	T2
		Finland	T2	T2	T2	T2
		France	T2, T3	T2	T2, T3	T2
		Germany	T3	T2	T2	T2
		Greece	T2	T2	T2	T2
		Hungary	T2	T2	T2	T2
		Iceland	T2	T2	T2	T2

(Continued)

Table 3 (*Continued*)

Annex	Continent	Nation	Dairy		Non-dairy	
			Enteric	Manure	Enteric	manure
		Italy	T2	T2	T2	T2
		Latvia	T2	T2	T2	T2
		Liechtenstein	T2	T2	T2	T2
		Lithuania	T2	T2	T2	T2
		Luxembourg	T2	T2	T2	T2
		Malta	T2	T2	T2	T2
		Monaco	NO	NO	NO	NO
		The Netherlands	T3	T2	T2	T2
		Norway	T2	T2	T2	T2
		Poland	T2	T2	T2	T2
		Portugal	T2	T2	T2	T2
		Romania	T2	T2	T2	T2
		Slovakia	T2	T2	T2	T2
		Slovenia	T2	T2	T2	T2
		Spain	T2	T2	T2	T2
		Sweden	T2	T2	T2	T2
		Switzerland	T3	T2	T2	T2
		Ukraine	T3	T2	T3	T2
		United Kingdom	T2	T2	T1, T2	T2

Annex 1	Oceania	Australia	T2	T2	T2	T2, T3
		New Zealand	T2	T2	T2	T2
Non-Annex 1	South America	Brazil	T2	T2	T2	T2
		Uruguay	T2	T2	T2	T2
	Africa	South Africa	T2	T2	T2	T2
	Asia	China	T2	T2	T2	T2
		India	T2	T1	T2	T1

[a] T1 = Tier 1, T2 = Tier 2, T3= Tier 3, NO = Not occurring, NE = Not estimated.

5.2 Sheep

Methane emissions from sheep are generally estimated by national inventories using tier 2 or tier 1 emissions factors (Table 4). The tier 2 method was used more often than the tier 1 approach. Generally, countries that applied the tier 2 method used the IPCC (2006) equations to estimate sheep feed requirements, manure excretion and methane emissions. France used a more advanced method to calculate enteric methane emission factors for sheep and New Zealand reported applying a country-specific tier 2 method. The French method for sheep is similar to the approach described earlier for French bovines, except for the calculation of methane from enteric fermentation. The New Zealand approach to calculate emissions from enteric fermentation and manure management in sheep was similar to that used for bovines. The tier 2 emission factors for both species were developed by Clark et al. (2003) and are regularly improved using new experimental research. The New Zealand computations for sheep enteric methane are carried out on a monthly basis and use country-specific data for sheep populations, pasture quality and productivity (e.g. milk yield and live weight). This data is generally available only at a national scale for sheep and beef cattle.

5.3 Pigs

The IPCC tier 1, enteric methane emission factors for pigs were used by 30 of the 47 nations reviewed (Table 4). The remaining countries used a tier 2 method for this source, except France which used tier 3. Tier 2 emission factors from the IPCC (2006) were primarily used to estimate methane from pig manure. Only 10 nations used a tier 1 method for this source. Most nations that used tier 2 emission factors entail computing pig's gross energy and protein intakes to meet feed requirements and estimating manure excretion from feed intake and diet digestibility. Methane emissions from manure are computed using survey data on manure storage systems and relevant IPCC conversion factors. This tier 2 method is slightly more advanced in some nations because manure excretion is based on national data instead of default excretion estimates provided by the IPCC.

The tier 2 and tier 3 methods that nations used to estimate enteric methane from pigs do not differ to that described for bovines and sheep. The Ym for pigs was normally derived from the references of the IPCC (2006) guidelines. Country-specific Ym values were also developed by a few nations, for example, Germany, but these estimates were similar to the IPCC estimate differing by <0.1 units.

5.4 Poultry

Nations are not required to estimate enteric methane emissions from poultry. This category is considered by the IPCC to emit negligible emissions. Thus, there are currently no tier 1 emission factors for this source. Nevertheless, 12 nations estimated enteric methane emissions from poultry and 3 reported using tier 2 emission factors (Table 5), because the method was developed from national research in their countries. The emission factors from such research, for example, Wang and Huang (2005) could potentially be applied to estimate enteric methane from poultry in other countries. A slightly higher number of nations estimated methane from poultry manure using tier 2 emission factors instead of tier 1. Australia was the only nation that reported a tier 3 method. The quantity of manure excreted by poultry was estimated using national data or IPCC default values.

5.5 Other livestock

The other livestock category enteric fermentation methane emissions were usually calculated by nations using tier 1 emission factors (Table 5). Nations seldom applied the tier 2 method to estimate enteric methane emissions from all species that are part of this livestock category. Most nations only used tier 2 emission factors to estimate methane emissions from a few economically important production species within the other livestock category, for example, for goat and deer. In general, the tier 2 approach in computing enteric methane emissions from other livestock species was very similar to the IPCC method for cattle and sheep, apart from the Ym parameter, which was not necessarily species-specific. For example, the South African Ym parameter for goats was based on sheep measurements. This approach to estimate Ym is generally acceptable as long as species have similar digestive systems. This approximation method is also recommended by the IPCC for other livestock species not listed in their guidelines and can be applied based on live weight for tier 1 emission factors.

Almost half of the nations reviewed used a tier 2 method to estimate methane from the manure of a species contained within the other livestock category. Of these nations, 13 used tier 2 methane emission factors for manure for all species within this category. The IPCC tier 2 equations and default manure excretion rates were generally used to estimate methane emissions from other livestock species manure.

5.6 Methodology conclusions

Livestock methane emissions factors can be computed more accurately using a tier 2 method instead of a default tier 1 approach because there is country-specific information available around methane emissions for the major production species. This approach is widely used by annex 1 nations to estimate methane from key livestock categories, for example, bovine, pigs and sheep. In addition, several annex 1 nations use this method for all species and integrate

Table 4 Method applied to quantify sheep and pigs methane emission factors for annex 1 and non-annex 1 nations

			Sheep		Pigs	
Annex	Continent	Nation	Enteric	Manure	Enteric	Manure
Annex 1	North America	Canada	T1	T2	T1	T2
		USA	T1	T2	T1	T2
Annex 1	Asia	Japan	T1	T1	T2	T2
		Kazakhstan	T2	T1	T1	T1
		Turkey	T1	T1	T1	T1
Annex 1	Asia/Europe	Russia	T1	T1	T2	T2
Annex 1	Europe	Austria	T1	T1	T1	T2
		Belarus	T1	T1	T1	T2
		Belgium	T1	T1	T1	T2
		Bulgaria	T2	T2	T1	T2
		Croatia	T2	T2	T2	T2
		Cyprus	T1	T1	T1	T1
		Czech Republic	T1	T1	T1	T1
		Denmark	T2	T2	T2	T2
		Estonia	T1	T1	T2	T2
		Finland	T2	T2	T2	T2
		France	T2, T3	T2	T2, T3	T2
		Germany	T1	T2	T2	T2
		Greece	T2	T2	T1	T1
		Hungary	T1	T2	T1	T2
		Iceland	T2	T2	T1	T1
		Italy	T2	T1	T1	T2

		Latvia	T1	T1	T1	T2
		Liechtenstein	T2	T2	T2	T2
		Lithuania	T2	T2	T2	T2
		Luxembourg	T2	T2	T1	T2
		Malta	T2	T1	T1	T2
		Monaco	NO	NO	NO	NO
		The Netherlands	T1	T1	T1	T2
		Norway	T2	T1	T1	T2
		Poland	T1	T1	T1	T1
		Portugal	T2	T2	T2	T2
		Romania	T2	T2	T2	T2
		Slovakia	T2	T2	T1	T1
		Slovenia	T1	T2	T1	T2
		Spain	T2	T1	T2	T2
		Sweden	T1	T1	T1	T2
		Switzerland	T2	T2	T2	T2
		Ukraine	T2	T2	T1	T2
		United Kingdom	T1	T2	T1	T2
Annex 1	Oceania	Australia	T2	T2	T2	T3
		New Zealand	T2	T2	T1	T2
Non-Annex 1	South America	Brazil	T1	T1	T1	T2
		Uruguay	T1	T1	T1	T1
	Africa	South Africa	T2	T2	T2	T2
	Asia	China	T2	T2	T1	T2
		India	T1	T1	T1	T1

Note: T1 = Tier 1, T2 = Tier 2, T3 = Tier 3, NO = Not occurring, NE = Not estimated.

with other methods to estimate air pollutants, for example, ammonia. The tier 2 method that nations applied were generally similar (i.e. emission factor and an estimate of intake to drive percentage losses), but the equations and parameters that nations used to calculate important inputs required to derive methane emissions, such as feed intakes, Ym and manure excretion were often based on national research. These country-specific equations are usually more accurate than the IPCC alternatives and should be used by the inventory where possible, for example, sheep. There may also be potential to develop more advanced methane emission factors for bovine and sheep by using national measurements of enteric methane emissions from these species.

6 Mitigation strategies

Whole-farm GHG models have assessed the effect of several farm practices on emissions. This section describes what can be captured, how they can be captured within models. Examples include selecting higher-yielding livestock, improving fecundity, reducing replacement rates, varying the grazing period, treatment of managed manure and improving nutrient management. The effect of breeding higher-yielding livestock on GHG emissions has varied between studies. Murphy et al. (2017) showed that a 13% increase in beef cattle live weight gain per day reduced GHG emission per unit of carcass weight by 19%. Similar results were reported by Bell et al. (2011) for milk production. In both studies, the reduction was explained by a decrease in methane emission due to improvements in feed conversion efficiency (kg of milk or meat/kg of feed) and decreasing maintenance costs. In contrast, O'Brien et al. (2011) reported Holstein-Friesian cows with higher milk yields increased emissions per unit of the product relative to lower-yielding cows. This was primarily caused by higher-yielding Holstein-Friesian cows having lower herd fertility rates and a shorter lifespan, which increased replacement heifer requirements and thus GHG emissions (O'Brien et al., 2011). The study indicates that it is important to consider a combination of genetic traits to reduce GHG emission intensity.

Beukes et al. (2010) modelled the effect of reducing herd replacement rates on GHG emissions from dairy systems. The analysis showed that a 10% reduction in replacement rate reduced GHG emission per unit of a product by an average of 14%. This decrease in GHG emissions occurred due to a decline in the number of non-productive animals. Bovine fertility is possible to improve through genetic selection and better reproductive management, that is, heat detection aids (Vellinga et al., 2011). Improving reproductive management can also reduce emissions from sheep farms. For instance, Ledgard et al. (2011) reported higher lambing percentages were partly responsible for a 22% reduction in GHG emission from New Zealand sheep production. Increased animal yields and farm profitability were the main drivers of this change.

The influence of the length of the grazing period on GHG emissions was evaluated by Schils et al. (2005) and O'Brien et al. (2015). Schils et al. (2005) investigated the effect of reducing the grazing period and reported that the strategy decreases nitrous oxide emissions from animal excreta deposited on pasture, but increases methane and nitrous oxide emissions from manure storage to a similar extent. Thus, the strategy had no net effect on GHG emissions. Furthermore, Schils et al. (2005) reported that the strategy has undesirable effects on other types of pollution, for example, ammonia. Thus, this demonstrates that models should also consider the undesirable secondary effects (pollution swapping) of GHG mitigation strategies (del Prado et al., 2013). The effect of extending the grazing period was examined by O'Brien et al. (2015). The study showed the strategy caused a greater reduction in methane and nitrous oxide emissions from manure storage than an increase in nitrous oxide emissions from animal excreta deposited on pasture. In addition, the strategy reduced dairy systems' GHG emission intensity by 0.17% per day and improved economic performance.

The potential for feed additives and growth promoters to mitigate methane emissions were assessed by Beauchemin et al. (2011) and Stackhouse et al. (2012). Beauchemin et al. (2011) reported that dietary oils mitigated the GHG emission intensity of western Canadian beef systems by 2–8%. However, Hristov et al. (2013) reported that this strategy was economically risky and highlighted that the long-term mitigation efficacy of oils and fats was uncertain. It also may be difficult to record this mitigation potential in national GHG emission inventories. Stackhouse et al. (2012) showed that antibiotic growth promoters reduced GHG emission intensity by 5–9% for California beef cattle, but these growth implants are proscribed in the EU. Few models directly assess strategies to mitigate methane from manure storage. McAuliffe et al. (2017) highlighted anaerobic digestion as a promising technology to mitigate the environmental impact of pig manure. The strategy can also capture bioenergy and produce electricity but requires more favourable tariffs.

6.1 Commercial farm

The environmental sustainability of commercial livestock farms often has been estimated in developed nations using whole farm or LCA models. O'Brien et al. (2015) estimated the milk carbon footprints of 221 Irish dairy farms using a life cycle approach and evaluated economic performance. The research showed a wide range in Irish dairy farms carbon footprints (0.50–1.97 kg of CO_2-equivalent/kg of fat and protein corrected milk). Generally, dairy farms with lower milk footprints were more profitable. For example, the milk carbon footprint of the top-third of farms in terms of net margin per hectare was 15% less than the bottom-third. Partial least square regression indicated that

Table 5 Method applied to quantify poultry and other livestock[b] methane emission factors for annex 1 and non-annex 1 nations

			Poultry		Other livestock	
Annex	Continent	Nation	Enteric	Manure	Enteric	manure
Annex 1	North America	Canada	NE	T2	T1	T1
		USA	NE	T2	T1	T2
Annex 1	Asia	Japan	NE	T2	T1	T1
		Kazakhstan	NE	T1	T1	T1
		Turkey	NE	T1	T1	T1
Annex 1	Asia/Europe	Russia	NE	T1	T1	T1
Annex 1	Europe	Austria	T2	T1	T1	T1
		Belarus	T1	T1	T1, T2	T1,T2
		Belgium	T1	T1	T1	T1
		Bulgaria	NA	T2	T1	T1
		Croatia	T1	T2	T1	T2
		Cyprus	NE	T1	T1	T1
		Czech Republic	NE	T1	T1	T1
		Denmark	T1	T2	T1	T2
		Estonia	NE	T1	T1	T1
		Finland	NE	T2	T1	T2
		France	NE	T2	T2	T2
		Germany	NE	T2	T1	T2
		Greece	T1	T1	T1	T1
		Hungary	T1	T2	T1	T2
		Iceland	T1	T1	T1	T1
		Italy	NE	T1	T1, T2	T1
		Latvia	NE	T1	T1	T1

		Liechtenstein	T2	T2	T2	T2
		Lithuania	NE	T1	T1	T1
		Luxembourg	NO	T2	T1	T2
		Malta	T1	T2	T1	T1
		Monaco	NO	NO	NO	NO
		The Netherlands	NE	T2	T1	T1
		Norway	T1	T2	T1, T2	T1
		Poland	NE	T1	T1	T1
		Portugal	NE	T2	T1	T1, T2
		Romania	NE	T2	T2	T2
		Slovakia	NE	T1	T1	T1
		Slovenia	NE	T2	T1	T2
		Spain	NE	T2	T1, T2	T1, T2
		Sweden	NE	T1	T1	T1
		Switzerland	T2	T2	T2	T2
		Ukraine	NE	T2	T1	T1, T2
		United Kingdom	NE	T2	T1	T1, T2
Annex 1	Oceania	Australia	NE	T3	T1	T2
		New Zealand	NE	T1	T1, T2	T1, T2
Non-Annex 1	South America	Brazil	NE	T1	T1	T1
		Uruguay	NE	T1	T1	T1
	Africa	South Africa	NE	T2	T1, T2	T1, T2
	Asia	China	NE	T1	T1, T2	T1, T2
		India	NE	T1	T1	T1

Note: T1 = Tier 1, T2 = Tier 2, T3= Tier 3, NO = Not occurring, NE = Not estimated.
Note: Other livestock includes goats, horses, mules, donkeys, camels, rabbits, deer, llamas, ostriches, reindeer and alpacas. Bovines, sheep, pigs and poultry were not part of this category.

farm practices which increase milk solids yield per hectare and the utilization of grazed grass mitigated GHG emission intensity and improved economic performance. The average methane emission factor for dairy cows (110 kg methane/cow) fell within the range of recent Irish inventory estimates.

Chobtang et al. (2016) and Del Prado et al. (2013) carried out similar research as O'Brien et al. (2015) for New Zealand and Spanish dairy farms. Del Prado et al. (2013) reported similar levels of variability among farms' carbon footprints as O'Brien et al. (2015), but Chobtang et al. (2016) found less variability for New Zealand dairy carbon footprints. This may be due to farms operating at a similar level of efficiency, but there was significant farm performance variability. Chobtang et al. (2016) also assessed other environmental measures such as acidification and health measures such as effects on human health. The variability across farms for these measures was generally much higher than the carbon footprint. Off-farm impacts were important for several environmental and health measures assessed and could be improved by increasing forage utilization. Veysset et al. (2014) also recommended the latter to mitigate GHG emissions from commercial French suckler beef farms. The authors highlighted that farms which produced the best forage and heaviest cattle mitigated GHG emission intensity by 50% compared to the least productive in the group. The least productive French farms were larger in the area and operated mixed crop-livestock systems.

Jones et al. (2014) estimated lamb carbon footprints for lowland, upland and hill sheep farms in the UK and reported lowland farms had the smallest mean footprint. There was substantial variability across lamb carbon footprints, and this was largely driven by animal and grassland productivity. The authors recommended a suite of management practices to reduce the carbon footprint of lamb, including improving lambing rates (lambs/ewe), increasing lamb growth rate and optimizing mating rates and concentrate usage. These improvements are likely to be problematic for hill farms where local conditions are much more challenging. Nevertheless, some improvement should be possible as Jones et al. (2014) reported top-performing hill farms outperformed the mean performance of lowland farms.

6.2 Modelling decisions

Quantifying GHG emissions from livestock systems involves making important modelling decisions such as the definition of system boundaries and allocation of emissions to co-products. O'Brien et al. (2011) examined the influence that different system boundaries have on GHG emissions from dairy farms varying in cow genotype and concentrate supplementation. This analysis showed that high-concentrate dairy systems reduced on-farm GHG emissions per kg of milk solids, but the opposite occurred for the New Zealand strain of Holstein Friesian when the system boundary was expanded to include off-farm GHG emissions. Further research by O'Brien et al. (2012) demonstrated similar results for grass-based and confinement dairy systems and highlighted that reducing national GHG emissions to comply with EU targets is likely to increase GHG emissions from dairy production through carbon leakage. This research demonstrates that livestock systems GHG emissions should be assessed using a complete approach like LCA to ensure changes to dairy systems reduce GHG emissions. O'Brien et al. (2011) recommended integrating the LCA method into the Irish inventory framework. This may be possible using datasets like the Teagasc national farm survey.

Various criteria or methods were evaluated for allocating GHG emissions to the products of sheep and dairy systems. Wiedemann et al. (2015) compared seven different criteria to allocate GHG emissions between sheep products, that is, live weight and greasy wool. The case study results were relatively similar when GHG emissions were expressed per kg of total products but varied widely when emissions were split between products based on their economic value. The study recommended allocating GHG emissions between sheep products according to their protein requirements or using the mass of protein sold. Dolle et al. (2009) and Mc Geough et al. (2012) showed similar variability in dairy systems GHG emission intensity due to the allocation method. The authors identified advantages and disadvantages with several methods. Mc Geough et al. (2012) recommended basing the allocation decision on the needs and availability of data, which agrees with the conclusions of Rice et al. (2017) for Irish dairy systems. Rice et al. (2017) reported simple mass allocation was the best approach in terms of the pedigree of data but advised using a range of allocation methods to understand the uncertainty associated with the decision.

7 Conclusion

Most environmental modelling studies use a life cycle approach to quantify livestock's GHG emissions and overall environmental performance. Generally, modelling studies estimated livestock methane emissions according to the calculations provided in national inventory reports or the IPCC guidelines, which ensure the compliance of approaches between different entities modelling at a national level. These calculations vary widely from country to country due to local conditions and livestock research capabilities, but normally show improvements in animal productivity and switching from suckler to dairy beef systems mitigate GHG emission intensity. There is significant potential to use models at a national level to estimate GHG emissions from commercial bovine systems to generate national footprints. Moreover, there is scope to use LCA and national inventory models to benchmark performance across countries and over time. Some national sustainability initiatives, for example, Origin Green are already employing these tools to improve farmers and the wider community understanding of GHG emissions from livestock production. Farm-scale models could be used to pin point the direction in relation to emissions reduction strategies, but it is important that models use an LCA approach to ensure that the analysis completed is robust and not facilitating carbon leakage. Models that are used in these situations should be peer reviewed, validated and evaluated against detailed experimental data.

8 Future research

The key future research required in the area of methane centres on biogenic methanes impact on global warming. Given the fact that methane is a short-lived gas there is increasing questions around its impact on climate change. There is a requirement for the IPCC and LCA methodologies to be relooked at in this context. It would be expected that the over-riding conclusions from this research will be that the impact of ruminant agriculture on climate change maybe over-estimated. Further information can be found at Cain et al. (2019).

9 References

Bannink, A., van Schijndel, M. W. and Dijkstra, J. 2011. A model of enteric fermentation in dairy cows to estimate methane emission for the Dutch National Inventory Report using the IPCC Tier 3 approach. *Animal Feed Science and Technology* 166-167, 603-618.

Basset-Mens, C., Ledgard, S. and Boyes, M. 2009. Eco-efficiency of intensification scenarios for milk production in New Zealand. *Ecological Economics* 68(6), 1615-1625.

Bauman, D. E., McCutcheon, S. N., Steinhour, W. D., Eppard, P. J. and Sechen, S. J. 1985. Sources of variation and prospects for improvement of productive efficiency in the dairy cow: a review. *Journal of Animal Science* 60(2), 583-592.

Beauchemin, K. A., Janzen, H. H., Little, S. M., McAllister, T. A. and McGinn, S. M. 2011. Mitigation of greenhouse gas emissions from beef production in western Canada - evaluation using farm-based life cycle assessment. *Animal Feed Science and Technology* 166-167, 663-677.

Bell, M. J., Wall, E., Russell, G., Simm, G. and Stott, A. W. 2011. The effect of improving cow productivity, fertility, and longevity on the global warming potential of dairy systems. *Journal of Dairy Science* 94(7), 3662-3678.

Berlin, J. 2002. Environmental life cycle assessment (LCA) of Swedish semi-hard cheese. *International Dairy Journal* 12(11), 939-953.

Beukes, P. C., Gregorini, P., Romera, A. J., Levy, G. and Waghorn, G. C. 2010. Improving production efficiency as a strategy to mitigate greenhouse gas emissions on pastoral dairy farms in New Zealand. *Agriculture, Ecosystems and Environment* 136(3-4), 358-365.

Cain, M., Lynch, J., Allen, M. J., Fuglestvedt, J. S., Frame, D. J. and Macey, A. H. 2019. Improved calculation of warming-equivalent emissions for short-lived climate pollutants. *Climate and Atmospheric Science* 2, 29. DOI: https://doi.org/10.1038/s41612-019-0086-4

Capper, J. L. 2011. The environmental impact of beef production in the United States: 1977 compared with 2007. *Journal of Animal Science* 89(12), 4249–4261.

Capper, J. L., Cady, R. A. and Bauman, D. E. 2009. The environmental impact of dairy production: 1944 compared with 2007. *Journal of Animal Science* 87(6), 2160–2167.

Cerri, C. C., Maia, S. M. F., Galdos, M. V., Cerri, C. E. P., Feigl, B. J. and Bernoux, M. 2009. Brazilian greenhouse gas emissions: the importance of agriculture and livestock. *Scientia Agricola* 66(6), 831–843.

Chobtang, J., Ledgard, S. F., McLaren, S. J., Zonderland-Thomassen, M. and Donaghy, D. J. 2016. Appraisal of environmental profiles of pasture-based milk production: a case study of dairy farms in the Waikato region, New Zealand. *The International Journal of Life Cycle Assessment* 21(3), 311–25.

Clark, H., Brookes, I. and Walcroft, A. 2003. *Enteric Methane Emissions from New Zealand Ruminants 1990–2001 Calculated Using an IPCC tier 2 Approach*. Report prepared for the Ministry of Agriculture and Forestry. Wellington, New Zealand.

Clarke, A. M., Brennan, P. and Crosson, P. 2013. Life-cycle assessment of the intensity of production on the greenhouse gas emissions and economics of grass-based suckler beef production systems. *The Journal of Agricultural Science* 151(5), 714–726.

de Vries, M. and de Boer, I. J. M. 2010. Comparing environmental impacts for livestock products: a review of life cycle assessments. *Livestock Science* 128(1–3), 1–11.

de Vries, M., van Middelaar, C. E. and de Boer, I. J. M. 2015. Comparing environmental impacts of beef production systems: a review of life cycle assessments. *Livestock Science* 178, 279–288.

del Prado, A., Mas, K., Pardo, G. and Gallejones, P. 2013. Modelling the interactions between C and N farm balances and GHG emissions from confinement dairy farms in northern Spain. *Science of the Total Environment* 465, 156–165.

Dolle, J. B., Gac, A. and Le Gall, A. 2009. L'empreinte carbone du lait et de la viande bovine. *Recherches sur les Ruminants* 16, 233–6.

Doreau, M., van der Werf, H. M. G., Micol, D., Dubroeucq, H., Agabriel, J., Rochette, Y. and Martin, C. 2011. Enteric methane production and greenhouse gases balance of diets differing in concentrate in the fattening phase of a beef production system. *Journal of Animal Science* 89(8), 2518–2528.

Ecoinvent 2010. Ecoinvent Centre. Ecoinvent 2.0 database. Swiss centre for life cycle inventories, Dübendorf. Accessed 22-02-2018. Available at: www.ecoinvent.ch.

Gavrilova, O. and Vilu, R. 2012. Production-based and consumption-based national greenhouse gas inventories: an implication for Estonia. *Ecological Economics* 75, 161–173.

Guinee, J. B., Gorree, M., Heijungs, R., Huppes, G., Kleijn, R., de Koning, A., van Oers, L., Wegener Sleeswijk, A., Suh, S., Udo de Haes, H. A., de Brujin, H., van Duin, R. and Huijbregts, M. A. J. 2002. *Handbook on Life Cycle Assessment. An Operational Guide to the ISO Standards*. Kluwar Academic Publishers, Dordrecht, The Netherlands.

Hessle, A., Bertilsson, J., Stenberg, B., Kumm, K.-I. and Sonesson, U. 2017. Combining environmentally and economically sustainable dairy and beef production in Sweden. *Agricultural Systems* 156, 105–114.

Hristov, A. N., Oh, J., Firkins, J. L., Dijkstra, J., Kebreab, E., Waghorn, G., Makkar, H. P. S., Adesogan, A. T., Yang, W., Lee, C., Gerber, P. J., Henderson, B. and Tricarico, J.

M. 2013. Special topics – mitigation of methane and nitrous oxide emissions from animal operations: I. A review of enteric methane mitigation options. *Journal of Animal Science* 91(11), 5045–5069.

Intergovernmental Panel on Climate Change (IPCC) 1997. *Revised 1996 IPCC Guidelines for National Greenhouse Gas Inventories*. Vol. 3 reference manual. Cambridge University Press, Cambridge, UK.

Intergovernmental Panel on Climate Change (IPCC) 2006. *2006 Intergovernmental Panel on Climate Change Guidelines for National Greenhouse Inventories. Prepared by the National Greenhouse Gas Inventories Program* Eggleston, H. S., Buendia, L., Miwa, K., Ngara, T. and Tanabe, K. (Eds). Institute for Global Environmental Strategies, Hayama, Japan.

International Organisation for Standardisation (ISO) 2006a. *Environmental Management - Life Cycle Assessment: Principles and Framework*. (ISO 14040:2006). European Committee for Standardization, Brussels, Belgium.

International Organisation for Standardisation (ISO) 2006b. *Environmental Management - Life Cycle Assessment: Requirements and Guidelines*. (ISO 14044:2006). European Committee for Standardization, Brussels, Belgium.

Jones, A. K., Jones, D. L. and Cross, P. 2014. The carbon footprint of lamb: sources of variation and opportunities for mitigation. *Agricultural Systems* 123, 97–107.

Kirchgessner, M., Windisch, W. and Muller, H. L. 1994. Methane release from dairy cows and pigs. In: Proceedings of the XIII Symposium on Energy Metabolism of Farm Animals, Aguilera (Ed.). EAAP Publication 76, 339–402.

Ledgard, S. F., Lieffering, M., Coup, D. and O'Brien, B. 2011. Carbon footprinting of New Zealand lamb from the perspective of an exporting nation. *Animal Frontiers* 1(1), 40–45.

McAuliffe, G. A., Takahashi, T., Mogensen, L., Hermansen, J. E., Sage, C. L., Chapman, D. V. and Lee, M. R. F. 2017. Environmental trade-offs of pig production systems under varied operational efficiencies. *Journal of Cleaner Production* 165, 1163–1173.

McGeough, E. J., Little, S. M., Janzen, H. H., McAllister, T. A., McGinn, S. M. and Beauchemin, K. A. 2012. Life-cycle assessment of greenhouse gas emissions from dairy production in Eastern Canada: A case study. *Journal of Dairy Science* 95(9), 5164–75.

Murphy, B., Crosson, P., Kelly, A. K. and Prendiville, R. 2017. An economic and greenhouse gas emissions evaluation of pasture-based dairy calf-to-beef production systems. *Agricultural Systems* 154, 124–132.

O'Brien, D., Bohan, A., McHugh, N. and Shalloo, L. 2016. A life cycle assessment of the effect of intensification on the environmental impacts and resource use of grass-based sheep farming. *Agricultural Systems* 148, 95–104.

O'Brien, D., Hennessy, T., Moran, B. and Shalloo, L. 2015. Relating the carbon footprint of milk from Irish dairy farms to economic performance. *Journal of Dairy Science* 98(10), 7394–7407.

O'Brien, D., Shalloo, L., Buckley, F., Horan, B., Grainger, C. and Wallace, M. 2011. The effect of methodology on estimates of greenhouse gas emissions from grass-based dairy systems. *Agriculture, Ecosystems and Environment* 141(1–2), 39–48.

O'Brien, D., Shalloo, L., Patton, J., Buckley, F., Grainger, C. and Wallace, M. 2012. Evaluation of the effect of accounting method, IPCC v. LCA, on grass-based and confinement

dairy systems' greenhouse gas emissions. *Animal: An International Journal of Animal Bioscience* 6(9), 1512-27.
Olesen, J. E., Schelde, K., Weiske, A., Weisbjerg, M. R., Asman, W. A. H. and Djurhuus, J. 2006. Modelling greenhouse gas emissions from European conventional and organic dairy farms. *Agriculture, Ecosystems and Environment* 112(2-3), 207-220.
Peters, G. P. 2008. From production-based to consumption-based national emission inventories. *Ecological Economics* 65(1), 13-23.
Peters, G. P. and Hertwich, E. G. 2008. Post-Kyoto greenhouse gas inventories: production versus consumption. *Climatic Change* 86(1-2), 51-66.
Rebitzer, G., Ekvall, T., Frischknecht, R., Hunkeler, D., Norris, G., Rydberg, T., Schmidt, W. P., Suh, S., Weidema, B. P. and Pennington, D. W. 2004. Life cycle assessment: Part 1: Framework, goal and scope definition, inventory analysis, and applications. *Environment International* 30(5), 701-720.
Reckmann, K., Traulsen, I. and Krieter, J. 2013. Life cycle Assessment of pork production: A data inventory for the case of Germany. *Livestock Science* 157(2-3), 586-96.
Redding, M. R., Devereux, J., Phillips, F., Lewis, R., Naylor, T., Kearton, T., Hill, C. J. and Weidemann, S. 2015. Field measurement of beef pen manure methane and nitrous oxide reveals a surprise for inventory calculations. *Journal of Environmental Quality* 44(3), 720-728.
Rice, P., O'Brien, D., Shalloo, L. and Holden, N. M. 2017. Evaluation of allocation methods for calculation of carbon footprint of grass-based dairy production. *Journal of Environmental Management* 202(1), 311-319.
Ridoutt, B. G., Page, G., Opie, K., Huang, J. and Bellotti, W. 2013. Carbon, water and land use footprints of beef cattle production systems in southern Australia. *Journal of Cleaner Production* 73, 24-30.
Rosemann, C., Haenel, H. D., Dammgen, U., Freibauer, A., Wulf, S., Eurich-Menden, B., Dohler, H. and Osterburg, B. 2017. Calculation of Gaseous and Particulate Emissions from German Agriculture 1990-2015. Report on Methods and Data (RMD). Thunen report 46. Johann Heinrich von Thünen-Institut. https://www.thuenen.de/media/publikation en/thuenen-report/Thuenen_Report_57.pdf. DOI:10.3220/REP1519913866000.
Rotz, C. A., Montes, F. and Chianese, D. S. 2010. The carbon footprint of dairy production systems through partial life cycle assessment. *Journal of Dairy Science* 93(3), 1266-1282.
Schils, R. L. M., Verhagen, A., Aarts, H. F. M., Kuikman, P. J. and Sebek, L. B. J. 2006. Effect of improved nitrogen management on greenhouse gas emissions from intensive dairy systems in the Netherlands. *Global Change Biology* 12(2), 382-391.
Schils, R. L. M., Verhagen, A., Aarts, H. F. M. and Šebek, L. B. J. 2005. A farm level approach to define successful mitigation strategies for GHG emissions from ruminant livestock systems. *Nutrient Cycling in Agroecosystems* 71(2), 163-175.
Shalloo, L., Dillon, P., Rath, M. and Wallace, M. 2004. Description and validation of the Moorepark Dairy System Model. *Journal of Dairy Science* 87(6), 1945-1959.
Soussana, J. F., Tallec, T. and Blanfort, V. 2010. Mitigating the greenhouse gas balance of ruminant production systems through carbon sequestration in grasslands. *Animal: an International Journal of Animal Bioscience* 4(3), 334-350.
Stackhouse, K. R., Rotz, C. A., Oltjen, J. W. and Mitloehner, F. M. 2012. Growth promoting technologies reduce the carbon footprint, ammonia emissions, and costs of California beef production systems. *Journal of Animal Science* 90(12), 4656-4665.

Thomassen, M. A., van Calker, K. J., Smits, M. C. J., Iepema, G. L. and de Boer, I. J. M. 2008. Life cycle assessment of conventional and organic milk production in the Netherlands. *Agricultural Systems* 96(1-3), 95-107.

United Nations Framework Convention on Climate Change (UNFCCC) 2017a. National inventory submissions 2017. Accessed 24-04-2020. Available at: https://unfccc.int/process-and-meetings/transparency-and-reporting/reporting-and-review-under-the-convention/greenhouse-gas-inventories-annex-i-parties/submissions/national-inventory-submissions-2017.

United Nations Framework Convention on Climate Change (UNFCCC) 2017b. Submitted biennial update reports (BURs) from non-annex 1 parties. Accessed 24-04-2020. Available at: https://unfccc.int/BURs.

van Middelaar, C. E., Berentsen, P. B. M., Dijkstra, J. and De Boer, I. J. M. 2013. Evaluation of a feeding strategy to reduce greenhouse gas emissions from dairy farming: the level of analysis matters. *Agricultural Systems* 121, 9-22.

Vellinga, T. V., de Haan, M. H. A., Schils, R. L. M., Evers, A. and van den Pol-van Dasselaar, A. 2011. Implementation of GHG mitigation on intensive dairy farms: farmers' preferences and variation in cost effectiveness. *Livestock Science* 137(1-3), 185-195.

Veysset, P., Lherm, M., Bébin, D., Roulenc, M. and Benoit, M. 2014. Variability in greenhouse gas emissions, fossil energy consumption and farm economics in suckler beef production in 59 French farms. *Agriculture, Ecosystems and Environment* 188, 180-91.

Wang, S. Y. and Huang, D.-J. 2005. Assessment of greenhouse gas emissions from poultry enteric fermentation. *Asian-Australasian Journal of Animal Sciences* 18(6), 873-878.

Wiedemann, S. G., Ledgard, S. F., Henry, B. K., Yan, M.-J., Mao, N. and Russell, S. J. 2015. Application of life cycle assessment to sheep production systems: investigating co-production of wool and meat using case studies from major global producers. *The International Journal of Life Cycle Assessment* 20(4), 463-76.

Zehetmeier, M., Baudracco, J., Hoffmann, H. and Heißenhuber, A. 2012. Does increasing milk yield per cow reduce greenhouse gas emissions? A system approach. *Animal: an International Journal of Animal Bioscience* 6(1), 154-166.

Part 2

Breeding, animal husbandry and manure management

Chapter 3

The contribution of animal breeding to reducing the environmental impact of livestock production

Yvette de Haas, Wageningen University and Research, The Netherlands; Marco C. A. M. Bink, Hendrix Genetics Research, Technology & Services B.V., The Netherlands; Randy Borg, Cobb Europe B.V., The Netherlands; Erwin P. C. Koenen, CRV, The Netherlands; Lisanne M. G. Verschuren, Topigs Norsvin Research Center B.V./Wageningen University and Research, The Netherlands; and Herman Mollenhorst, Wageningen University and Research, The Netherlands

1 Introduction

The world human population is expanding, and the demand for food in the year 2050 is expected to be approximately 70% greater than in 2010 (FAO, 2009). Demand for meat and other livestock products is highly elastic to income (Delgado et al., 1999), and therefore as population affluence improves, the demand for livestock products will increase further. Also, the global human population is ageing, and older people typically consume larger quantities of animal-derived protein than children (Steinfeld et al., 2006). The expected 70% increase in food demand requires an annual increase in food production of

http://dx.doi.org/10.19103/AS.2020.0077.05

1.3% per annum. This increase in food demand requires an increased efficiency of food production, both animal and crop derived. Moreover, competition for land from other industries (e.g. biofuels) implies this increased animal and crop production must be achieved from an ever-declining land base. Although feed efficiency, as currently defined, is not synonymous with production efficiency, it undoubtedly will be a major contributor to increasing production from an ever-decreasing food-producing land base. The global production of red meat is expected to increase from 229 million tonnes in 1999–2001 to 465 million tonnes in 2050, while global milk production is expected to increase from 580 to 1043 million tonnes in the same period (Steinfeld et al., 2006). This increased production must, however, be achieved in an environmentally responsible and sustainable manner.

There is considerable commentary nowadays on climate change and its implications, as well as possible mitigation strategies. Animal production generates greenhouse gas (GHG) emissions as methane (CH_4) from enteric fermentation and manure, nitrous oxide (N_2O) from the widespread use of nitrogenous fertilizers and animal manure and carbon dioxide (CO_2) from the fossil fuels for energy usage plus land-use change. Methane, however, is not only an environmental hazard but is also associated with a loss of carbon from the rumen and therefore an unproductive use of energy (Johnson and Johnson, 1995). There is a wide variation in documented calculations of animal agriculture contributions to GHG (Herrero et al., 2011). O'Mara (2011) stated that animal agriculture is responsible for 8.0% to 10.8% of global GHG emissions based on calculations from the Intergovernmental Panel on Climate Change (IPCC), but if complete lifecycle analysis (i.e. accounting for the production of inputs to animal agriculture as well as the change in land use such as deforestation) is undertaken, this figure can be up to 18%. Cattle are the largest contributors to the global emission of CH_4 from enteric fermentation (O'Mara, 2011).

In December 2015, 195 nations signed the Paris Agreement. The Paris Agreement's central aim is to strengthen the global response to the threat of climate change. Therefore, in this century, the global temperature rise should at least be kept well below 2°C above pre-industrial levels. To pursue efforts to limit the temperature increase even further, a more stringent aim is set to a threshold of 1.5°C, instead of 2°C. Additionally, the agreement aims to increase the ability of countries to deal with the impacts of climate change and to make finance flows consistent with low GHG emission and climate-resilient pathways. All sectors have to reduce their environmental impact, including the livestock sector. In this chapter, we will review the current environmental impact of livestock species, show historical trends and quantify the contribution of animal breeding in reducing the environmental impact of livestock species.

2 The environmental impact of livestock production

Animal production is responsible for 14.5% of total anthropogenic GHG emissions (Gerber et al., 2013). Approximately half of these emissions originate directly from animal production, whereas the other half comes from feed production. Of the total of 7.1 gigatonnes CO_2-equivalents produced by the global livestock sector, beef production is responsible for 41%, cattle milk production for 20%, pig meat production for 10%, chicken meat production for 6% and chicken egg production for 3%. The remaining 21% is produced by other species and purposes, like buffalo and small ruminants for milk and meat, and by draft power, manure as fuel, and so on (Gerber et al., 2013).

Several tools have been developed to assess the environmental impacts of animal agriculture on a product basis. One of these tools is FeedPrint (Vellinga et al., 2013; Wageningen Livestock Research, 2015). FeedPrint is a life cycle analyses (LCA) tool for the assessment of environmental impacts of animal production, especially GHG emissions, developed primarily for the Dutch animal feed industry. FeedPrint, therefore, contains a full database of animal feed ingredients and their respective impacts from cradle to farmgate, including, for example, fertilizer production and application, field emissions, energy use for field work, transport and feed production. Based on data available in this tool, we could show that the GHG emissions for the production of 1 kg of pig meat are higher than for 1 kg of broiler meat, followed by chicken eggs and dairy production (Fig. 1a). Although results differ between studies, this ranking is in line with, for example, De Vries and De Boer (2010) or Williams et al. (2006). The relatively low numbers of GHG emissions for the production of milk and eggs are mainly due to the high water content of these products compared to meat and the fact that animals do not need to be slaughtered to acquire the product. When values of GHG emissions are recalculated to per kg protein basis, numbers come much closer together, with broiler meat being among the best-performing products (De Vries and De Boer, 2010; Williams et al., 2006). Beef production is outside the scope of this chapter, but, in general, calculated GHG emissions per kg product and kg protein are much higher for beef than for the other animal products.

There are different sources of GHG emissions related to animal production. The best-known and most-investigated one is enteric methane (CH_4) emission from ruminants. For dairy production, animal-related CH_4 emissions account for over one-third of total GHG emissions (Fig. 1b). However, in general, about half of the GHG emissions related to animal production come from feed-production-related processes (Gerber et al., 2013). For poultry in The Netherlands, even more than 80% of GHG emissions are related to feed production (Fig. 1b), with CO_2 being the predominant contributor, followed by nitrous oxide (N_2O).

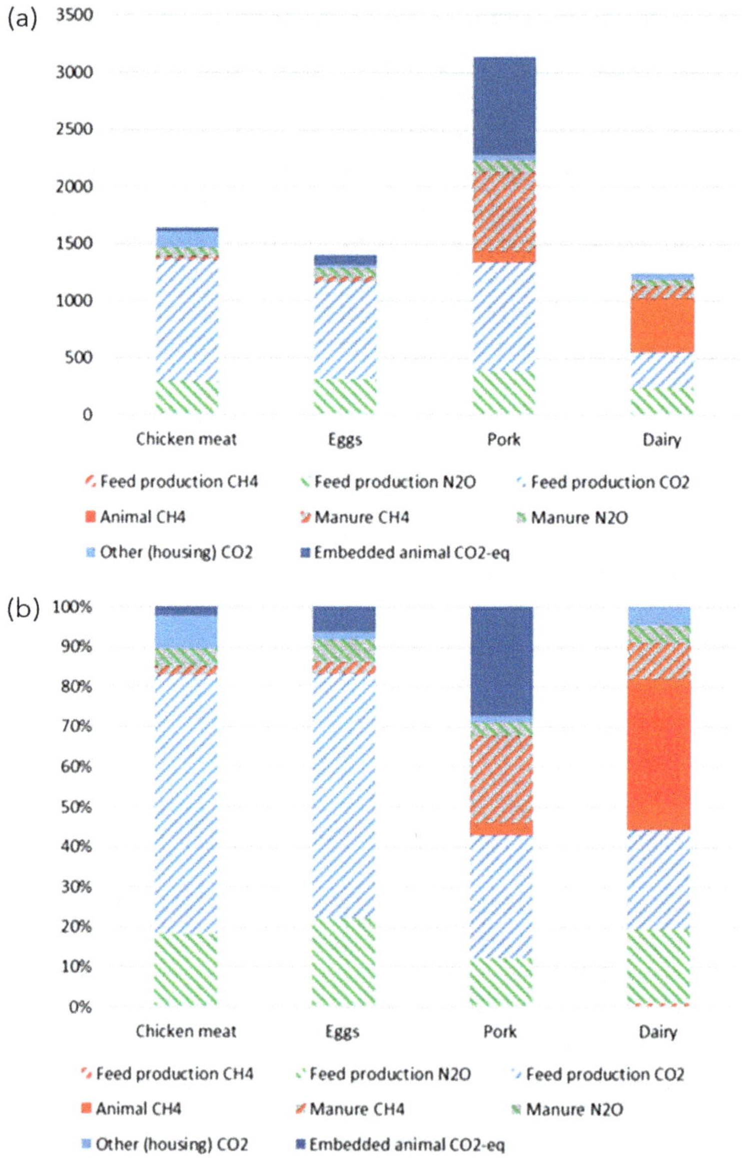

Figure 1 Greenhouse gas emissions (kg CO_2-equivalents) per tonne product (a) and relative contribution (%) to total greenhouse gas emissions (b) for four main animal products in The Netherlands based on FeedPrint (Vellinga et al., 2013; Wageningen Livestock Research, 2015). (Embedded animal = emissions caused by parent and young stock which are accounted to the final product).

Next to the emission of GHG, nitrogen (N)- and phosphorus (P)-related environmental impacts, like acidification, eutrophication and depletion of P resources, are important when animal products are concerned (e.g. Thomassen and de Boer, 2005). These impacts could be assessed using lifecycle analysis but more straightforward by nutrient balances and derived efficiency parameters (Mu et al., 2016). The efficiency of N and P usage at animal level seems to be a relevant indicator for animal breeding, as animal breeding aims at improving animal production and efficiently using resources, resulting in a reduction of the environmental impact. The objective of this study was to quantify the contribution of animal breeding to reducing the environmental impact of the four major livestock species in The Netherlands, namely broilers (meat), laying hens (eggs), pigs (meat) and dairy cattle (milk).

In this chapter, we will show both the historical trends and the quantification of the contribution of animal breeding to the environmental impact of several livestock species: broilers, layers, pigs and dairy cattle. At the end of the chapter is a summarizing conclusion with recommendations.

3 Broilers: environmental impact and the contribution of breeding

3.1 Historical trends

Havenstein et al. (2003) performed a study where they compared the feed conversion of broilers genetically representative of animals for the years 1957 versus 2001 when fed broiler diets representative for diets fed in 1957 and 2001. Average body weights on 42 and 84 days of age for the 1957 broiler on the 1957 diets were 539 and 1430 g versus 2672 and 5520 g for the 2001 broiler on the 2001 diets, respectively. The 42-day feed conversion (FC) for the 1957 broiler on the 1957 diets was 2.34, and for the 2001 broiler on the 2001 diet was 1.63. The 2001 broiler on the 2001 feed was estimated to have reached 1815 g BW at 32 days of age with an FC of 1.47, whereas the 1957 broiler on the 1957 feed would not have reached that BW until 101 days of age with an FC of 4.42. Their conclusion was that 85-90% of the improvement in the performance of broilers is due to genetic selection, and 10-15% due to nutrition (Havenstein et al., 2003).

Havenstein et al. (2003) also showed that the slaughter weight increases over the years. This distorts the trend of the feed conversion ratio (FCR) over the years. Therefore, LEI (currently Wageningen Economic Research) calculated the trend in FCR corrected to a slaughter weight of 2.15 kg based on Dutch data, which showed that the FCR decreases with almost 1% per year (Fig. 2).

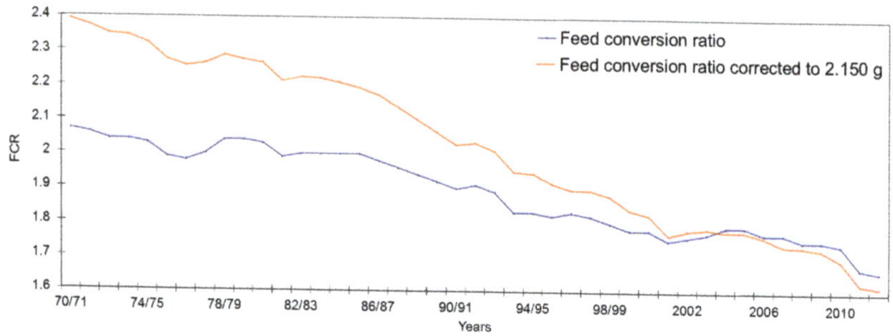

Figure 2 Trend in the feed conversion ratio of Dutch broilers (based on data from LEI (currently Wageningen Economic Research)).

Table 1 Summary data of broiler dataset with reference year, date of first and last placement, number of flocks, average number of animals per flock (Avg. numb.), average age at slaughter (Avg. age), average final body weight (after 1 day fasting) (Avg. weight) and average feed conversion ratio (Avg. FCR)

Year	First	Last	# flocks	Avg. numb.	Avg. age	Avg. weight	Avg. FCR
2014	07-05-13	28-08-14	12	33 105	37.2	2.2	1.66
2018	16-06-17	07-08-18	10	63 716	40.3	2.7	1.56

3.2 Quantification of contribution of animal breeding

3.2.1 Materials and methods

Data on the performance of Dutch commercial flocks were obtained from a broiler breeding company (Cobb Europe B.V., Boxmeer, The Netherlands) and contained 12 flocks with placement date in the period from May 2013 to August 2014, referred to as 2014, and 10 flocks from the period June 2017 to August 2018, referred to as 2018 (Table 1).

Feed composition was assumed to be equal in both periods and was derived from FeedPrint 2015.03 (Vellinga et al., 2013; Wageningen Livestock Research, 2015). Emissions of GHG related to the production of feed ingredients, and their N and P content, were collected from the database of FeedPrint 2018 (Wageningen Livestock Research, 2018).

Emissions of GHG are expressed in CO_2 equivalents (CO_2-eq.), which is a unit to express the contribution of different GHGs to their global warming potential (GWP), relative to CO_2. Methane (CH_4) and nitrous oxide (N_2O) have a GWP of respectively 36.75 (34 for biogenic CH_4) and 298 CO_2-eq (Myhre et al., 2013).

Efficiencies of N and P are expressed in percentages and calculated as N and P in output, for broilers in live animals, over input with feed. To express the output in kg N and P, body composition after 1 day fasting was based on Caldas (2015).

Published by Burleigh Dodds Science Publishing Limited, 2021.

3.2.2 Results

The environmental impact caused by GHG emissions from broiler production decreased with 1.6% per year, and N and P efficiency increased with 1.7% per year (Table 2). This decrease in environmental impact was fully caused by a decrease in FCR (Table 1, 1.5%), which was larger than expected based on a literature overview of the development of FCR, where FCR decreased with about 1%.

3.2.3 Discussion

As data were available only for the broiler growing phase, data analysis focused on this phase. This phase, however, is responsible for the vast majority of the GHG emissions of the whole broiler production chain (Fig. 1), with feed production in the broiler growing phase being responsible for 83% of total GHG emissions. Results, therefore, could be considered representative for the whole broiler production process.

The dataset used for this study was based on commercial flocks, instead of a research dataset. Therefore, it showed some peculiarities, for example, a considerable increase in flock size and age at slaughter, the latter also causing an increase in final body weight. These differences represent a global shift in slaughter weights for the broiler industry. However, for purposes of comparing GHG emissions, slaughter age and flock size differences could have contributed to an underestimation of the decrease in FCR, resulting in high reduction percentages. However, even with a more conservative estimation of 1% decrease in FCR, we could conclude that GHG emissions decrease and N and P efficiency increases with the current breeding goal for broilers.

4 Layers: environmental impact and the contribution of breeding

4.1 Historical trends

Pelletier (2018) provided an evaluation on the extent of and plausible reasons for the change in the life cycle environmental footprint of conventional Canadian

Table 2 Emissions of GHG and N and P efficiency of broiler production in 2014 and 2018

Year	GHG emissions (kg CO_2-eq/kg BW)	N efficiency (%)	P efficiency (%)
2014	1.43	53.2	46.0
2018	1.34	56.7	49.1

egg production over a 50-year interval spanning from 1962 to 2012. For this, the historical trajectory of the industry toward more sustainable practices was elucidated. Supply chain acidifying emissions declined by 61%, eutrophying emissions by 68% and GHG emissions by 72%. Despite the >50% increase in annual egg production volumes - from 43.4 million dozen to 65.7 million dozen eggs in 1962 and 2012, respectively, the industry's overall environmental footprint actually decreased across all emissions and resource use domains considered. These observed changes are attributable to a combination of factors, including improved feed efficiency, changes in diet composition and manure management.

A similar study was performed by Pelletier et al. (2014) on the comparison of the environmental footprint of the egg industry in the United States in 1960 and 2010. They showed a similar reduction in GHG emissions by 71%.

The increase in egg production and the decrease in FCR were also present in Dutch data (Fig. 3), based on national data collated yearly (KWIN, 2011, 2013, 2017).

4.2 Quantification of contribution of animal breeding

4.2.1 Materials and methods

For laying hens a full LCA model was available (Van Winkoop, 2013), which takes into account parent stock and layers, both including their rearing phase.

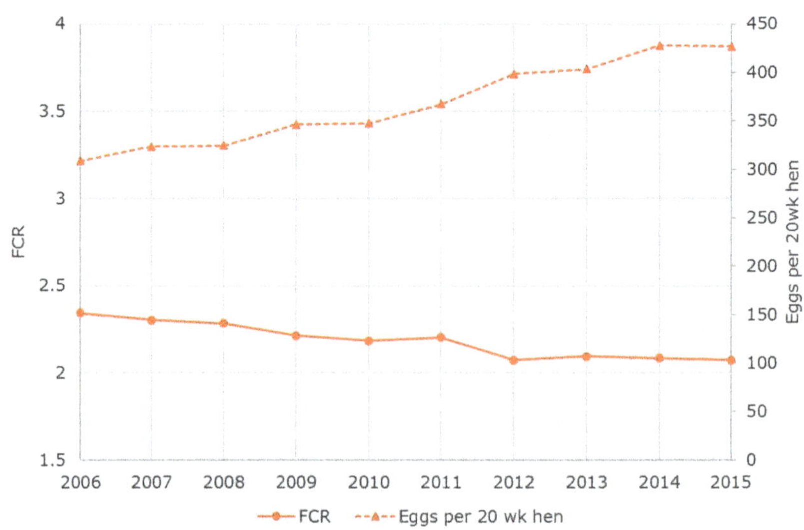

Figure 3 The trends in egg production and feed conversion ratios per Dutch laying hen from 2006 to 2015.

Published by Burleigh Dodds Science Publishing Limited, 2021.

For the calculation of genetic progress, however, only changes in the laying period were incorporated in the calculations.

Data on the development of modern commercial brown and white layer lines were obtained from a breeding company (Institut de Sélection Animale B.V. a Hendrix Genetics Company, Boxmeer, The Netherlands) and contained for the brown layers data for the years 2008 and 2017 (Table 3) and for the white layers for the years 2009 and 2017 (Table 3). These data should be considered to be from top-performing flocks and were available for the egg production phase only. These data were used to assess the rate of improvement, as this was anticipated to be representative for top-performing and average-performing commercial flocks. For a full LCA assessment, however, more data were needed and were derived from the 'commercial product' guides (Institut de Sélection Animale B.V. a Hendrix Genetics Company, Boxmeer, The Netherlands; Table 4). These data were also used for calculating environmental impacts for the commercial situation. It should be noted that these guides should hold for a broad range of commercial settings, including more challenging environments.

Feed composition was assumed to be equal in all periods and was derived from FeedPrint 2015.03 (Vellinga et al., 2013; Wageningen Livestock Research,

Table 3 Summary data of top-performing brown and white layers with reference year, total life time, mortality rate during laying period (from 18 weeks onwards), number of eggs per housed hen, total egg mass and average feed conversion ratio (Avg. FCR) during the laying period

Line	Year	Lifetime (weeks)	Mortality rate (%)	Eggs per housed hen	Total egg mass (kg)	Avg. FCR
Brown	2008	75	6	324	20.6	2.25
	2017	90	5	429	27.0	2.14
White	2009	75	6	329	20.7	2.16
	2017	90	5	433	27.3	2.05

Source: Institut de Sélection Animale B.V. a Hendrix Genetics Company, Boxmeer, The Netherlands.

Table 4 Summary data of commercial brown and white layers, total life time, mortality rate during laying period (from 18 weeks onwards), number of eggs per housed hen, total egg mass and average feed conversion ratio (FCR) during the laying period

Line	Lifetime (weeks)	Mortality rate (%)	Eggs per housed hen	Total egg mass (kg)	Avg. FCR
Brown	80	7.8	353	22.1	2.29
White	90	7.5	411	25.9	2.24

Source: Product guides, Institut de Sélection Animale B.V. a Hendrix Genetics Company, Boxmeer, The Netherlands.

2015). Emissions of GHG related to the production of feed ingredients, and their N and P content, were collected from the database of FeedPrint 2018 (Wageningen Livestock Research, 2018).

Emissions of GHG are expressed in CO_2 equivalents (CO_2-eq.), which is a unit to express the contribution of different GHG to global warming, their global warming potential (GWP), relative to CO_2. Methane (CH_4) and nitrous oxide (N_2O) have a GWP of, respectively, 36.75 (34 for biogenic CH_4) and 298 CO_2-eq (Myhre et al., 2013).

Efficiencies of N and P are expressed in percentages and calculated as N and P in output over input with feed. For laying hens, only eggs were considered as output and N and P contents were calculated with N and P content of raw egg (edible part; Finglas et al., 2015) applied to total egg mass (Tables 3 and 4) corrected for 15% shells (pers. Comm., Institut de Sélection Animale B.V. a Hendrix Genetics Company).

4.2.2 Results

Results for commercial brown and white layers (Table 5) show that impacts do not differ much.

The environmental impact caused by GHG emissions from egg production decreased by 0.7% per year for brown layers and by 0.9% per year for white layers (Table 6). N and P efficiency increased with 0.5% per year for brown layers and with 0.7% per year for white layers. The decrease in environmental impact was only partly caused by a decrease in FCR as also the production period was extended, due to genetic progress (Table 3). Especially for the assessment of GHG emissions, for which a full LCA including parent stock and rearing was used, the extended production period considerably contributed to the reduction of environmental impacts.

4.2.3 Discussion

As data for calculating genetic progress were only available for the laying period, total improvements could be expected to be larger when also improvements

Table 5 Emissions of GHG and N and P efficiency of egg production by commercial brown and white laying hens

Line	GHG emissions (kg CO_2-eq/kg egg)	N efficiency (%)	P efficiency (%)
Brown	2.18	30.2	15.5
White	2.09	30.9	15.8

Table 6 Emissions of GHG and N and P efficiency of egg production by top-performing brown laying hens in 2008 and 2017

Line	Year	GHG emissions (kg CO_2-eq/kg egg)	N efficiency (%)	P efficiency (%)
Brown	2008	2.18	30.6	15.7
	2017	2.03	32.0	16.4
White	2009	2.10	31.9	16.3
	2017	1.95	33.5	17.1

in parent stock and rearing would be taken into account. The laying period, however, is responsible for the vast majority of the GHG emissions of egg production (Fig. 1), with feed production in the laying period being responsible for 83% of total GHG emissions.

Based on this analysis we conclude that genetic progress is considerable in both brown and white layers, where white hens currently perform better and also improve faster than brown hens with respect to the environmental impact of production. As most brown hens produce brown eggs and most white hens produce white eggs and consumers in some countries prefer brown over white eggs, both types of layers still exist.

5 Pigs: environmental impact and the contribution of breeding

5.1 Historical trends

The feed efficiency of growing pigs has been a matter of serious commercial and scientific interest since at least 1970, but early recording technology made it difficult to produce accurate feed intake data at the individual level (Knap, 2009). Since electronic feeders were introduced, the pig breeding industry has been making good genetic improvement in feed conversion ratio (FCR), but this has been mainly due to genetic improvement of growth and body composition traits. A 35-year time trend illustrated by Knap and Wang (2012) shows very clearly that the average commercial FCR has come down from 3.3 to 2.6, with a quite considerable bandwidth among terminal crosses which does not show any signs of narrowing over time (Fig. 4).

5.2 Quantification of contribution of animal breeding

5.2.1 Materials and methods

Data from an experiment, which is described in Sevillano et al. (2018), were obtained from a pig breeding company (Topigs Norsvin Research Center B.V.,

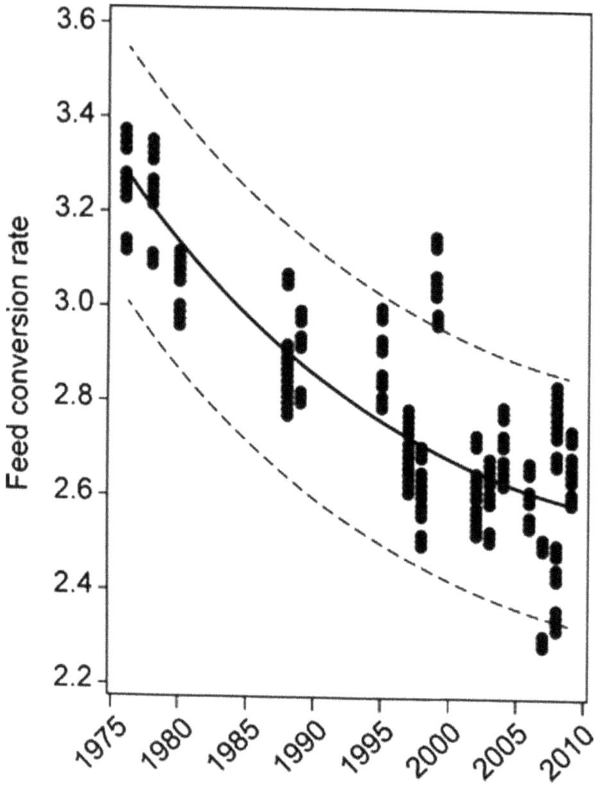

Figure 4 Time trends of feed conversion ratio in grower-finisher pigs of 103 terminal crosses, as recorded in public commercial product evaluation trials in Denmark, France, Germany, The Netherlands, UK and the United States. Unadjusted phenotypic population means, data from 18 literature and internet sources. The trend line is a spline interpolation plot through the data, with its 95% confidence limits. Figure copied from Knap and Wang (2012).

Beuningen, The Netherlands). In this experiment a corn/soybean meal (CS) diet and a cereals/alternative ingredient (CA) diet was fed to intact boars and gilts. The CS diet resembles American practice, but the impact of feed ingredients was calculated as fed in The Netherlands, which means, for example, that soybean meal mainly originated from Argentina and Brazil. The CA diet resembles Dutch practice, with many by-products in the diet. For calculating genetic progress, data from 400 pigs in 2014 (December 2013–May 2014) and 401 pigs in 2016 (November 2015–March 2016) were used (Table 7). Data only considered the growing-finishing phase (i.e. from 22 kg to approximately 120 kg of live weight). Feed composition was derived from Sevillano et al. (2018).

Emissions of GHG related to the production of feed ingredients, and their N and P content, were collected from the database of FeedPrint 2018

(Wageningen Livestock Research, 2018). Emissions of GHG are expressed in CO_2 equivalents (CO_2-eq.), which is a unit to express the contribution of different GHG to global warming, their global warming potential (GWP), relative to CO_2. Methane (CH_4) and nitrous oxide (N_2O) have a GWP of respectively 36.75 (34 for biogenic CH_4) and 298 CO_2-eq (Myhre et al., 2013). Efficiencies of N and P are expressed in percentages and calculated as N and P in output over input with feed. Protein deposition was calculated as described by Sevillano et al. (2018) and was used to calculate N deposition, whereas P deposition was calculated based on Pettey et al. (2015).

5.2.2 Results

The environmental impact caused by GHG emissions from pig production decreased by 0.6-0.7% per year dependent on the diet (Table 8). Nitrogen efficiency increased by 1.6–1.7%, whereas P efficiency increased by 0.4–0.6% over the two years. On all environmental indicators, boars performed slightly better than gilts. The N efficiency could be calculated more precisely than P efficiency because back fat thickness was measured in the experiment and used to calculate protein deposition. Therefore, the decrease in environmental impact was not only caused by a decrease in FCR (lower feed intake at same growth rate; Table 7) but also by higher protein deposition at similar growth.

5.2.3 Discussion

Data analysis focused on the grower-finisher phase because good-quality, detailed information was available for this phase. Furthermore, the chosen method for GHG emission calculation accounted for the effect of feed production on GHG emissions only. The full analysis of the whole production cycle could have given different results, as Groen et al. (2016) showed that CH_4 emissions from manure management, crop yields and reproduction performance are important processes determining whole chain GHG emissions from pig production. These results correspond well with data, shown in Fig. 1, where reproduction and rearing phase (27%) and emissions from manure (25%) contribute considerably to the total impact of pig production. Groen et al. (2016), however, also showed that FCR is the most important factor, with the highest influence on whole chain GHG emission from pig production. This is in agreement with the data shown in Fig. 1, where feed production alone explains more than 40% of GHG emissions from pig production.

Differences in the environmental impact of pigs, fed either a CS or CA diet, were most clear in P efficiency, caused by low digestibility of P in some by-products in the CA diet (e.g., rapeseed and sunflower meal). When the CS diet would have been calculated as fed in the country where corn and soybean

Table 7 Summary of pig dataset on corn/soybean meal (CS) and cereals/alternative ingredients (CA) diet with sex (Male = M, Female = F), reference year, average daily feed intake (ADFI) on starter, grower and finisher feed, average daily gain during the growing-finishing phase (ADG), average final empty body weight at slaughter and protein deposition during the growing-finishing phase

Diet	Sex	Year	ADFI starter (kg/d)	ADFI grower (kg/d)	ADFI finisher (kg/d)	ADG (g/d)	Empty body weight (kg)	Protein deposition (kg)
CS	M	2014	1.33	2.18	2.94	955	92.7	16.4
CS	M	2016	1.22	2.16	3.01	960	91.7	16.5
CS	F	2014	1.38	2.21	2.92	938	95.4	16.3
CS	F	2016	1.27	2.19	3.00	943	94.4	16.4
CA	M	2014	1.36	2.21	3.04	961	92.4	17.1
CA	M	2016	1.25	2.19	3.12	967	91.4	17.2
CA	F	2014	1.43	2.24	3.02	929	93.7	16.6
CA	F	2016	1.33	2.22	3.09	934	92.7	16.7

Table 8 Emissions of GHG and N and P efficiency of male (M) and female (F) pigs on corn/soybean meal (CS) and cereals/alternative ingredients (CA) diet in 2014 and 2016

Diet	Sex	Year	GHG emissions (kg CO_2-eq/ kg BW gain)	N efficiency (%)	P efficiency (%)
CS	M	2014	1.93	44.7	36.8
CS	M	2016	1.90	46.2	37.3
CS	F	2014	1.99	43.2	36.9
CS	F	2016	1.96	44.7	37.3
CA	M	2014	1.70	43.7	25.8
CA	M	2016	1.68	45.1	26.1
CA	F	2014	1.78	41.2	25.3
CA	F	2016	1.76	42.5	25.5

were grown, GHG impacts of pigs fed the CS diets probably would have been lower.

From this analysis we could conclude that current breeding goals decrease the environmental impact of producing pig meat and that boars are more efficient and, therefore, have a lower environmental impact than gilts.

6 Dairy cattle: environmental impact and the contribution of breeding

6.1 Historical trends

Over the past 100 years, the range of traits considered for genetic selection in dairy cattle populations has progressed to meet the demands of both industry and society (Miglior et al., 2017). At the turn of the twentieth century, dairy farmers were interested in increasing milk production; however, a systematic strategy for selection was not available. Organized milk performance recording took shape, followed quickly by conformation scoring. Methodological advances in both genetic theory and statistics around the middle of the century, together with technological innovations in computing, paved the way for powerful multi-trait analyses. As more sophisticated analytical techniques for traits were developed and incorporated into selection programs, production began to increase rapidly, and the wheels of genetic progress began to turn. By the end of the century, the focus of selection had moved away from being purely production oriented toward a more balanced breeding goal. This shift occurred partly due to increasing health and fertility issues and partly due to societal pressure and welfare concerns. Traits encompassing longevity, fertility, calving, and health, have now been integrated into national selection indices.

With these indices, milk production is still increasing per year. As shown in Fig. 5, milk production in The Netherlands has increased by 46% between 1990 and 2017. Because of the increased milk production, the feed intake has increased but to a lesser extent than milk production; therefore, the efficiency of dairy production (kg milk/kg feed) has improved over the years.

6.2 Quantification of contribution of animal breeding

6.2.1 Selection indices

Selection indexes are utilized by livestock breeders of many species around the world and aid in the selection of animals for use within a breeding program where there are several traits of economic or functional importance. Selection indexes provide an overall 'score' of an animal's genetic value for a specific purpose and are calculated based on weightings placed on individual traits that are deemed to be important for that purpose. Selection indexes assist producers in making 'balanced' selection decisions. The derivation of a selection index starts with the definition of the overall breeding objective.

The next stage in developing a selection index is to calculate economic values for each trait, generally with a bio-economic model, where the economic value is the increase in revenue from a unit change of a trait while everything else is held constant. Then, selection index theory (Hazel, 1943) is commonly used to calculate the most appropriate index weights and responses to

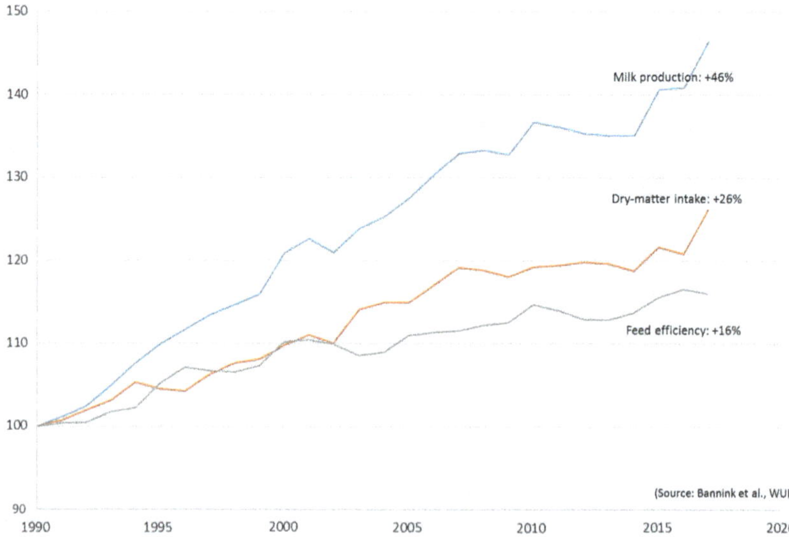

Figure 5 Trends in milk yield, feed intake and feed efficiency of the Dutch dairy cattle population between 1995 and 2017 (extended and based on Bannink et al. (2011)).

selection for a set of traits given the genetic and phenotypic (co)variances and the economic values of traits in the index. The resulting selection index is the sum of n estimated breeding values (EBVi) for each trait multiplied by their respective index weights b_i.

$$\text{Index} = b_1\text{EBV1} + b_2\text{EBV2} + \ldots + b_n\text{EBVn}$$

6.2.2 Quantification effect of breeding

In our case, we started with a selection index representing the national breeding goal for dairy cattle of The Netherlands (CRV, 2018). The Dutch national breeding goal consists of milk component traits, longevity, health traits (udder health, claw health), fertility traits (interval first-last insemination, calving interval), conformation traits (for udder and for feed and legs), calving traits (calving ease and vitality of calves) and feed efficiency (Table 9). We added enteric CH_4 emissions to this index as a correlated trait. Genetic parameters were obtained from the literature (Lassen and Lovendahl, 2016). The heritability for enteric CH_4 production is 0.21, and genetic correlations with milk lactose, protein, fat and dry matter intake are 0.43, 0.37, 0.77 and 0.42 (−0.42 for feed saved), respectively. Correlations between enteric CH_4 production and other traits in the breeding goal were set to zero. All phenotypic correlations were also set to zero.

Selection index calculations show how much the traits are predicted to change per year. This is plotted in Fig. 6 for both CH_4 production (g/d) and for CH_4 intensity (CH_4 production expressed per kg of milk). It shows that CH_4 production per cow will steadily increase as a correlated response to selection for the current breeding goal. However, the methane intensity drops. Further reductions could be achieved when actively selecting on lower methane-emitting cows, by adding more weight on CH_4 in the national breeding goal. Selecting actively against methane would result in healthy, fertile, long-living cows that emit less CH_4. Actively selecting against CH_4 emission, however, requires large-scale recording of individual CH_4 emissions.

6.2.3 Discussion

The predicted future trends in enteric methane production are based on the genetic parameters used in the selection indices. The correlation of 0.77 between protein yield and enteric methane production is strong and impacts the results. Further research is needed to estimate reliable genetic parameters between enteric methane production and other traits of interest (e.g., the traits in the breeding goal). Estimating these genetic parameters requires that a large enough dataset is built, which includes records of enteric CH_4 emission of many individual cows.

Table 9 Heritabilities of (on diagonal) and genetic correlations (below diagonal) between the 15 traits in the Dutch national breeding goal (CRV, 2018)

	1	2	3	4	5	6	7	8	9	10	11	12	13	14	15
1. Lactose	**0.43**														
2. Fat	0.38	**0.58**													
3. Protein	0.88	0.58	**0.50**												
4. Longevity	0.36	0.35	0.42	**0.12**											
5. Udder health	−0.03	−0.02	−0.06	0.36	**0.09**										
6. Interval first-last insemination	−0.34	−0.24	−0.29	0.25	0.27	**0.08**									
7. Calving interval	−0.44	−0.33	−0.37	0.11	0.21	0.85	**0.15**								
8. Udder conformation	−0.08	−0.04	−0.10	0.11	0.27	−0.05	0.00	**0.34**							
9. Feet and leg conformation	0.02	0.04	0.05	0.25	0.21	0.00	0.00	0.35	**0.17**						
10. Direct calving ease	0.07	0.15	0.11	0.24	0.15	0.20	0.24	0.00	0.00	**0.07**					
11. Maternal calving ease	0.00	0.00	0.00	0.16	0.09	0.25	0.24	0.10	0.10	0.19	**0.05**				
12. Direct vitality	0.05	0.09	0.02	0.14	0.05	0.10	0.14	0.00	0.00	0.60	0.14	**0.04**			
13. Maternal vitality	−0.04	−0.07	0.03	0.16	0.07	0.32	0.24	0.00	0.00	0.11	0.34	−0.16	**0.09**		
14. Claw health	0.00	0.15	0.07	0.33	0.09	0.10	0.14	0.15	0.65	0.16	0.06	0.03	0.10	**0.18**	
15. Saved feed costs	0.20	0.35	0.30	0.50	−0.03	−0.10	−0.30	−0.09	−0.29	0.41	−0.20	0.17	−0.05	0.11	**0.25**

Published by Burleigh Dodds Science Publishing Limited, 2021.

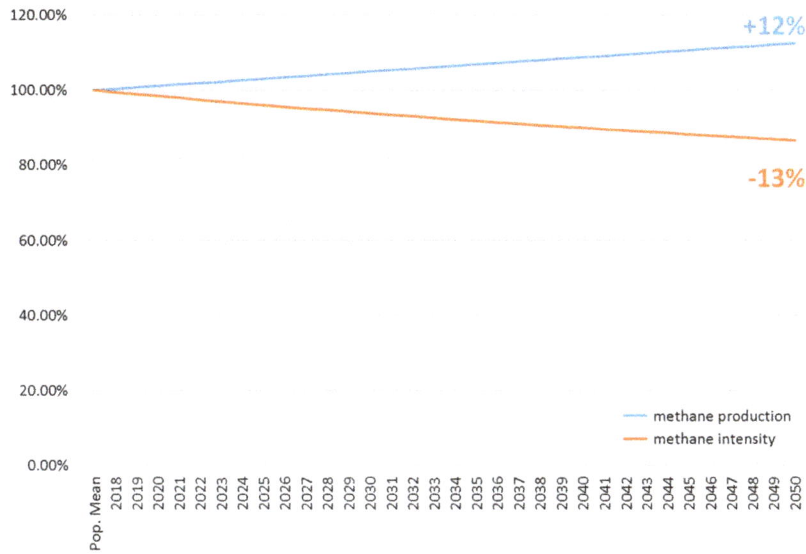

Figure 6 Expected future trends in CH_4 production (g/d) and CH_4 intensity (g/kg milk) for the Dutch dairy cattle population with breeding on the current national breeding goal.

7 Conclusion

Animal production is responsible for 14.5% of total anthropogenic GHG emissions. Approximately half of these emissions originate directly from animal production, whereas the other half comes from feed production. Animal breeding aims at improving animal production and efficient use of resources, which results in a reduction of the environmental impact. The objective of this study was to quantify the contribution of animal breeding in reducing the environmental impact of the four major livestock species in The Netherlands, namely broilers (meat), laying hens (eggs), pigs (meat) and dairy cattle (milk).

A literature review was performed to assess the current status of and historical trends in environmental impact, mainly focused on GHG emissions, and general performance criteria, like feed efficiency and lifetime production. Emissions related to the feed production dominate the impacts by broilers and laying hens. For pigs, the emissions during feed production and from manure are important contributors. For dairy cattle, as being ruminants, enteric methane emission is a large contributor to total GHG emissions. Historical trends show considerable improvements in efficiency over the last decades, in which breeding has an important role. The literature review showed that the contribution of breeding to reducing the environmental impact of animal production is led by an indirect response through selection on increased efficiency.

Also a quantitative assessment was made on the current environmental impact of the four animal products and the effect of recent genetic improvements. For broiler meat, chicken eggs and pig meat the focus was on GHG emissions, and nitrogen (N) and phosphorus (P) efficiency, whereas for dairy the focus was on enteric methane emissions, an important contributor to GHG emissions. Data were partly provided by breeding organizations, that is, the partners in the Breed4Food consortium (www.breed4food.com).

The analyses in this chapter demonstrate that animal breeding can provide a mitigation tool to lower the environmental impact of livestock species. Genetic improvement of livestock is a particularly cost-effective technology, producing permanent and cumulative changes in performance:

- For broilers, it was shown that GHG emissions decreased with 1.7% and N and P efficiency increased by 1.6% per year due to the current breeding goals.
- For laying hens, white and brown hens were considered and it was concluded that white hens currently have a lower GHG impact and better N and P efficiency than brown hens and that improvements over the past 10 years went faster for white hens.
- For pigs, data were available from a well-controlled study with two diets and animals divided by sex over a time frame of two years. Results showed that in the growing-fattening phase of pigs, GHG emissions decrease and N and P efficiency increase with the current breeding goal. Furthermore, boars had a lower environmental impact than gilts.
- For dairy cattle, results showed that with the current breeding goal, CH_4 production per cow per day increases but CH_4 intensity (i.e. CH_4 production per kg milk) decreases.

All reported results are achieved without specific selection on environmental traits, but as an indirect response of the current breeding goals for each species, which is a combination of health, growth and (feed) efficiency. If direct selection of environmental traits is desired, recording of new traits is required, for example, N and P contents of meat and eggs and methane emission of individual dairy cows.

Direct measurement of GHG impact of animal production is difficult, but not impossible, which hampers active selection against GHG emissions of animals. In the short run, indirect selection against GHG emissions could be further optimized by putting more selection pressure on efficiency traits, while accounting for the effects on other important traits, for example, health, longevity and reproduction. In the long run, recording schemes could be set up to either record the desired traits on commercial farms (for dairy) or in parental stock (for pigs and poultry).

The LCA analyses performed in this study could be further improved by also including information of the parent stock and rearing phases. It is expected that when including genetic progress in parent stock and rearing phases of parents and commercials, the contribution of genetics to reduce GHG emissions per kg product has an even bigger impact.

8 Where to look for further information

8.1 Further reading

Selection index theory: Hazel, L. N. (1943) The genetic basis for constructing selection indexes. *Genetics* 28, 476-490. & Lush, J. L. (1960) Improving dairy cattle by breeding. I. Current status and outlook. *Journal of Dairy Science* 43, 702-706.

LCA analyses: Thomassen, M. A. and De Boer, I. J. M. (2005) Evaluation of indicators to assess the environmental impact of dairy production systems. *Agriculture, Ecosystems & Environment* 111: 185-199. & De Vries, M. and De Boer, I. J. M. (2010) Comparing environmental impacts for livestock products: A review of life cycle assessments. *Livestock Science* 128, 1-11.

Report on 'The contribution of breeding to reducing environmental impact of animal production', https://library.wur.nl/WebQuery/wurpubs/549934.

8.2 Key conferences

WCGALP (world conference of genetics applied to animal production) is well attended by members of the animal breeding community (industry and scientists).

GGAA (greenhouse gas of agriculture and animal) is well attended by scientists in all disciplines (nutrition, breeding, microbiology, etc.) working on the reduction of environmental impact of livestock production.

9 Acknowledgements

This study was financially supported by the Dutch Ministry of Economic Affairs (TKI Agri & Food project 16022) and the Breed4Food Partners Cobb Europe, CRV, Hendrix Genetics and Topigs Norsvin.

10 References

Bannink, A., Van Schijndel, M. W. and Dijkstra, J. (2011). A model of enteric fermentation in dairy cows to estimate methane emission for the Dutch National Inventory Report using the IPCC Tier 3 approach. *Anim. Feed Sci. Technol.*, 166-167: 603-618.

Caldas, J. (2015). *Calorimetry and Body Composition Research in Broilers and Broiler Breeders*. Fayetteville, AR: University of Arkansas, 243.

CRV. (2018). Statistical indicators, E-20. *NVI*. Available at: https://www.cooperatie-crv.nl/wp-content/uploads/2018/10/E_20-NVI_aug2018_en.pdf.

De Vries, M. and De Boer, I. J. M. (2010). Comparing environmental impacts for livestock products: A review of life cycle assessments. *Livest. Sci.*, 128(1-3): 1-11.

Delgado, C., Rosegrant, M., Steinfeld, H., Ehui, S. and Courbois, C. (1999). *Livestock to 2020 - The Next Food Revolution*. Food, Agriculture, and the Environment Discussion. Washington, D.C.; Rome, Italy; Nairobi, Kenya: International Food Policy Research Institute; Food and Agriculture Organization of the United Nations; International Livestock Research Institute.

FAO. (2009). How to feed the world in 2050. Available at: http://www.fao.org/fileadmin/templates/wsfs/docs/expert_paper/How:to_Feed_the_World_in_2050.pdf.

Finglas, P., Roe, M., Pinchen, H., Berry, R., Church, S. and Dodhia, S. (2015). *McCance and Widdowson's the Composition of Foods Integrated Dataset 2015 - User Guide*. London, UK: Public Health England. PHE publications gateway number: 2014822.

Gerber, P. J., Steinfeld, H., Henderson, B., Mottet, A., Opio, C., Dijkman, J., Falcucci, A. and Tempio, G. (2013). *Tackling Climate Change through Livestock - A Global Assessment of Emissions and Mitigation Opportunities*. Rome, Italy: Food and Agriculture Organization of the United Nations (FAO).

Groen, E. A., van Zanten, H. H. E., Heijungs, R., Bokkers, E. A. M. and de Boer, I. J. M. (2016). Sensitivity analysis of greenhouse gas emissions from a pork production chain. *J. Clean. Prod.*, 129: 202-211.

Havenstein, G. B., Ferket, P. R. and Qureshi, M. A. (2003). Growth, livability, and feed conversion of 1957 versus 2001 broilers when fed representative 1957 and 2001 broiler diets. *Poult. Sci.*, 82(10): 1500-1508.

Hazel, L. N. (1943). The genetic basis for constructing selection indexes. *Genetics*, 28(6): 476-490.

Herrero, M., Gerber, P., Vellinga, T., Garnett, T., Leip, A., Opio, C., Westhoek, H. J., Thornton, P. K., Olesen, J., Hutchings, N., Montgomery, H., Soussana, J.-F., Steinfeld, H. and McAllister, T. A. (2011). Livestock and greenhouse gas emissions: The importance of getting the numbers right. *Anim. Feed Sci. Technol.*, 166-167: 779-782.

Johnson, K. A. and Johnson, D. E. (1995). Methane emissions from cattle. *J. Anim. Sci.*, 73(8): 2483-2492.

Knap, P. W. (2009). Voluntary feed intake and pig breeding. In: Torrallardona, D. and Roura, E. (Eds) *Voluntary Feed Intake in Pigs*. Wageningen, the Netherlands: Wageningen Academic Publishers.

Knap, P. W. and Wang, L. (2012). Pig breeding for improved feed efficiency. In: Patience, J. F. (Ed.) *Feed Efficiency in Swine*. Wageningen, the Netherlands: Wageningen Academic Publishers.

KWIN. (2011). *Kwantitatieve Informatie Veehouderij 2011-2012*. Wageningen, the Netherlands: Wageningen Livestock Research.

KWIN. (2013). *Kwantitatieve Informatie Veehouderij 2013-2014*. Wageningen, the Netherlands: Wageningen Livestock Research.

KWIN. (2017). *Kwantitatieve Informatie Veehouderij 2017-2018*. Wageningen, the Netherlands: Wageningen Livestock Research.

Lassen, J. and Lovendahl, P. (2016). Heritability estimates for enteric methane emissions from Holstein cattle measured using noninvasive methods. *J. Dairy Sci.*, 99(3): 1959-1967.

Miglior, F., Fleming, A., Malchiodi, F., Brito, L. F., Martin, P. and Baes, C. F. (2017). A 100-year review: Identification and genetic selection of economically important traits in dairy cattle. *J. Dairy Sci.*, 100(12): 10251–10271.

Mu, W., van Middelaar, C. E., Bloemhof, J. M., Oenema, J. and de Boer, I. J. M. (2016). Nutrient balance at chain level: A valuable approach to benchmark nutrient losses of milk production systems. *J. Clean. Prod.*, 112: 2419–2428.

Myhre, G., Shindell, D., Bréon, F.-M., Collins, W., Fuglestvedt, J., Huang, J., Koch, D., Lamarque, J.-F., Lee, D., Mendoza, B., Nakajima, T., Robock, A., Stephens, G., Takemura, T. and Zhang, H. (2013) Anthropogenic and natural radiative forcing. In: Stocker, T. F., Qin, D., Plattner, G.-K., Tignor, M., Allen, S. K., Boschung, J., Nauels, A., Xia, Y., Bex, V. and Midgley, P. M. (Eds) *Climate Change 2013: The Physical Science Basis. Contribution of Working Group I to the Fifth Assessment Report of the Intergovernmental Panel on Climate Change*. Cambridge, UK and New York, NY: Cambridge University Press.

O'Mara, F. P. (2011). The significance of livestock as a contributor to global greenhouse gas emissions today and in the near future. *Anim. Feed Sci. Technol.*, 166–167: 7–15.

Pelletier, N. (2018). Changes in the life cycle environmental footprint of egg production in Canada from 1962 to 2012. *J. Clean. Prod.*, 176: 1144–1153.

Pelletier, N., Ibarburu, M. and Xin, H. (2014). Comparison of the environmental footprint of the egg industry in the United States in 1960 and 2010. *Poult. Sci.*, 93(2): 241–255.

Pettey, L. A., Cromwell, G. L., Jang, Y. D. and Lindemann, M. D. (2015). Estimation of calcium and phosphorus content in growing and finishing pigs: Whole empty body components and relative accretion rates. *J. Anim. Sci.*, 93(1): 158–167.

Sevillano, C. A., Nicolaiciuc, C. V., Molist, F., Pijlman, J. and Bergsma, R. (2018). Effect of feeding cereals-alternative ingredients diets or corn-soybean meal diets on performance and carcass characteristics of growing-finishing gilts and boars. *J. Anim. Sci.*, 96(11): 4780–4788.

Steinfeld, H., Gerber, P., Wassenaar, T., Castel, V., Rosales, M. and De Haan, C. (2006). *Livestock's Long Shadow: Environmental Issues and Options*. Rome, Italy: Food and Agriculture Organization of the United Nations.

Thomassen, M. A. and de Boer, I. J. M. (2005). Evaluation of indicators to assess the environmental impact of dairy production systems. *Agric. Ecosyst. Environ.*, 111(1-4): 185–199.

Van Winkoop, C. (2013). The contribution of layer chicken breeding to the reduction of climate change. Wageningen, the Netherlands: Wageningen University; Animal Breeding and Genomics Centre, 44.

Vellinga, T. V., Blonk, H., Marinussen, M., Van Zeist, W. J., De Boer, I. J. M. and Starmans, D. (2013). *Methodology Used in FeedPrint: A Tool Quantifying Greenhouse Gas Emissions of Feed Production and Utilization*. Wageningen, the Netherlands: Wageningen UR Livestock Research, 121. http://edepot.wur.nl/254098.

Wageningen Livestock Research. (2015). *FeedPrint 2015.03*. Wageningen, the Netherlands: Wageningen University and Research.

Wageningen Livestock Research. (2018). *FeedPrint 2018.01*. Wageningen, the Netherlands: Wageningen University and Research.

Williams, A. G., Audsley, E. and Sandars, D. L. (2006). *Determining the Environmental Burdens and Resource Use in the Production of Agricultural and Horticultural Commodities*. Bedford: Cranfield University and Defra.

Chapter 4

Quantifying the contribution of livestock health issues to the environmental impact of their production systems

Stephen G. Mackenzie, Trinity College Dublin, Ireland; and Ilias Kyriazakis, Queen's University of Belfast, UK

1 Introduction

There is an increased interest about the consequences of animal health on the environmental impact of livestock systems. This should not be surprising given the association between animal health and the efficiency with which they utilise the important resource inputs to livestock production, which are key factors in determining the environmental impact of these systems (de Vries and de Boer, 2010; McAuliffe et al., 2016; Tallentire et al., 2018). Animal health has major consequences on how animals utilise their resources, including feed, increasing inputs, including medication, and reducing outputs, such as amount of milk or meat produced per animal, or per unit of input (Perry et al., 2018).

While the consequences of animal health for the environmental impacts of livestock systems have been addressed to a limited extent, for example (ADAS, 2015; Mostert et al., 2018; Skuce et al., 2016), to date the focus has been almost exclusively on the implications of conditions which impact on ruminant systems and the implications of these for greenhouse gases (GHGs). This focus is natural, due to the high profile nature of both the issue of anthropogenic GHGs and the large contribution of ruminant systems to this issue (Gerber

http://dx.doi.org/10.19103/AS.2020.0077.07

et al., 2013). However, the environmental issues facing the livestock industry are much broader than this: livestock systems have a major impact on global land use (Weindl et al., 2017), water use (Doreau et al., 2012) and the acidification of soils and waterways (Leip et al., 2015), to name just some of the issues. In addition, the majority of meat consumed globally is from pigs and poultry (FAO, 2019), but these production systems cause less GHGs than ruminants (Gerber et al., 2013). With consumption of non-ruminant livestock products expected to increase dramatically to 2050 (FAO, 2011), non-ruminant animal farming faces concerns around the impact of feed production (particularly the associated land and water use), and the pollution caused by their manure (de Vries and de Boer, 2010; Poore and Nemecek, 2018). These issues represent important future challenges that the industry must overcome to maintain its public acceptability.

Climate change is expected to exacerbate animal health issues, which in turn would affect the environmental impact of livestock systems, thus creating a vicious circle that links animal health, environmental impact and climate change. A combination of changes in air temperature, precipitation, as well as the frequency and magnitude of extreme weather events are all expected to have a negative impact on outcomes for livestock health and welfare (Lacetera, 2019). Climate change may also affect the quantity, quality and composition of animal feed, both directly and indirectly. The increase in the incidence of fungal contamination of animal feedstuffs is one such indirect effect on animal health due to increased concentration of mycotoxins in the feed (Bernabucci et al., 2011).

The focus of our chapter is on the environmental impact of endemic livestock health challenges that lead to deterioration in animal health, and on the potential impacts arising from their mitigations. This is because, while epidemics may have devastating effects on livestock systems, these would be usually controlled through stamping out by eradication of affected herds and flocks (Geering et al., 1999). In some instances, where large-scale eradication is unacceptable, other control measures, such as vaccination programmes supplemented by other disease control measures, can be used to eliminate many epidemic health and welfare challenges (Roth, 2011). The dearth of information on this issue in relation to the environmental impacts of epidemic health challenges for livestock means we are unable to give the latter issue detailed consideration in our chapter.

The first part of our chapter concentrates on the potential of animal health to affect the environmental impact of livestock systems. Because in several ways animal health affects environmental impact of livestock, that is, through decreases in productivity and efficiency and increases in culling, applies to all livestock systems, we review the consequences of animal health on resource inputs and outputs across livestock systems. This includes the implications for a variety of important variables for environmental impacts, such as the utilisation of nutrients in feed, mortality/culling rates, breeding performance and also milk quality in the case of dairy systems (ADAS, 2015). Subsequently, we review the literature to date which has quantified the impact of health challenges for the environmental impacts of livestock systems.

The potential of successful health interventions to mitigate negative environmental impacts represents a point of synergy between concerns around environmental sustainability and animal welfare, both of which represent 'hot topics' in the discourse surrounding the livestock industry and its sustainability. This is a topic that has been under-represented in the literature, both because of the difficulty to address it in practical terms, and because of a lack of framework that will allow us to do so (Rushton, 2017). The challenges associated with modelling health interventions and their potential to mitigate environmental impacts constitute the last section in our chapter. This part has a heuristic value, given the current lack of information in the literature.

2 Consequences of health challenges on resource inputs and outputs of the animal and production system

Given that a large part of the environmental impact assessment of livestock systems is done on the basis of the principle of mass flows (FAO, 2016a, 2018), below we review the consequences of health challenges on how an animal uses the nutrient resources it consumes. One of the most profound consequences of health challenges is a decrease in the efficiency of how nutrient resources are used (Kyriazakis and Houdijk, 2007; Sandberg et al., 2007), which in turn will have consequences on the fate of nutrient excretion. However, in addition to this, health challenges may have indirect consequences on the resource inputs required to produce the key outputs of livestock production systems, such as milk, eggs and meat. These indirect consequences would arise from the fact that system outputs may need to be discarded because of their quality or for safety reasons; individual animals may need to be culled because their performance is no longer considered economically viable by their keepers; and ultimately some animals may succumb to the consequences of the health challenges. Alternatively, some health challenges would require inputs from pharmaceuticals or specialist management, which in turn utilise resources for their production. For these reasons, both the direct and indirect consequences of health challenges on resource inputs into animals and their production system are considered below.

2.1 Anorexia during health challenges

A reduction in the voluntary food intake, henceforth called anorexia, is associated with most challenges that affect livestock health (Kyriazakis, 2014). During subclinical infections, anorexia manifests as a ~20% reduction in voluntary food intake and is frequently the only indication that an animal's health is being challenged (Hite et al., 2020; Kyriazakis et al., 1998). In more severe infections, there may be a complete cessation of eating, but this is usually associated with severe infection outcomes, such as death (Lough et al., 2015). Anorexia is also associated with metabolic diseases, such as acidosis or ketosis in ruminants, and is often used as an indicator that the management of the animals has not been appropriate and requires attention (González et al., 2008). Ketosis in lactating cows, which results from a severe energy gap between intake and output, may lead to complete cessation of eating, but intake returns almost immediate to its previous levels after appropriate treatment (Andersson, 1988; Berge and Vertenten, 2014).

With all other things being equal, anorexia during health challenges would be expected to lead to a reduction in animal performance, and therefore, the animal would need to take more time to achieve the same production output. This in itself would imply that the animal would need to divert proportionally more resources in maintaining itself for a longer period of time, and therefore the efficiency with which its food is being utilised would be reduced. However, studies that compared the performance of infected animals with that of uninfected animals that have been offered the same amount of food (usually called pair-fed animals), suggest that the former utilise the same amount of food less efficiently than the latter (Holmes, 1993). A frequent observation is that infected animals grow more slowly and deposit less protein and fat in their bodies than the uninfected pair-fed animals (Holmes, 1993; Escobar et al., 2004). This implies that infection per se is associated with other effects in feed utilisation, which will be detailed below.

2.2 The effects of infectious challenges on digestion, absorption and utilisation of nutrient resources

Due to their significance, the effects of infectious challenges on the digestion, absorption and utilisation of resources has been studied most extensively during gastrointestinal parasitism. Depending on the site of infection, digestion and absorption of nutrients may be significantly affected (Sandberg et al., 2007). In ruminant animals parasitized with gastrointestinal parasites, a reduction in nutrient digestion has been observed, especially when the parasites affect the organs of digestion. This is the case of parasitism with the abomasal parasites, *Ostertagia ostertagi* in cattle and *Teladorsagia circumcincta* in sheep, as the gastric glands of the abomasum are damaged (Holmes, 1993; Ceï et al., 2018). Similarly, whether the process of nutrient absorption is affected would depend on the site of gastrointestinal parasitism and its effect on organ integrity. Several gastrointestinal parasites in ruminants are associated with gastrointestinal damage, such as epithelial and mucus loss (Coop and Kyriazakis, 1999; Mavrot et al., 2015). In the case of blood-sucking parasites, such as *Haemonchus contortus*, blood and plasma loss is also the outcome of the infection. All these effects are associated with the loss of considerable quantities of protein into the digestive tract of the host animal (Holmes, 1993).

Similar observations have been reported in non-ruminant animals parasitized by gastrointestinal parasites. The apparent digestibility of protein was reduced during pathogen challenges with parasitic worms in pigs and was smaller for pathogens that affected mainly the stomach, compared with those affecting latter parts of the gastrointestinal tract (Hale et al., 1985; Midha et al., 2018) . For instance, in broilers infected with the small intestine coccidia *Eimeria acervulina*, the apparent ileal digestibility of every single amino acid in the food was reduced by ~5% (Rochell et al., 2016). As well as damaging parasitized enterocytes, *Eimeria* parasitism increases plasma protein leakage and mucogenesis, thus increasing endogenous amino acid flow.

Infectious challenges are expected to increase the (maintenance) requirements for nutrient resources through a variety of routes: (1) animals need to repair damage to its tissues or replenish lost fluids as a direct consequence of infection. (2) Fever, which accompanies several infections would increase energy expenditure and hence requirements. (3) Infected animals need to mount an immune response in order to cope with the infection and eventually overcome it. Hosts may respond with innate or acquired immunity, or a combination of these, depending on the type of challenge and the stage of a particular infection (Sandberg et al., 2007).

The Sandberg et al. (2007) investigations were the most comprehensive attempt to quantify such effects on nutrient resource requirements. It is difficult to make generalisations about these quantitative effects, given that different infectious challenges would increase the different sources of increased requirements to different extents, but some suggestions can be made. As far increases in energy requirements are concerned, it has been suggested that maintenance requirements to be 1.7–2.2 times as great in challenged as opposed to non-challenged animals. There appears to be highly specific pathogen differences, with the greatest energetic cost occurring for infectious challenges associated with fever, especially during the stages of acute infection (Sandberg et al., 2006). Otesile et al. (1991) challenged pigs with *Trypanosoma brucei* and found that they had a significant increase (1.7 times) in energy maintenance requirements than healthy controls.

In the case of increased protein requirements due to infection challenges, it has been estimated that protein maintenance animals challenged by a varied of pathogenic and non-pathogenic challenges varied from 1.3 times to 4 times that in the healthy. Webel et al. (1998) found that chicks challenged by a non-pathogenic antigen had increased lysine maintenance requirements, tended to have increased threonine maintenance requirements, but there was no effect for arginine requirements. Different effects on maintenance may suggest that amino acid requirements are affected to different extents and that this needs to be accounted for the changes in their utilisation during infectious challenges.

2.3 The effects of non-infectious challenges on digestion, absorption and utilisation of nutrient resources

Production diseases, which arise from the management of usually high-producing animals, are a good case to consider the effect of non-infectious challenges on digestion, absorption and utilisation of nutrient resources. They include the group of diseases usually referred to as 'metabolic diseases', such as rumen acidosis and ketosis in dairy cows, pregnancy toxaemia in ewes and ascites in broilers. Most of these conditions, at least in their subclinical form, may predispose animals to susceptibility to infectious challenges, and therefore, it is difficult to disassociate their consequences from them. For example, while it would be expected that the nutrient resource requirements which are associated with lameness would be very small, several metabolic diseases with significant resource requirements are a predisposing factor for lameness. Their subclinical forms are also very widespread, for example, sub-clinical ketosis was found to have an average prevalence of 24.1% in a global study of the issue in dairy herds (Brunner et al., 2019).

Sub-acute ruminal acidosis (SARA) in dairy cows is a good case in point for its effects on nutrient digestion and use. SARA is estimated to have an overall prevalence in dairy herds of 11–33% in studies that have investigated the issue (Kleen and Cannizzo, 2012). It arises from feeding of high-energy density diets to meet the energy requirements of high-yielding cows, which leads to reduced rumen pH and increased risk of acidosis. Ruminal acidosis can cause erratic fluctuations in feed intake, and low rumen pH causes the cow to go 'off-feed', which reduces the production of fermentation acids, allowing the pH to recover. Changes in the rumen environment lead to reduced digestion of fibrous ingredients and lowers the efficiency of microbial protein production in the rumen (Beauchemin, 2007), thus affecting directly digestion and absorption of nutrient resources. At the same time, low rumen pH leads to inflammatory changes both locally, including damage to the gastrointestinal tract wall, and systemically, associated with an acute phase response (Bertoni et al., 2008). This acute phase response may not only be associated with increased nutrient resource requirements, but also with significant changes in the energy and lipid metabolism in different body tissues. Due to the nature and the focus of the condition, and the difficulty in disassociating these effects, they have not been quantified separately.

Because most production diseases are associated with negative nutrient resource balance and the animal needs to sacrifice its body or tissue mastitis reserves for the sake of continuing the production of milk or eggs, there would be less nutrient resources available for the maintenance of 'integrity' functions, such as defences to pathogens. This would make the animal more vulnerable to challenges, previously easy to control. This explanation has been put forward for the occurrence of mastitis in high-yielding dairy cows (O'Rourke, 2009), which are less able to direct resources towards the functions of innate and acquired immunity.

2.4 Indirect consequences of health challenges on system inputs and outputs

In growing animals, the usual consequences of health challenges are increases in their mortality and decreases in their growth rate and feed efficiency. This is especially the case with endemic viral diseases, affecting non-vaccinated animals. For example, infection with the porcine reproductive and respiratory syndrome virus (PRRSv) resulted in an additional 11% increase in mortality among nursery pigs and 6% increase among growing pigs in the United States pig herds (Kliebenstein et al., 2004). The same reports suggest increased use of medication among PRRSv-affected herds, although the increase is not sufficiently quantified.

In reproducing animals, health challenges lead directly or indirectly to reproductive failure/fertility and culling. In herds endemically infected by PRRSv, there is a substantial increase in the percentage of abortions per sow year, and decreases in weaned pigs produced per sow year and number of farrowings per sow year (Valdes-Donoso et al., 2018). Similarly, it has long been established that there is a clear association between health and fertility for cows affected by conditions such as mastitis, lameness and ketosis (Bertoni et al., 2008). In all these cases, the percentage of animals which are prematurely culled or die from the conditions may be substantial.

Finally, health challenges may lead to product loss from affected animals. Such losses may be associated with the condemnation or discard of milk, and the partial or whole condemnation of carcasses from normal or emergency slaughter (Garcia et al., 2019). Specifically, during mastitis the condemnation of milk is associated with increased inflammatory cell content, which affects milk quality, and antibiotic residues in the milk due to treatment of infected udders. This is on top of the reduction in milk yield and any culling of cows.

Above, we developed a framework which enables to view the effects of health challenges on nutrient resource requirements, which in turn enables to quantify their effects on mass flows. The framework should allow us to review and account for the effect of health challenges on environmental impact, and this is done below.

3 Quantifying the environmental impact of health challenges

In this section, we review the literature published to date presenting quantitative results on the environmental impacts of health issues in livestock production systems. The life cycle assessment (LCA) framework is the widely accepted way to holistically assess the environmental impact of livestock systems (FAO, 2016b, 2016c, 2018), although there are no specific guidelines when applying the approach to health issues. While there have been a large number of environmental impact studies of livestock systems as reviewed by Poore and Nemecek (2018), a relatively small proportion of these focus on the implications of animal health issues for the environmental impacts of livestock production. These studies have begun to quantify the link between animal health issues and the environmental impacts of livestock systems, and they can broadly be grouped into two categories:

1. Studies that model the environmental impacts caused by the existence of a health issue.
2. Studies that investigate the potential of specific treatment strategies to mitigate a health issue and the potential implications for the environmental impacts of the production system.

Both approaches can be useful exercises when used appropriately. In the first category, approaches can be entirely hypothetical and take a broad view across multiple different diseases to understand where the greatest potential benefits for targeted treatment strategies lie, both in terms of environmental impact mitigation, and in traditional economic terms. Using the second approach, studies can provide a realistic picture of the real-world benefits in terms of impact mitigation that specific treatment or prevention strategies can have. In this section, we focus on studies that have taken the first approach.

Table 1 lists the studies found to date that have quantified the environmental impacts caused by specific health issues in livestock production systems. Almost all studies to date that have investigated the impact of health issues are on ruminant production systems, the exception being Li et al. (2015), who investigated the effect of vaccination for PRRSv on key sources of GHG emissions for pig production (see Section 4 for further details on this). Only two studies considered environmental impact categories beyond GHGs in relation to health issues (Chen et al., 2016; Hospido and Sonesson, 2005). The most wide-ranging and comprehensive study found was the ADAS (2015) report on the impact of endemic diseases and conditions for the GHGs and economic performance across the UK sectors of beef, dairy and dual purpose cattle systems. The ADAS study quantified both the GHGs caused by 10 important

Table 1 A summary of quantitative results reported in literature on the environmental impact of livestock health issues. Where a study addressed more than one disease, its results are reported separately in the table below

Study	Study type	Species (system)	Disease/s studied	Impact of health issue for GHGs (%)	Functional units and other methodological details	Other notes
(Skuce et al., 2016)	Modelling	Sheep	Gastrointestinal nematodes	+10	1 kg carcass weight	Theoretical calculation based on three levels of infection. +10% is based on going from 20% of flock infected to 0.
(Fox et al., 2018)	Experimental trial	Sheep (Lambs)	Parasitic worm infection (Teladorsagia circumcincta)	+33 methane emissions	Methane per dry matter intake	No overall GHG assessment.
(Skuce et al., 2016)	Modelling	Cattle (beef)	Neosporosis	+2.2–4.7	1 kg carcass weight	Theoretical calculation based on three levels of infection. Highest increase is based on going from 20% of flock infected to 0.
(Macleod et al., 2018)	Modelling	Cattle (dairy and beef)	Trypanosomiasis	+9	1 kg of edible protein at the farm gate	Milk and meat protein are equivalent in this calculation.
(Chen et al., 2016)	Modelling	Cattle (dairy)	Lameness	+8	1 kg of fat and protein-corrected milk (FPCM). FPCM milk = milk volume × (0.25 + 0.122 × fat% + 0.077 × protein%)	Study considered other impact categories, namely eutrophication potential, acidification potential and abiotic depletion.

(*Continued*)

Table 1 (*Continued*)

Study	Study type	Species (system)	Disease/s studied	Impact of health issue for GHGs (%)	Functional units and other methodological details	Other notes
(Mostert et al., 2018)	Modelling	Cattle (dairy)	Subclinical ketosis and related conditions	+2.3 per case	1 ton of FPCM milk	Calculation includes assumptions of probabilities for associated conditions including mastitis, metritis, displaced abomasum, lameness and clinical ketosis.
(Özkan et al., 2015)	Modelling	Cattle (dairy)	Mastitis	+2	1 L milk produced	
(ADAS, 2015)	Modelling	Cattle (dairy and beef)	Overall impact of disease on UK herd GHGs	+6 dairy and +6.6 beef	1000 L FPCM milk and 1 tonne edible carcass weight	The overall impact of all health issues considered for the UK cattle sector compared to theoretical healthy herd.
(ADAS, 2015)	Modelling	Cattle (dairy and beef)	Bovine viral diarrhoea (BVD)	+20 dairy and +130 beef	1000 L FPCM milk and 1 tonne edible carcass weight	Increase for cases where farms impacted by the condition compared to a theoretical healthy herd.
(ADAS, 2015)	Modelling	Cattle (dairy and beef)	Johne's disease	+25 dairy and +40% beef	1000 L FPCM milk and 1 tonne edible carcass weight	Increase for cases where farms impacted by the condition compared to a theoretical healthy herd.
(ADAS, 2015)	Modelling	Cattle (dairy and beef)	Salmonella	+20 dairy and +30 beef	1000 L FPCM milk and 1 tonne edible carcass weight	Increase for cases where farms impacted by the condition compared to a theoretical healthy herd.

(ADAS, 2015)	Modelling	Cattle (dairy and beef)	Liver flukes	+10 dairy and +10 beef	1000 L FPCM milk and 1 tonne edible carcass weight	Increase for cases where farms impacted by the condition compared to a theoretical healthy herd.
(ADAS, 2015)	Modelling	Cattle (dairy and beef)	IBR	+8 dairy and +20 beef	1000 L FPCM milk and 1 tonne edible carcass weight	Increase for cases where farms impacted by the condition compared to a theoretical healthy herd.
(ADAS, 2015)	Modelling	Cattle (dairy and beef)	Lameness	+7.5 dairy only	1000 L FPCM milk and 1 tonne edible carcass weight	Increase for cases where farms impacted by the condition compared to a theoretical healthy herd.
(ADAS, 2015)	Modelling	Cattle (dairy and beef)	Mastitis	+7 dairy only	1000 L FPCM milk and 1 tonne edible carcass weight	Increase for cases where farms impacted by the condition compared to a theoretical healthy herd.
(ADAS, 2015)	Modelling	Cattle (dairy and beef)	Infertility	+16 dairy and +21% beef	1000 L FPCM milk and 1 tonne edible carcass weight	Increase for cases where farms impacted by the condition compared to a theoretical healthy herd.
(ADAS, 2015)	Modelling	Cattle (dairy and beef)	Calf pneumonia	+1 dairy and +4 beef	1000 L FPCM milk and 1 tonne edible carcass weight	Increase for cases where farms impacted by the condition compared to a theoretical healthy herd.
(ADAS, 2015)	Modelling	Cattle (dairy and beef)	Calf diarrhoea	<1 dairy and +4 beef	1000 L FPCM milk and 1 tonne edible carcass weight	Increase for cases where farms impacted by the condition compared to a theoretical healthy herd.

health issues in UK systems and the potential of associated treatments and prevention strategies to mitigate them.

The 10 diseases and conditions investigated by ADAS were: infectious bovine rhinotracheitis (IBR), bovine viral diarrhoea (BVD), calf pneumonia, calf diarrhoea, liver fluke, mastitis, lameness, infertility, paratuberculosis and salmonella (ADAS, 2015). Other health issues for cattle considered in literature to date include: subclinical ketosis (Mostert et al., 2018), trypanosomiasis (Macleod et al., 2018) and neosporosis disease (Skuce et al., 2016). The studies show a wide range in the GHG emissions attributable to different health issues, the most dramatic results including an estimated 130% increase in GHG emissions from beef farms affected by bovine viral diarrhoea (BVD), and an increase of 25% in GHGs for dairy farms impacted by Johne's disease (paratuberculosis) compared to a 'healthy farm' scenario (ADAS, 2015).

The underlying data behind these models are able to shed light on why certain health issues cause greater GHG emission increases than others. Previously in this chapter we listed various pathways through which health issues can negatively affect animal performance in ways likely to increase GHGs from livestock production, but the studies in Table 1 quantify this link directly. Table 2 (reproduced from ADAS (2015)) summarises the major assumptions used in that study regarding changes in dairy cow performance when affected by different health issues compared to a 'healthy herd'. For example, on average, BVD increases mortality from 2% to nearly 5%, increases levels of abortion/still birth rates and reduces conception rates (Gunn et al., 2004; Houe, 1999). In addition to the increased mortality risk, the issues with reduced productivity mean that on average culling rate increases from 17% to 20%. All of these changes are likely to cause an increase in the GHGs of the farming system by decreasing the productive outputs. As can be seen from Table 2, the presence of Johne's disease (+20%) and *Salmonella* (+16%) were deemed to result in the largest increases in GHGs per cow per year compared to a healthy herd scenario. In the ADAS report calculations, BVD only produced a 10% increase in GHGs per cow per year, but due to losses in productive outputs (particularly reduced milk yield) its overall impact on GHGs per 1000 litres of milk produced was similar to that of salmonella. The differences between the data for Johne's disease and salmonella/BVD demonstrate how different mechanisms can drive GHG increases from health conditions. The increase in GHGs from salmonella and BVD are driven by increase in mortality, decrease in milk yield (this is more dramatic for BVD) and a reduction in the calving rate. However, for Johne's disease it is a dramatic increase in the culling rate from 17% to 33% along with some losses in milk yield which drive the increase in GHGs. This demonstrates that how important management responses can be for determining the impact of health conditions on GHGs, and increased GHGs do not just emanate from direct impacts on animal performance.

While conditions such as BVD and Johne's disease have been identified as causing the highest increases in GHGs for UK cattle in beef and dairy systems respectively, the literature beyond the ADAS report has tended to focus on conditions that the industry considers the most important from an economic perspective. As mastitis and lameness have previously been identified as some of the most economically important diseases in dairy herds (ADAS, 2015), they are associated with some of the highest losses in cattle welfare (Bennett and IJpelaar, 2003), and mastitis is the most likely condition to cause an animal to be culled (Liang, 2013; Ózsvári, 2017). It is thus unsurprising that these health issues are the subject of multiple studies with respect to their environmental impacts. The increase in GHGs caused by widespread problems such as mastitis, lameness and subclinical ketosis are calculated to be in the range of 2–8% in several studies (ADAS, 2015; Chen et al., 2016; Hospido and Sonesson, 2005; Mostert et al., 2018; Özkan et al., 2015). As can be seen in Table 2, instances of mastitis are linked to increased mortality, while also decreasing lactation yield, with the complications associated with the condition leading farmers to increase the culling rate (Rollin et al., 2015). The condition also leads to milk being discarded, due to treatment of the condition and instances of raised somatic cell counts. All of these factors along with its very high prevalence in dairy herds (estimated at 37% of all cows over 1 year in the ADAS report), mean the condition inevitably has a large effect on animal performance in dairy herds and results in GHG increases of around 7% (ADAS, 2015). Both Chen et al. (2016) and the ADAS report attributed an increase of around 8% in GHGs to scenarios of increased lameness on farm. These increases are caused by reduced feed intake resulting in weight loss and reduction in milk production, and lameness experienced early in lactation is also associated with reduced fertility, which further increases GHG intensity of the milk produced (Collick et al., 1989; Lucey et al., 1986). Chen et al. (2016) also reported increases of around 10% in acidification, eutrophication and fossil resource depletion from dairy production due to lameness. They concluded that for lameness, the same mechanisms of reduced feed intake and milk production were driving the increase in all impact categories considered. The lack of consideration of impacts beyond GHGs in this area of study to date means it is difficult to state if this is the case for other health issues beyond lameness.

Another interesting case study is that of subclinical ketosis (SCK), the implications of which for GHGs was investigated in a detailed LCA study by Mostert et al. (2018). This study took into account the full impact of SCK on Dutch dairy herds including the likelihood of complications such as mastitis, metritis and lameness, as well as progression to clinical ketosis. While SCK is prevalent in up to 49% of European dairy herds causing significant economic losses, the study suggested that overall the impact of SCK in terms of increased

Table 2 A summary of the estimated impact of various health issues on livestock performance parameters and greenhouse gas (GHG) emissions in UK dairy systems (adapted from ADAS (2015)).

	Mortality (%)	Lactation yield (litres)	Cull rate (%)	Calving interval (days)	Productive life (lactations)	Calves per cow per year	GHG emissions per cow
Healthy	2.0	7831	17	368	6.1	0.89	6995
Industrial baseline	3.0	7539	26	362	3.9	0.86	7105
Bovine viral diarrhoea (BVD)	4.9	7204	20	451	5.0	0.67	7670
Calf pneumonia	2.0	7831	17	368	6.1	0.89	7039
Calf scour	2.0	7831	17	368	6.1	0.89	7003
Infectious bovine rhinotracheitis	4.0	7560	17	401	6.0	0.78	7326
Infertility	2.0	7194	20	449	5.2	0.73	7485
Lameness	2.0	7534	19	387	3.2	0.85	7245
Liver fluke	3.9	7804	17	376	6.0	0.87	7693
Mastitis	3.2	7766	19	368	5.2	0.89	7415
Paratuberculosis	2.0	7590	33	405	3.1	0.81	8423
Salmonella	5.0	7584	17	402	6.0	0.77	8104

GHGs is relatively small at 0.9% per case of SCK, increasing to 2.3% when predicted complications are accounted for (Mostert et al., 2018). Importantly, however, the Mostert study presented detailed data on exactly what caused the predicted increases in GHGs, which should serve as a template for other studies in this area. For cases of SCK only, with no further complications, 72% of the predicted increase in GHGs is caused by a prolonged calving interval with the rest caused by reduced milk production. When individual cows experience complications relating to SCK then the potential increases in GHGs increase and the source of these increases change. In those cases involving mastitis-displaced abomasum and clinical ketosis, the increased removal of heifers for culling was the second largest contributing factor to increased GHGs. Contrastingly, in cases involving lameness, reduced milk production was the largest contributing factor to increased GHGs followed by prolonged interval between calving. The Mostert et al. (2018) study predicted that individual cases of culled cows with SCK increased GHG emissions by around 20%, whereas any cases of cows which died on farm increased GHGs by 50% per case compared to healthy cows. This was because meat from culled cows was considered to be viable and replace other low-value meat products, thus off-setting the need to produce these, whereas meat from cows that died on farm was assumed to be discarded. While important queries may be raised about the validity of this methodological approach, the detail in such calculations opens up a potential future pathway in bringing environmental impact considerations into the existing economic models which optimise culling decisions in dairy herds based on economic outcomes (Liang, 2013).

4 A framework to evaluate the environmental impact of health interventions

A more interesting and challenging question than what impacts are caused by the existence of health issues, is how to quantify the potential of prevention and treatment strategies to mitigate environmental impacts in livestock systems. The added interest here is that rather than just stating the extra environmental burden caused by certain health conditions, researchers can use modelling to identify the most effective strategies that can reduce this. The challenges they face in doing so are actually similar to those faced by farmers, vets and economists when trying to establish which health interventions are actually cost-effective. As described by Rushton (2017) there are still huge issues with the availability of baseline level data on the prevalence of endemic diseases and welfare issues in livestock populations, even in OECD countries. This hampers our ability to understand how cost-effective health interventions are likely to be at the scale of a national or transnational population.

4.1 Studies that have modelled the environmental impact of health interventions

A key difference between quantifying the potential environmental impact mitigation from health interventions and the economic consequences of doing so is that the environmental impact of several interventions, for example, producing vaccines or medicines, is likely to be small in comparison to any significant projected benefits through improved animal performance. The relatively small environmental impact of interventions means there is unlikely to be any meaningful trade-off in terms of environmental impacts if a treatment or prevention strategy has been shown to be effective (ADAS, 2015). This completely contrasts with the economic modelling where there is often a fine balance between such costs and benefits. For example, vaccinations only become cost-effective when the risk from disease is deemed to be above a certain threshold (Linhares et al., 2015). In reality, economic and animal welfare considerations still drive decisions around whether to implement health interventions on farm. As such, the question of how any cost-effective interventions are in terms of environmental impact reduction is of great interest to researchers in this area looking to incorporate multiple objectives in their modelling to establish why particular health interventions should be prioritised.

This is reflected in work such as the ADAS (2015) report where marginal abatement cost curve (MACC) analysis was performed to identify the most cost-effective strategies to reduce GHGs in UK cattle through health interventions (ADAS, 2015). As such, it was modelled on the national scale and linked epidemiological, LCA and economic modelling. Their use of MACC analysis to relate the treatment/prevention of health issues, GHGs and economic in this way was innovative and to our knowledge has not been replicated elsewhere for livestock systems. However, the MACC methodology is known to have important limitations. For example, it is not able to properly rank cost-negative mitigation measures as traditional MACC methodology will tend to favour scenarios that actually produce lower emission savings (Taylor, 2012; Ward, 2014). More generally, the methodology is viewed as lacking transparency with respect to the calculations in some cases, with limited ability to interpret uncertainty in the underlying data and being unable to account for interactions between individual treatment options (Kesicki and Ekins, 2012). Despite these criticisms it will be important that economic considerations are integrated into future work on health issues and environmental impacts in livestock systems, whether using MACCs or other relevant methods to account for environmental externalities in economic models of livestock systems (e.g. Jongeneel et al., 2014).

Table 3 summarises the studies we found that assessed the impact of specific treatment and prevention strategies to reduce GHGs caused by health issues in animal production systems. Fewer studies have addressed the more difficult task of assessing the impact of treatment and prevention strategies, than those quantifying the impacts caused by health conditions discussed in Section 3. The majority of the studies were modelling exercises applying empirical evidence within an LCA modelling framework. However, some experimental studies, which directly measured GHG emissions were also identified (Fox et al., 2018; Houdijk et al., 2017; Li et al., 2015). The predicted benefits for GHGs from specific treatment/prevention strategies based on the evidence of their effectiveness are naturally more conservative than the theoretical reductions calculated based on eliminating health issues. This is because they need to be based on evidence that the predicted improvements to system performance can be achieved through these treatment and prevention strategies.

The shared challenges between modelling the potential environmental impact mitigation from health interventions and their economic consequences, combined with increased interest in understanding the potential environmental mitigation from health interventions, mean a guideline framework for researchers constructing such models would be an important resource.

4.2 Integrating epidemiological data in environmental impact assessment

Figure 1 sets out the important components of modelling the environmental impacts of health issues and health interventions in livestock systems. An essential component for understanding the potential of interventions to mitigate health issues is to model the expected changes to production parameters and animal health outcomes from any individual or combination of interventions. This includes mortality rates, fertility rates, the utilisation of feed and quality characteristics of products from the farming system. This is consistent with the frameworks currently used by LCAs to assess the consequences of management changes on the environmental impact of a livestock system. The level of data required for the epidemiological component of the exercise is entirely dependent on the scale at which researchers model on, and in many cases appropriate data may only be available at the farm or herd level, which can be very difficult to scale-up to model interventions at the regional or national level (Garner and Hamilton, 2011; Rushton, 2017). This is a problem for LCA models, too, which often set out to present the impact of health conditions at the national or regional level. Work to date modelling the

Table 3 A summary of quantitative results reported in literature on animal health issues and associated treatment/prevention strategies. Each reported outcome represents the potential mitigation of GHGs shown using the treatment/prevention strategies. Where a study addressed more than one disease, its results are reported separately in the table below.

Study	Study type	Species (System)	Disease/s studied	Health intervention/s	Reported GHGs outcome	Functional units and other methodological details	Other notes
(Houdijk et al., 2017)	Experimental trial	Sheep (Ewes)	Parasitic worm infection (*Teladorsagia circum- cincta*)	Anthelminthics	−14%	1 kg Live weight gain	Increases and decreases shown for different sources of GHG production.
(Kenyon et al., 2013)	Experimental trial	Sheep (Lambs)	Gastrointestinal nematodes	Anthelminthics (four strategies – 'monthly' (NST), 'strategic' (SPT), 'targeted' (TST) or based on 'clinical signs' (MT)	−9% for best treatment case compared to worst case	1 kg Live weight gain	All groups in trial treated with Anthelminthics just using different strategies.
(Li et al., 2015)	Experimental trial	Pigs	Porcine reproductive and respiratory syndrome virus (PRRSv)	Vaccination for PRRSv	+11% CO_2 +300% N_2O NH_3 CH_4 and H_2S all unchanged	Figures represent increase in emissions from manure storage when unvaccinated pigs infected with PRRSv	No overall GHG assessment.
(Skuce et al., 2016)	Modelling	Cattle (dairy)	Infectious bovine rhinotracheitis (IBR) disease		−1.5 to 3%	1 kg carcass weight	Reduction based on any one of control measures listed being implemented.

(Hospido and Sonesson, 2005)	Modelling	Cattle (dairy)	Mastitis	Scenarios for blanket and selective dry cow therapy	−2.5% GWP in 'improved herd' scenario	1 L of milk	Improved scenario was based on a reduction in discarded milk and culling. Higher reductions reported for other impact categories which included eutrophication, acidification and ozone depletion.
(ADAS, 2015)	Modelling	Cattle (dairy and beef)	Bovine viral diarrhoea (BVD)	Vaccination, double fencing, identification of PI animal	−14% dairy and −50% beef	1 L FPCM milk and 1 tonne edible carcass weight	Values represent the impact of implementing all treatments in model compared to no treatment.
(ADAS, 2015)	Modelling	Cattle (dairy and beef)	Johne's disease	Vaccination, colostrum management and hygiene, buying policy, test and cull	−13% dairy and −21% beef	1000 L FPCM milk and 1 tonne edible carcass weight	Values represent the impact of implementing all treatments in model compared to no treatment.
(ADAS, 2015)	Modelling	Cattle (dairy and beef)	Salmonella	Vaccination, vector control, hygiene	−15% dairy and −13% beef	1000 L FPCM milk and 1 tonne edible carcass weight	Values represent the impact of implementing all treatments in model compared to no treatment.
(ADAS, 2015)	Modelling	Cattle (dairy and beef)	Liver flukes	Strategic treatment, grazing management	−7% dairy and −5% beef	1000 L FPCM milk and 1 tonne edible carcass weight	Values represent the impact of implementing all treatments in model compared to no treatment.
(ADAS, 2015)	Modelling	Cattle (dairy and beef)	IBR	Vaccination, double fencing, identification of latently infected carrier animal	−6% dairy and −13% beef	1000 L FPCM milk and 1 tonne edible carcass weight	Values represent the impact of implementing all treatments in model compared to no treatment.

(Continued)

Table 3 *(Continued)*

Study	Study type	Species (System)	Disease/s studied	Health intervention/s	Reported GHGs outcome	Functional units and other methodological details	Other notes
(ADAS, 2015)	Modelling	Cattle (dairy and beef)	Lameness	Change in cow 'hardiness', mobility management, cow comfort	−4% dairy only	1000 L FPCM milk and 1 tonne edible carcass weight	Values represent the impact of implementing all treatments in model compared to no treatment.
(ADAS, 2015)	Modelling	Cattle (dairy and beef)	Mastitis	Milking routine, dry cow therapy, nutrition, housing and milking machine maintenance	−1% dairy only	1000 L FPCM milk and 1 tonne edible carcass weight	Values represent the impact of implementing all treatments in model compared to no treatment.
(ADAS, 2015)	Modelling	Cattle (dairy and beef)	Infertility	Fixed time AI, tail paint/kamar and activity metres, nutrition	−10% dairy and −12% beef	1000 L FPCM milk and 1 tonne edible carcass weight	Values represent the impact of implementing all treatments in model compared to no treatment.
(ADAS, 2015)	Modelling	Cattle (dairy and beef)	Calf pneumonia	Vaccination, colostrum intake, buildings, stocking density and buying policy	−<1% dairy and −1% beef	1000 L FPCM milk and 1 tonne edible carcass weight	Values represent the impact of implementing all treatments in model compared to no treatment.
(ADAS, 2015)	Modelling	Cattle (dairy and beef)	Calf diarrhoea	Housing and calf comfort, prophylactic therapy, vaccination and colostrum screening	−<1% dairy and −1% beef	1000 L FPCM milk and 1 tonne edible carcass weight	Values represent the impact of implementing all treatments in model compared to no treatment.

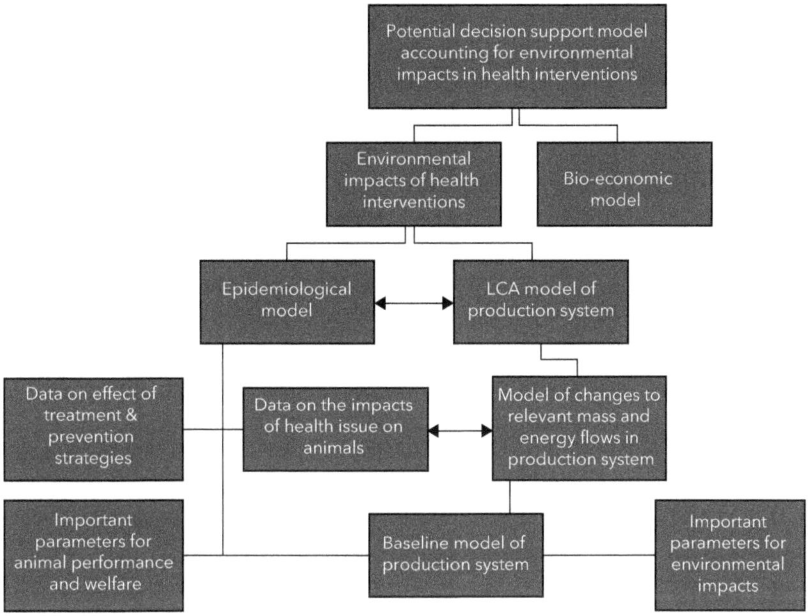

Figure 1 The important components for modelling the environmental impacts of health issues and health interventions in livestock systems through Life Cycle Assessment (LCA).

environmental impact of health interventions has acknowledged its limitations with regard to the robustness of the scenarios presented for animal health outcomes (e.g. ADAS, 2015). However, this may be an unfairly high standard to judge early work in this area, which can only work with the best data available to further our understanding of the potential for successful health interventions to reduce environmental impacts.

More challenging, when dealing with health issues, are the spatial and temporal scales that need to be taken into account to assess their effectiveness. Setting boundaries for the technical system being modelled is necessary in LCA. Commonly for LCA of livestock systems the model may be set up to represent a representative farming scenario within a region, without any precise definition of its location or spatial properties (McAuliffe et al., 2016). However, the effectiveness of many treatment and prevention strategies can only be considered at larger regional or national scales. In general, epidemic diseases affect multiple farms, and therefore have a strong spatial component. For example, it may be necessary to assume sufficient uptake of a preventative vaccine within a region to establish herd immunity without the need for extra controls on travelling animals, for example, in the case of bovine TB (Smith and Delahay, 2018). The same applies to endemic diseases whose transmission depends on a biological vector, such as trypanosomiasis. For this reason, several researchers have modelled the eradication of a disease as opposed to controlling a health challenge, as discussed above (Macleod et al., 2018; Skuce et al., 2016). This approach also comes with complexity, an example being the models of preventing IBR in cattle in both the ADAS (2015) report and later by Skuce et al. (2016). The latter study acknowledged that it may not be possible to achieve the eradication of the condition in the UK as has been shown to be possible elsewhere (Ackermann et al., 1990), without a full national programme of compulsory control including testing, culling vaccination and restriction of movement. As such, the implementation of control strategies on individual farms, such as vaccinations or double fencing, is predicted to have a reduced effect in disease reduction when introduced individually. In contrast, some treatment and prevention strategies can be pursued successfully in individual farms and can thus have 'isolated' consequences. For example, such is the case of hygiene measures: the washing of udders and general improvements to cleanliness of housing to reduce instances of mastitis (Lam et al., 2013). The challenges here are faced in all epidemiological modelling of transboundary diseases (FAO, 2016d): researchers need to justify why the scale at which they are modelling the treatment and prevention of health issues is appropriate and ensure that where possible they have considered that any issues they face with upscaling data from trials done at the farm level. They should also consider relevant interactions between different treatment strategies where possible.

Similarly, there are key temporal considerations in epidemiological modelling, with temporal scale being key to understand the effectiveness of prevention and treatment strategies (Onstad, 1992). For example, if one is interested in quantifying the environmental impact of a health challenge control over time, then the temporal effects on the epidemiology of the challenge will need to be taken into account. Currently, there is an increased interest in showing the reductions in environmental impact of a livestock sector achieved through management or breeding (Tallentire et al., 2018; Ottosen et al. 2020), and this exercise has analogies with what we are discussing here in relation to health challenge controls.

4.3 Identifying the impact of health issues and interventions on important mass and energy flows

Translating the outcomes of any epidemiological model of a health intervention into a model of the mass and energy flows for an animal production system is a significant challenge. One of the particular challenges here is that sensitive aspects of the LCA model may not be routinely recorded, even in experimental trials measuring the effectiveness of treatment and prevention strategies in reducing specific health conditions. For example, while aspects of animal performance such as feed intake, growth rate and milk or egg production would be routinely measured, the effect of health conditions on enteric methane production or the amount of volatile solids contained in manure would not be. While predictions of the environmental impact mitigation achieved from prevention and treatment strategies can be made just from traditional performance data, factors such as enteric methane and volatile solid excretion are extremely sensitive in models of environmental impacts in animal production (Opio et al., 2013). Most calculations relating GHGs and health issues in animal production use a fixed equation based on the energy and dry matter intake of animals to calculate enteric methane emissions. The methods either directly follow, or are derived from the tier 2 IPCC protocols for calculating methane emissions from livestock (IPCC, 2006). Several studies reference the fact that this is a sensitive assumption within their models which could be improved (Macleod et al., 2018; Mostert et al., 2018). Importantly, these emission factors and the predicted enteric methane emissions are not altered when considering the scenarios impacted by health conditions. This may actually not be true. Evidence of this can be seen from studies that have the relevant measurements in experiments (Fox et al., 2018; Houdijk et al., 2017). In the Houdijk et al. (2017) study, peri-parturient parasitism reduced feed intake by 9% and litter weight gain by 7% for rearing ewes, while also doubling maternal weight loss, and these were, at least partly, caused by parasite-induced anorexia. While net methane production was actually reduced, the methane yield per kg of digestible organic matter intake increased by 14%. Infection caused reduced feed intake depriving the animal of needed resources, reduced the efficiency with which feed was being used and increased the proportion of organic material being released as volatile organic compounds. The manure methane and nitrous oxide yields also increased 46% and 31% respectively per kg of digestible organic matter intake. These increased yields were a result of reduced nutrient absorption, which is a symptom of parasite infection (Coop and Kyriazakis, 1999; Shea-Donohue et al., 2017), and also contribute to the increased GHGs caused by untreated infection in this study.

These recorded increases in the yield of methane and nitrous oxide are important points for considering the environmental impact of health issues. Models that assume a constant relationship between feed intake and enteric methane emissions, as well as emissions from manure, may underestimate the increase in GHGs and other important emissions caused by health issues affecting the digestive system. To date these detailed experimental studies in relation to health issues and the production of GHGs have been conducted on sheep systems (Fox et al., 2018; Houdijk et al., 2017). However, similar findings could be reasonably predicted in dairy and beef systems for conditions involving parasitism and/or instances of diarrhoea. Increases in the methane and nitrous oxide yields per kg of digestible organic matter intake would increase the predicted increase in GHGs caused by such health issues produced by studies such as by ADAS (2015) on dairy and beef systems. Notably, Houdijk et al (2017) also found that release of respiratory CO_2 increased per kg intake of digestible organic matter by 17% and in the LCA results overall by 15%. CO_2 released this way is almost never taken into account in LCA models of livestock systems and certainly not modelled to increase in any of the other studies on health issues found in this review. These results suggest that the impact of some health issues on GHGs in livestock systems may be underestimated in many of the current LCA models produced. If possible, LCA models of health issues in livestock and potential mitigations should identify the relevant mechanisms within animals by which health issues may increase GHGs produced on farm from livestock systems, as well as accounting for expected changes to herd performance as is already typical.

Another important consideration in terms of mass and energy flows when modelling the environmental impact of health issues in livestock impact is any impact they may have on the quality of the output products. This is a different issue to some of the examples discussed earlier where milk may fail quality tests or for other reasons have to be discarded as a result of health issues for this is easily accounted for as waste within an LCA model. A relevant example in this case would be an alteration in carcass composition that reduced the quality of meat produced by sheep or beef cattle as a result of being challenged with gastrointestinal parasites (Holmes, 1993). Accounting for such issues is linked to a wider methodological challenge that researchers face when constructing LCA models of livestock systems that can account for the quality of the product produced. The key issue being the importance of the chosen functional unit in LCA modelling (Hospido et al., 2010).

The currently accepted convention, and standard way of representing functional unit in LCA of meat production systems is to use a variation on edible meat or carcass weight with no further properties describing the nutritional quality of the meat produced (e.g. ADAS, 2015; FAO, 2016c, 2018). Essentially its chemical composition and physical characteristics are usually assumed to be either unrelated to whatever treatments (changes to diet, health treatment, management changes) a livestock LCA has considered or it is acknowledged (not always explicitly) that the model was simply not able to take the relationship between these two things into account. This situation contrasts slightly with the standard functional unit used in dairy LCA of fat and protein-corrected milk (FPCM) (FAO, 2016a) using methods such as that set out by Bartl et al. (2011):

For conversion of kg milk to kg FPCM the yield was adjusted for energy to a standard milk with 4% fat, 3.2% protein and 4.8% lactose using the formula FPCM (kg/day) ¼ milk (kg/day) [0.038 fat (g/kg) þ 0.024 protein (g/kg) þ 0.017 lactose (g/kg)]/3.14 (Bartl et al., 2011).

A few studies of dairy systems build further on this to include either FPCM produced per land occupied, or to include the meat produced and sold from dairy systems in the functional unit, but the vast majority use a mass or volume measurement of FPCM (FAO, 2016a). More broadly in LCA of food systems there has been a trend towards developing nutritional functional units based on a set of nutritional properties for food products, which match guidelines for human dietary requirements (Saarinen et al., 2017). For LCA of livestock production systems McAuliffe et al. (2018) presented such a methodology based on using important nutritional properties for high-protein foods adapted from the methods of Saarinen et al. (2017) for Finnish food systems. To adopt such an approach requires confidently linking the consequences of production diseases for the chemical properties of livestock tissue. If such data is available for the health issues and treatments/prevention strategies being considered in LCA models, then it should be accounted for using an appropriate functional unit as discussed above, but to date this has not been the case for any studies of health issues in meat production systems.

4.4 Beyond health issues for ruminants and greenhouse gas emissions

Work to date on the environmental impacts of health issues in livestock has focussed almost exclusively on ruminant systems. We found only one study that specifically focussed on the environmental impact of health issues in non-ruminant systems. Li et al. (2015) assessed the implications of PRRSv infection on growth, dietary nutrient utilisation efficiency, manure output, as well as emissions of CO_2, CH_4, H_2S, N_2O and NH_3 gases from stored manure in both vaccinated and unvaccinated pigs. This study was able to show that in addition to the expected reduction in growth performance, exposure to PRRSv significantly increased CO_2 and N_2O released both directly from pigs and their manure when stored, with no observed impact on NH_3 or CH_4 emissions. Manure from infected pigs had a greater C:N ratio, and infection with PRRSv-increased manure output did not have an observed impact on the utilisation and retention of N in infected animals. These increased levels of organic matter excreted explained the increased CO_2 emissions from their manure. The increased C:N ratio was also observed to stimulate more fermentation and microbial activity during manure storage. Reaction routes as part of this increased microbial activity during storage were thought to cause the discrepancy between N_2O (increased) and NH_3 (no change) emissions from the manure of infected pigs.

Ultimately, the study was unable to demonstrate that prior vaccination reduced the negative impact of PRRSv on animal performance or GHG emissions (Li et al., 2015). While no overall GHG assessment was performed, the results of their work show that PRRSv caused an increase in the GHGs (increased CO_2 and N_2O) from pig production systems beyond the expected drops in animal performance. As with the case of parasite infection in sheep production discussed above, failure to take into account the link between infection and increase in on-farm GHG emissions may lead to researchers underestimating the impact of these conditions in LCA studies. Moreover, while their results were not fed into an LCA model, it was notable that neither NH_3 nor H_2S emissions from the pigs and during manure storage increased when animals were infected with PRRSv nor did infection have any detectable impact on nutrient utilisation. Nutrient utilisation efficiency and NH_3 emissions in particular, both in barn and during manure storage, are a major factor in determining the acidification and eutrophication caused by pig farming systems (Mackenzie et al., 2015; McAuliffe et al., 2016).

Most studies on the environmental impact of health issues (with exceptions, Chen et al., 2016; Hospido and Sonesson, 2005) did not consider other forms of environmental impact beyond GHGs, such as eutrophication, acidification and abiotic resource use. The focus on ruminant systems to date in this area is understandable as discussed above. However, when considering the environmental impacts of livestock production in broader terms beyond just GHGs, it becomes much more important to account for the role of non-ruminant production systems (de Vries and de Boer, 2010). Focussing on a single environmental impact does allow researchers to concentrate on translating the outputs of epidemiological models into fewer important mass flows. For example, LCA models which consider only GHGs will account solely for CH_4, N_2O, CO_2 and any hydrofluorocarbon emissions released during by the production system considered (Opio et al., 2013). Accounting for further impact categories such as acidification potential or water use, as recommended in most of the FAO technical guidelines on livestock LCA (FAO, 2016c, 2016b, 2018), would mean that researchers are more likely to capture potential trade-offs in the environmental impact benefits of health interventions. While each impact category added to the model increases uncertainty, as researchers have to establish the potential impact of health issues and treatment strategies on a greater number of physical flows in the system, it will be important to overcome these issues in future work in this area. As a case in point, the results from the Li et al. (2015) study are indicative that certain health conditions may have greater implications for some forms of environmental impact over others. Depending on the context that studies are carried out in, researchers may deem health issues that decrease nutrient utilisation efficiency, thus increasing NH_3 and other emissions are more problematic from an environmental impact perspective than those which do not in pig and poultry farming systems. For example, within the EU pig production represents around 25% of NH_3 emissions (Philippe et al., 2011) and EU has now defined standards which dictate limits for NH_3 emissions to air in housing as well as N and P excretions in manure (EU Science Hub, 2017). As such those looking to quantify which treatment and prevention strategies could yield the most important mitigation of environmental impacts in EU pig and poultry systems may wish to prioritise the reduction of nutrient excretion in their models. They would need to be able to identify the relevant mechanisms for health issues in pig and poultry systems in relation to nutrient utilisation and in barn emissions set out for PRRSv in the Li et al. (2015) study. The LCA models would also need to account for multiple impact categories in order to interpret these data into useful outputs for decision-making.

5 Conclusions

We reviewed important concepts for quantifying the relationship between health issues in livestock systems and their environmental impacts, as well as the studies published on this issue to date. In doing so we outlined a general framework and guidelines for constructing such models, highlighted a number of important methodological considerations for modelling in this area and outlined some significant gaps in current knowledge of this topic that should be addressed in future research. As this area of modelling requires the integration of epidemiological models with environmental impact models, there are unique temporal and spatial considerations to account for when modelling livestock health issues.

It is also important when integrating these two areas of modelling that the impact of health issues on sensitive mass flows within the animal and on farm, such as methane or ammonia emissions are taken into account. In most cases, researchers assume a constant relationship between the feed intake or energy intake of animals and on-farm emissions in their models of health issues and GHGs (Macleod et al., 2018; Mostert et al., 2018). However, while initial experimental work in this area to date suggests this may not always be the case, studies that have measured the relationships between on-farm emissions and health issues are sparse, presenting a challenge in developing effective models on this subject. Wherever possible they should be accounted for in models of the environmental impact of health issues, otherwise this will likely lead to inaccuracies in reporting the environmental impact of health issues in livestock. Likewise, no studies have thus far utilised relevant methods to account for the implications of health issues for product quality with respect to their environmental impacts. New methods being developed in this area for nutritional functional units in LCA studies of food systems (McAuliffe et al., 2018; Sonesson et al., 2017) should allow researchers to account for this in future where relevant data is available, for example, with respect to health challenges and product composition.

In most cases, the implications of health issues for environmental impacts beyond GHGs of livestock systems have not been considered or quantified to date. This contradicts several FAO guideline documents on how to quantify the environmental impact of livestock systems and represents a significant gap in our knowledge of the issue. Surprisingly few studies have considered the impact of health issues in pig and poultry production systems at all, with these sectors making up an ever-increasing proportion of livestock products consumed around the world, and this is an area that will warrant future exploration by researchers.

Ultimately, in order to actually be integrated into decision-making processes, environmental impact assessment models of health interventions will need to integrate economic considerations to produce recommendations for the most cost-effective mitigation measures. The ADAS (2015) report demonstrated the possibilities for this while modelling at a national scale. Work at this level could be used to justify funding or subsidies for particular health interventions at the national level on the combination of animal welfare and environmental grounds. However, there may also be scope for impactful work in this area to be done through farm level modelling. This could integrate health interventions and their potential benefits for environmental impacts into decision support tools as is being currently being developed in other areas of livestock husbandry such as feeding (Garcia-Launay et al., 2018; Mackenzie et al., 2016), breeding (Ottosen et al., 2020; Tallentire et al., 2018) and environmental management (Pexas et al., 2020; Ten Hoeve et al., 2014).

6 Where to look for further information

- The Oxford University Research Archive (https://doi.org/10.5287/bodleian: 0z9MYbMyZ) stores a streamlined global database of GHGs associated with food and beverage production systems.
- For interesting discussion on the challenges of economics in animal health modelling – Rushton, J. 2017a. Improving the use of economics in animal health – Challenges in research, policy and education. Prev. Vet. Med. 137, 130–139. https://doi.org/10.1016/J.PREVETMED.2016.11.020.
- The latest FAO technical guidelines published on developing LCA models of livestock systems can be found here – http://www.fao.org/partnerships/leap/publications/guidelines/en/.
- For an overview of basic principles and important methodological issues in LCA modelling – Curran, M. A. 2012. Life cycle assessment handbook: a guide for environmentally sustainable products. Scrivener Publishing LLC, Beverley, MA.
- ADAS, 2015. Study to Model the Impact of Controlling Endemic Cattle Diseases and Conditions on National Cattle Productivity, Agricultural Performance and Greenhouse Gas Emissions. Final Report to Defra/AHVLA on Project FFG1016. 210.

7 References

Ackermann, M., Müller, H. K., Bruckner, L. and Kihm, U. 1990. Eradication of infectious bovine rhinotracheitis in Switzerland: review and prospects. *Vet. Microbiol.* 23(1-4), 365-70. doi:10.1016/0378-1135(90)90168-U.

ADAS 2015. *Study to Model the Impact of Controlling Endemic Cattle Diseases and Conditions on National Cattle Productivity, Agricultural Performance and Greenhouse Gas Emissions.* Final Report to Defra/ AHVLA on Project FFG1016 (vol. 210).

Andersson, L. 1988. Subclinical ketosis in dairy cows. *Vet. Clin. North Am. Food Anim. Pract.* 4(2), 233-51. doi:10.1016/S0749-0720(15)31046-X.

Bartl, K., Gómez, C. A. and Nemecek, T. 2011. Life cycle assessment of milk produced in two smallholder dairy systems in the highlands and the coast of Peru. *J. Clean. Prod.* 19(13), 1494-505. doi:10.1016/j.jclepro.2011.04.010.

Beauchemin, K. A. 2007. Ruminal acidosis in dairy cows: balancing physically effective fiber with starch availability. In: *Florida Ruminant Nutrition Symposium,* 30-31 January 2007. Gainesville, FL, pp. 16-27.

Berge, A. C. and Vertenten, G. 2014. A field study to determine the prevalence, dairy herd management systems, and fresh cow clinical conditions associated with ketosis in Western European dairy herds. *J. Dairy Sci.* 97(4), 2145-54. doi:10.3168/jds.2013-7163.

Bernabucci, U., Colavecchia, L., Danieli, P. P., Basiricò, L., Lacetera, N., Nardone, A. and Ronchi, B. 2011. Aflatoxin b1 and fumonisin B1 affect the oxidative status of bovine peripheral blood mononuclear cells. *Toxicol. Vitr.* 25(3), 684-91. doi:10.1016/j.tiv.2011.01.009.

Bennett, R., IJpelaar, J., 2005. Updated Estimates of the Costs Associated with Thirty Four Endemic Livestock Diseases in Great Britain: *A Note. J. Agric. Econ.* 56, 135-144. https://doi.org/10.1111/j.1477-9552.2005.tb00126.x.

Bertoni, G., Trevisi, E., Han, X. and Bionaz, M. 2008. Effects of inflammatory conditions on liver activity in puerperium period and consequences for performance in dairy cows. *J. Dairy Sci.* 91(9), 3300-10. doi:10.3168/jds.2008-0995.

Brunner, N., Groeger, S., Canelas Raposo, J., Bruckmaier, R. M. and Gross, J. J. 2019. Prevalence of subclinical ketosis and production diseases in dairy cows in Central and South America, Africa, Asia, Australia, New Zealand, and Eastern Europe. *Transl. Anim. Sci.* 3(1), 84-92. doi:10.1093/tas/txy102.

Ceï, W., Salah, N., Alexandre, G., Bambou, J. C. and Archimède, H. 2018. Impact of energy and protein on the gastro-intestinal parasitism of small ruminants: a meta-analysis. *Livest. Sci.* 212, 34-44. doi:10.1016/j.livsci.2018.03.015.

Chen, W., White, E. and Holden, N. M. 2016. The effect of lameness on the environmental performance of milk production by rotational grazing. *J. Environ. Manage.* 172, 143-50. doi:10.1016/j.jenvman.2016.02.030.

Collick, D. W., Ward, W. R. and Dobson, H. 1989. Associations between types of lameness and fertility. *Vet. Rec.* 125(5), 103-6. doi:10.1136/vr.125.5.103.

Coop, R. L. and Kyriazakis, I. 1999. Nutrition-parasite interaction. In: *Vet. Parasitol.* 84(3-4), 187-204. doi:10.1016/S0304-4017(99)00070-9.

de Vries, M. and de Boer, I. J. M. 2010. Comparing environmental impacts for livestock products: a review of life cycle assessments. *Livest. Sci.* 128(1-3), 1-11. doi:10.1016/J.LIVSCI.2009.11.007.

Doreau, M., Corson, M. S. and Wiedemann, S. G. 2012. Water use by livestock: a global perspective for a regional issue? *Anim. Front.* 2(2), 9-16. doi:10.2527/af.2012-0036.

Escobar, J., Van Alstine, W. G., Baker, D. H. and Johnson, R. W. 2004. Decreased protein accretion in pigs with viral and bacterial pneumonia is associated with increased myostatin expression in muscle. *J. Nutr.* 134(11), 3047-53. doi:10.1093/jn/134.11.3047.

EU Science Hub 2017. *New EU Environmental Standards for Large Poultry and Pig Farms* [WWW Document]. Available at: https://ec.europa.eu/jrc/en/news/new-eu-environmental-standards-large-poultry-and-pig-farms (Accessed on 9 November 2019).

FAO 2011. *World Livestock 2011: Livestock in Food Security*. Rome, Italy.

FAO 2016a. *Environmental Performance of Large Ruminant Supply Chains Guidelines for Assessment. Livestock Environmental Assessment and Performance Partnership*. Rome, Italy.

FAO 2016b. *Environmental Performance of Large Ruminant Supply Chains Guidelines for Assessment*. Rome, Italy.

FAO 2016c. *Greenhouse Gas Emissions and Fossil Energy Use from Small Ruminant Supply Chains - Guidelines for Assessment. Livestock Environmental Assessment and Performance Partnership*. Rome, Italy.

FAO 2016d. *Economic Analysis of Animal Diseases (No. 18), Animal Production and Health Guidelines*. Rome, Italy.

FAO 2018. *Environmental Performance of Pig Supply Chains Guidelines for Assessment*. Rome, Italy.

FAO 2019. FAOSTAT [WWW Document]. Available at: http://www.fao.org/faostat/en/?#data/QL (Accessed on 30 August 2019).

Fox, N. J., Smith, L. A., Houdijk, J. G. M., Athanasiadou, S. and Hutchings, M. R. 2018. Ubiquitous parasites drive a 33% increase in methane yield from livestock. *Int. J. Parasitol.* 48(13), 1017-21. doi:10.1016/j.ijpara.2018.06.001.

Garcia, S. N., Osburn, B. I. and Cullor, J. S. 2019. A one health perspective on dairy production and dairy food safety. *One Heal.* 7, 100086. doi:10.1016/j.onehlt.2019.100086.

Garcia-Launay, F., Dusart, L., Espagnol, S., Laisse-Redoux, S., Gaudré, D., Méda, B. and Wilfart, A. 2018. Multiobjective formulation is an effective method to reduce environmental impacts of livestock feeds. *Br. J. Nutr.* 120(11), 1298-309. doi:10.1017/S0007114518002672.

Garner, M. G. and Hamilton, S. A. 2011. Principles of epidemiological modelling. *Rev. - Off. Int. Epizoot.* 30(2), 407-16. doi:10.20506/rst.30.2.2045.

Geering, W. A., Roeder, P. L., Obi, T. U. and Food and Agriculture Organization of the United Nations 1999. Manual on the preparation of national animal disease emergency preparedness plans [WWW Document]. Available at: http://www.fao.org/3/X2096E/X2096E06.htm (Accessed on 17 October 2019).

Gerber, P. J., Steinfeld, H., Henderson, B., Mottet, A., Opio, C., Dijkman, J., Falcucci, A. and Tempio, G. 2013. *Tackling Climate Change through Livestock - a Global Assessment of Emissions and Mitigation Opportunities*. Food and Agriculture Organization, Rome.

González, L. A., Tolkamp, B. J., Coffey, M. P., Ferret, A. and Kyriazakis, I. 2008. Changes in feeding behavior as possible indicators for the automatic monitoring of health disorders in dairy cows. *J. Dairy Sci.* 91(3), 1017-28. doi:10.3168/jds.2007-0530.

Gunn, G. J., Stott, A. W. and Humphry, R. W. 2004. Modelling and costing BVD outbreaks in beef herds. *Vet. J.* 167(2), 143-9. doi:10.1016/S1090-0233(03)00112-6.

Hale, O. M., Stewart, T. B. and Marti, O. G. 1985. Influence of an experimental infection of Ascaris suum on performance of pigs. *J. Anim. Sci.* 60(1), 220-5. doi:10.2527/jas1985.601220x.

Hite, J. L., Pfenning, A. C. and Cressler, C. E. 2020. Starving the enemy? Feeding behavior shapes Host-Parasite Interactions. *Trends Ecol. Evol. (Amst.)* 35(1), 68-80. doi:10.1016/j.tree.2019.08.004.

Holmes, P. H. 1993. Interactions between parasites and animal nutrition: the veterinary consequences. *Proc. Nutr. Soc.* 52(1), 113-20. doi:10.1079/pns19930043.

Hospido, A. and Sonesson, U. 2005. The environmental impact of mastitis: a case study of dairy herds. *Sci. Total Environ.* 343(1-3), 71-82. doi:10.1016/j.scitotenv.2004.10.006.

Hospido, A., Davis, J., Berlin, J. and Sonesson, U. 2010. A review of methodological issues affecting LCA of novel food products. *Int. J. Life Cycle Assess.* 15(1), 44-52. doi:10.1007/s11367-009-0130-4.

Houdijk, J. G. M., Tolkamp, B. J., Rooke, J. A. and Hutchings, M. R. 2017. Animal health and greenhouse gas intensity: the paradox of periparturient parasitism. *Int. J. Parasitol.* 47(10-11), 633-41. doi:10.1016/j.ijpara.2017.03.006.

Houe, H. 1999. Epidemiological features and economical importance of bovine virus diarrhoea virus (BVDV) infections. In: *Vet. Microbiol.* 64(2-3), 89-107. doi:10.1016/S0378-1135(98)00262-4.

IPCC 2006. Chapter 10: Emissions From Livestock and Manure Management. In: *Agriculture, Forestry and Other Land Use*. https://doi.org/http://www.ipcc-nggip.iges.or.jp/public/2006gl/index.html.

Jongeneel, R., Polman, N. and Van Der Ham, C. 2014. *Costs and Benefits Associated with the Externalities Generated by Dutch Agriculture*. In: *14th EAAE Congress*, 26-29 August 2014. Ljubljana, Slovenia.

Kenyon, F., Dick, J. M., Smith, R. I., Coulter, D. G., McBean, D. and Skuce, P. J. 2013. Reduction in greenhouse gas emissions associated with worm control in lambs. *Agriculture* 3(2), 271-84. doi:10.3390/agriculture3020271.

Kesicki, F. and Ekins, P. 2012. Marginal abatement cost curves: A call for caution. *Clim. Policy* 12(2), 219-36. doi:10.1080/14693062.2011.582347.

Kleen, J. L. and Cannizzo, C. 2012. Incidence, prevalence and impact of SARA in dairy herds. *Anim. Feed Sci. Technol.* 172(1-2), 4-8. doi:10.1016/j.anifeedsci.2011.12.003.

Kliebenstein, J. B., Johnson, C., Neumann, E., Marby, J., Bush, E., Seitzinger, A. and Green, A. 2004. *Economic Impact of PRRS on the Cost of Pork Production*. Iowa State University, https://www.pork.org/wp-content/uploads/2006/07/02-223-KLIEBENSTEIN.3-5-04.pdf

Kyriazakis, I., Tolkamp, B. J. and Hutchings, M. R. 1998. Towards a functional explanation for the occurrence of anorexia during parasitic infections. *Anim. Behav.* 56(2), 265-74 doi:10.1006/anbe.1998.0761.

Kyriazakis, I. and Houdijk, J. G. M. 2007. Food intake and performance of pigs during health, disease and recovery. In: Wiseman, J., Varley, M. A., McOrist, S. and Kemp, B. (Eds), *Paradigms in Pig Science*. Nottingham University Press, Nottingham, UK, pp. 493-513.

Kyriazakis, I. 2014. Pathogen-induced anorexia: a herbivore strategy or an unavoidable consequence of infection? *Anim. Prod. Sci.* 54(9). 1190-7. doi:10.1071/AN14431.

Lacetera, N. 2019. Impact of climate change on animal health and welfare. *Anim. Front.* 9(1), 26–31. doi:10.1093/af/vfy030.

Lam, T. J. G. M., Van Den Borne, B. H. P., Jansen, J., Huijps, K., Van Veersen, J. C. L., Van Schaik, G. and Hogeveen, H. 2013. Improving bovine udder health: a national mastitis control program in the Netherlands. *J. Dairy Sci.* 96(2), 1301–11. doi:10.3168/jds.2012-5958.

Leip, A., Billen, G., Garnier, J., Grizzetti, B., Lassaletta, L., Reis, S., Simpson, D., Sutton, M. A., de Vries, W., Weiss, F. and Westhoek, H. 2015. Impacts of European livestock production: nitrogen, sulphur, phosphorus and greenhouse gas emissions, land-use, water eutrophication and biodiversity. *Environ. Res. Lett.* 10(11), 115004. doi:10.1088/1748-9326/10/11/115004.

Li, M. M., Seelenbinder, K. M., Ponder, M. A., Deng, L., Rhoads, R. P., Pelzer, K. D., Radcliffe, J. S., Maxwell, C. V., Ogejo, J. A. and Hanigan, M. D. 2015. Effects of porcine reproductive and respiratory syndrome virus on pig growth, diet utilization efficiency, and gas release from stored manure. *J. Anim. Sci.* 93(9), 4424–35. doi:10.2527/jas.2015-8872.

Liang, D. 2013. *Estimating the Economic Losses from Diseases and Extended Days Open with a Farm-Level Stochastic Model*. University of Kentucky, Lexington, Kentucky. https://uknowledge.uky.edu/cgi/viewcontent.cgi?article=1022&context=animalsci_etds

Linhares, D. C. L., Johnson, C. and Morrison, R. B. 2015. Economic analysis of vaccination strategies for PRRS control. *PLoS ONE* 10(12), e0144265. doi:10.1371/journal.pone.0144265.

Lough, G., Kyriazakis, I., Bergmann, S., Lengeling, A. and Doeschl-Wilson, A. B. 2015. Health trajectories reveal the dynamic contributions of host genetic resistance and tolerance to infection outcome. *Proc. Biol. Sci.* 282 (1819), 2151–61. doi:10.1098/rspb.2015.2151.

Lucey, S., Rowlands, G. J. and Russell, A. M. 1986. The association between lameness and fertility in dairy cows. *Vet. Rec.* 118(23), 628–31. doi:10.1136/vr.118.23.628.

Mackenzie, S. G., Leinonen, I., Ferguson, N. and Kyriazakis, I. 2015. Accounting for uncertainty in the quantification of the environmental impacts of Canadian pig farming systems. *J. Anim. Sci.* 93(6), 3130–43. doi:10.2527/jas.2014-8403.

Mackenzie, S. G., Leinonen, I., Ferguson, N. and Kyriazakis, I. 2016. Towards a methodology to formulate sustainable diets for livestock: accounting for environmental impact in diet formulation. *Br. J. Nutr.* 115(10), 1860–74. doi:10.1017/S0007114516000763.

Macleod, M., Id, V. E., Wint, W., Shaw, A., Gerber, P. J., Id, G. C., Mattioli, R., Sykes, A. and Robinson, T. 2018. Assessing the greenhouse gas mitigation effect of removing bovine trypanosomiasis in eastern Africa. *Sustainability*, 1–19. doi:10.3390/su10051633.

Mavrot, F., Hertzberg, H. and Torgerson, P. 2015. Effect of gastro-intestinal nematode infection on sheep performance: A systematic review and meta-analysis. *Parasites Vectors* 8, 557. doi:10.1186/s13071-015-1164-z.

McAuliffe, G. A., Chapman, D. V. and Sage, C. L. 2016. A thematic review of life cycle assessment (LCA) applied to pig production. *Environ. Impact Assess. Rev.* 56, 12–22. doi:10.1016/j.eiar.2015.08.008.

McAuliffe, G. A., Takahashi, T. and Lee, M. R. F. 2018. Framework for life cycle assessment of livestock production systems to account for the nutritional quality of final products. *Food Energy Secur.* 7(3), e00143. doi:10.1002/fes3.143.

Midha, A., Janek, K., Niewienda, A., Henklein, P., Guenther, S., Serra, D. O., Schlosser, J., Hengge, R. and Hartmann, S. 2018. The intestinal roundworm Ascaris suum releases antimicrobial factors which interfere with bacterial growth and biofilm formation. *Front. Cell. Infect. Microbiol.* 8, 271. doi:10.3389/fcimb.2018.00271.

Mostert, P. F., van Middelaar, C. E., Bokkers, E. A. M. and de Boer, I. J. M. 2018. The impact of subclinical ketosis in dairy cows on greenhouse gas emissions of milk production. *J. Clean. Prod.* 171, 773–82. doi:10.1016/j.jclepro.2017.10.019.

O'Rourke, D. 2009. Nutrition and udder health in dairy cows: a review. *Ir. Vet. J.* 62 (Suppl. 4), S15–20. doi:10.1186/2046-0481-62-S4-S15.

Onstad, D. W. 1992. Temporal and spatial scales in epidemiological concepts. *J. Theor. Biol.* 158(4), 495–515. doi:10.1016/s0022-5193(05)80712-x.

Opio, C., Gerber, P., Mottet, A., Falcucci, A., Tempio, G., MacLeod, M., Vellinga, T., Henderson, B. and Steinfeld, H. 2013. *Greenhouse Gas Emissions from Ruminant Supply Chains: A Global Life Cycle Assessment*. Food and Agriculture Organization of the United Nations (FAO), Rome, Italy.

Otesile, E. B., Fagbemi, B. O. and Adeyemo, O. 1991. The effect of Trypanosoma brucei infection on serum biochemical parameters in boars on different planes of dietary energy. *Vet. Parasitol.* 40(3–4), 207–16. doi:10.1016/0304-4017(91)90101-z.

Ottosen, M., Mackenzie, S. G. and Kyriazakis, I. 2020 A method to account for the effect of genetic traits of pigs on the environmental impacts from their production system. *Int. J. Life Cycle Assess.* 25, 523–37. doi: 10.1007/s11367-019-01686-8

Özkan, Ş., Ahmadi, B. V., Bonesmo, H., Østerås, O., Stott, A. and Harstad, O. M. 2015. Impact of animal health on greenhouse gas emissions. *Adv. Anim. Biosci.* 6(1), 24–5. doi:10.1017/S2040470014000454.

Ózsvári, L. 2017. Economic cost of lameness in dairy cattle herds. *Anim. Res. Vet, Dairy* 6(2), 6. doi:10.15406/jdvar.2017.06.00176.

Perry, B. D., Robinson, T. P. and Grace, D. C. 2018. Review: animal health and sustainable global livestock systems. *Animal* 12(8), 1699–708. doi:10.1017/S1751731118000630.

Pexas, G., Mackenzie, S. G., Wallace, M. and Kyriazakis, I. 2020. Environmental impacts of housing and manure management in European pig production systems through a life cycle perspective: a case study in Denmark. *J. Clean. Prod.* 253, e120005

Philippe, F. X., Cabaraux, J. F. and Nicks, B. 2011. Ammonia emissions from pig houses: influencing factors and mitigation techniques. *Agric. Ecosyst. Environ.* 141(3–4), 245–60. doi:10.1016/j.agee.2011.03.012.

Poore, J. and Nemecek, T. 2018. Reducing food's environmental impacts through producers and consumers. *Science* 360(6392), 987–92. doi:10.1126/SCIENCE.AAQ0216.

Rochell, S. J., Parsons, C. M. and Dilger, R. N. 2016. Effects of Eimeria Acervulina infection severity on growth performance, apparent ileal amino acid digestibility, and plasma concentrations of amino acids, carotenoids, and α1-acid glycoprotein in broilers. *Poult. Sci.* 95(7), 1573–81. doi:10.3382/ps/pew035.

Rollin, E., Dhuyvetter, K. C. and Overton, M. W. 2015. The cost of clinical mastitis in the first 30 days of lactation: an economic modeling tool. *Prev. Vet. Med.* 122(3), 257–64. doi:10.1016/j.prevetmed.2015.11.006.

Roth, J. A. 2011. Veterinary vaccines and their importance to animal health and public health. *Procedia Vaccinol* 5, 127–36. doi:10.1016/j.provac.2011.10.009.

Rushton, J. 2017. Improving the use of economics in animal health – challenges in research, policy and education. *Prev. Vet. Med.* 137(B), 130-9. doi:10.1016/J.PREVETMED.2016.11.020.

Saarinen, M., Fogelholm, M., Tahvonen, R. and Kurppa, S. 2017. Taking nutrition into account within the life cycle assessment of food products. *J. Clean. Prod.* 149, 828-44. doi:10.1016/J.JCLEPRO.2017.02.062.

Sandberg, F. B., Emmans, G. C. and Kyriazakis, I. 2006. A model for predicting feed intake of growing animals during exposure to pathogens. *J. Anim. Sci.* 84(6), 1552-66. doi:10.2527/2006.8461552x.

Sandberg, F. B., Emmans, G. C. and Kyriazakis, I. 2007. The effects of pathogen challenges on the performance of naïve and immune animals: the problem of prediction. *Animal* 1(1), 67-86. doi:10.1017/S175173110765784X.

Shea-Donohue, T., Qin, B. and Smith, A. 2017. Parasites, nutrition, immune responses and biology of metabolic tissues. *Parasite Immunol.* 39(5). doi:10.1111/pim.12422.

Skuce, P. J., Bartley, D. J., Zadoks, R. N. and Macleod, M. 2016. *Livestock Health & Greenhouse Gas Emissions.* ClimateXChange Report for Gordon Struth & Keith McWhinnie, Scottish Government, Climate Change & Agriculture, 1-74.

Smith, G. C. and Delahay, R. J. 2018. Modeling as a decision support tool for bovine TB control programs in wildlife. *Front. Vet. Sci.* 5, 276. doi:10.3389/fvets.2018.00276.

Sonesson, U., Davis, J., Flysjö, A., Gustavsson, J. and Witthöft, C. 2017. Protein quality as functional unit - A methodological framework for inclusion in life cycle assessment of food. *J. Clean. Prod.* 140, 470-8. doi:10.1016/j.jclepro.2016.06.115.

Tallentire, C. W., Mackenzie, S. G. and Kyriazakis, I. 2018. Can novel ingredients replace soybeans and reduce the environmental burdens of European livestock systems in the future? *J. Clean. Prod.* 187, 338-47. doi:10.1016/J.JCLEPRO.2018.03.212.

Taylor, S. 2012. The ranking of negative-cost emissions reduction measures. *Energy Policy* 48, 430-8. doi:10.1016/j.enpol.2012.05.071.

Ten Hoeve, M., Hutchings, N. J., Peters, G. M., Svanström, M., Jensen, L. S. and Bruun, S. 2014. Life cycle assessment of pig slurry treatment technologies for nutrient redistribution in Denmark. *J. Environ. Manage.* 132, 60-70. doi:10.1016/j.jenvman.2013.10.023.

Valdes-Donoso, P., Alvarez, J., Jarvis, L. S., Morrison, R. B. and Perez, A. M. 2018. Production losses from an endemic animal disease: porcine reproductive and respiratory syndrome (PRRS) in selected Midwest US Sow Farms. *Front. Vet. Sci.* 5, 102. doi:10.3389/fvets.2018.00102.

Ward, D. J. 2014. The failure of marginal abatement cost curves in optimising a transition to a low carbon energy supply. *Energy Policy* 73, 820-2. doi:10.1016/j.enpol.2014.03.008.

Webel, D. M., Johnson, R. W. and Baker, D. H. 1998. Lipopolysaccharide-induced reductions in food intake do not decrease the efficiency of lysine and threonine utilization for protein accretion in chickens. *J. Nutr.* 128(10), 1760-6. doi:10.1093/jn/128.10.1760.

Weindl, I., Popp, A., Bodirsky, B. L., Rolinski, S., Lotze-Campen, H., Biewald, A., Humpenöder, F., Dietrich, J. P. and Stevanović, M. 2017. Livestock and human use of land: productivity trends and dietary choices as drivers of future land and carbon dynamics. *Glob. Planet. Change* 159, 1-10. doi:10.1016/J.GLOPLACHA.2017.10.002.

Chapter 5

Sustainable nitrogen management for housed livestock, manure storage and manure processing

Barbara Amon, Leibniz Institute for Agricultural Engineering and Bioeconomy (ATB), Germany and University of Zielona Góra, Poland; Lars Stouman Jensen, University of Copenhagen, Denmark; Karin Groenestein, Wageningen Livestock Research, The Netherlands; and Mark Sutton, UK Centre for Ecology & Hydrology (UKCEH), UK

1 Introduction

Nitrogen can take various forms (Fig. 1), including atmospheric di-nitrogen (N_2) and a wide range of reactive nitrogen (N_r) compounds, including all forms of nitrogen that are biologically, photochemically and radiatively active. Compounds of nitrogen that are reactive include ammonia (NH_3) and ammonium (NH_4^+), nitrous oxide (N_2O), nitrogen oxides (NO_x), nitrite (NO_2^-), nitrate (NO_3^-), nitric acid (HNO^-_3) and a wide range of organic nitrogen compounds ($R\text{-}NH_2$). Reactive forms of nitrogen are capable of cascading through the environment and causing an impact through smog, acid rain, biodiversity loss and so on as well as affecting climate (Butterbach-Bahl et al., 2011b). The design of abatement/mitigation measures requires a sound knowledge of the processes that influence formation and emission of all N_r compounds and N_2 into the environment, where nitrogen is lost to a wide range of atmospheric and aquatic pathways.

1.1 Ammonia

The principles of ammonia formation and its influencing factors are well known. Degradation of nitrogen-containing organic substance results in

http://dx.doi.org/10.19103/AS.2020.0077.08

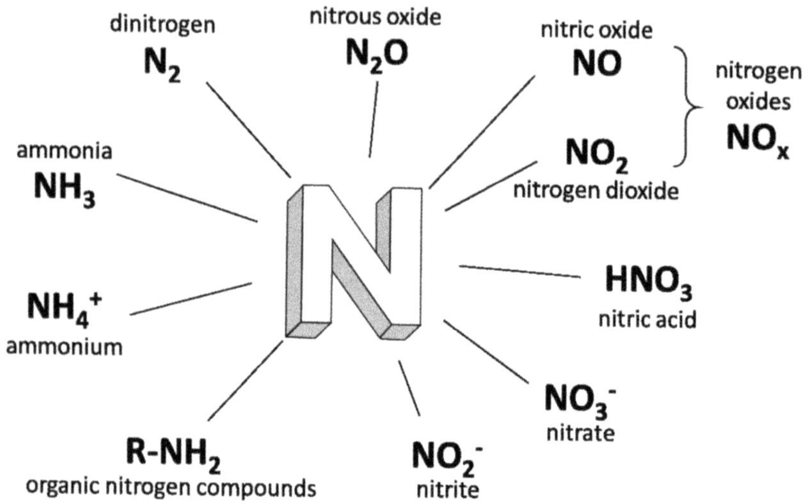

Figure 1 Major forms of nitrogen occurring in the environment. The sum of all forms except N_2 is often termed fixed or reactive nitrogen (Nr).

ammonium formation ($NH4^+$). There is an equilibrium between ammonium and ammonia:

$$H_2O + NH_3 \rightleftharpoons OH^- + NH_4^+$$

The degree to which ammonia forms the ammonium ion depends on the pH of the solution. If the pH is low, the equilibrium shifts to the right: more ammonia molecules are converted into ammonium ions. If the pH is high, the equilibrium shifts to the left: the hydroxide ion abstracts a proton from the ammonium ion, generating ammonia.

Ammonia emissions are governed by the difference between solution and atmosphere NH_3 partial pressure. High NH_3 concentrations in the solution and low NH_3 concentrations in the surrounding atmosphere increase NH_3 emissions. According to Henry's Law, ammonia emissions are also temperature dependent with rising temperatures increasing emissions (Fig. 2). Denmead et al. (1982) give the following equation:

$$NH_{3(\text{solution})} = \left(NH_{3(\text{solution})} + NH_{4(\text{solution})}\right) \Big/ \left(1 + 10^{0.09018 + (2729.92/T) - \text{pH}}\right)$$

where

$NH_{3(solution)}$ = NH_3 concentration in the solution

$NH_{3(solution)} + NH_{4\,(solution)}$ = The sum NH_3 and NH_4^+ in the solution

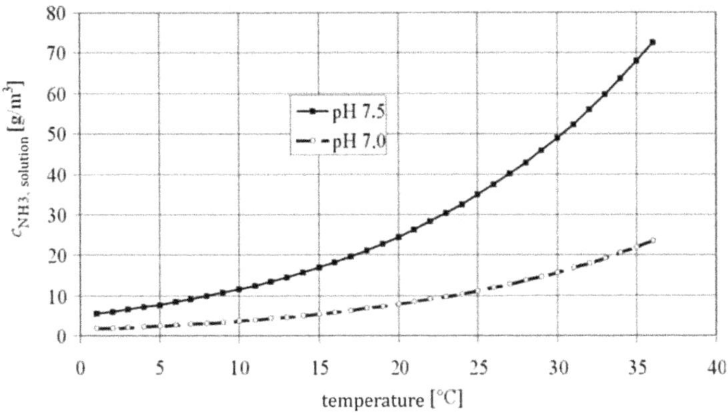

Figure 2 NH_3 concentration in the solution depending on temperature for pH 7.0 and pH 7.5 (Denmead at al., 1982).

T = Temperature in the solution [K]
pH = pH value in the solution

Ammonia emissions associated with livestock housing, manure storage, management and processing result from the degradation of urea by the ubiquitous enzyme urease which results in $NH4^+$ formation. Urea is mainly excreted in the urine and once it is hydrolysed it is much more prone to ammonia losses than organic nitrogen excreted in the faeces. In the case of poultry nitrogen is excreted largely in the form of uric acid, which hydrolyses like urea to produce ammonia. Where it is possible to dry excreta (e.g. in poultry litter), strategies may focus on reducing the hydrolysis rate of uric acid and urea. Once ammoniacal nitrogen (the sum of NH_3 + NH_4^+) is formed, strategies in animal housing and manure management focus on avoiding its volatilization to the atmosphere, for example, by reducing access to air, by reducing pH or by keeping the manure surface cool (cf. Fig. 2).

1.2 Nitrous oxide and di-nitrogen

The gases N_2O, NO_x and N_2 are formed both during the nitrification and the denitrification processes in the environment. The 'Leakage' model developed by Firestone and Davidson (1989) shows N_2O and NO_x losses as leakage flows during nitrification and denitrification (Fig. 3).

Nitrification oxidizes ammonia via nitrite to nitrate. This process is strictly aerobic. Autotrophic-nitrifying bacteria belong to the widespread group of nitrosomonas, nitrospira and nitrobacter, which are capable of growing on CO_2, O_2 and NH_4^+. Availability of NH_4^+ is mostly the limiting factor as CO_2 and O_2

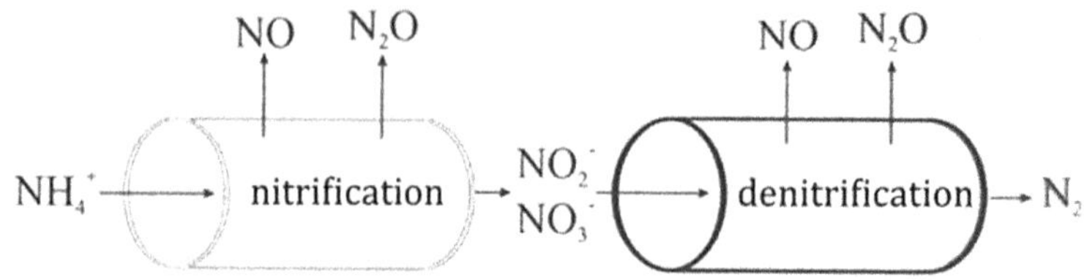

Figure 3 Leakage model for N_2O and NO_x losses during nitrification and denitrification (Firestone and Davidson, 1989).

are available in abundance. Low pH, lack of P and temperatures below 5°C or above 40°C lead to a reduction in nitrification activities. A water content of 60% of soil water-holding capacity is optimal for the nitrification process.

At low pH values, nitrification is carried out by bacteria and fungi. In contrast to the autotrophic nitrifiers, they need carbon sources for their growth. Their turnover rate is much lower compared to the autotrophic nitrifiers, but a substantial total turnover can still be achieved as a wider range of species have the ability for heterotrophic nitrification. N_2O production during nitrification is around 1%, NO_x production ranges between 1% and 4% of inputs N (Butterbach-Bahl et al., 2011a)

Denitrification reduces nitrate to N_2O, NO or N_2 when oxygen availability is low. NO_3^-, NO and N_2O serve as alternative electron acceptors when O_2 is lacking, and hence the denitrification occurs only under strictly anaerobic conditions. Molecular N_2 is the last part of the denitrification reaction chain and it is the only biological process that can turn reactive nitrogen into molecular N_2. Denitrifying bacteria are heterotrophic and facultative anaerobic. This means that they use O_2 as electron acceptor and switch to alternative electron acceptors (NO_3^-, NO and N_2O) when oxygen availability is low. Denitrifying bacteria are wide spread and show a high biodiversity.

Controlling factors for denitrification have been extensively investigated, mainly under lab conditions. Complex interactions exist between the various influencing factors which make a prediction of N_2O emissions difficult under practical conditions.

Denitrification is mainly governed by oxygen availability. Denitrification starts when the O_2 concentration decreases to below 5% (e.g. Hutchinson and Davidson, 1993). This may be the case in poorly aerated soils (e.g. high water content, in excess of 80% water-filled pore space), but also in soils where a high biological turnover consumes the oxygen faster than the supply. Easily degradable carbon sources and high nitrate concentrations also enhance the denitrification rate, while low temperature and low pH limit denitrification activity.

The relationship between N_2 and N_2O formation is mainly governed by the relationship between electron acceptor and reducing agent, and by the

O_2 concentration in the substrate. N_2 is only formed under strictly anaerobic conditions and a wide C: NO_3^- ratio. High nitrate concentrations increase the rate of N_2O production. These differences have effects in practice concerning N losses from housed livestock and manure storage, according to the extent of oxygen and carbon availability in different systems.

1.3 Nitrate and other nitrogen leaching and run-off

Diffuse pollution of groundwater and surface waters with N (and P) is a problem in many regions of the world, especially in areas with high livestock production. Animal manures contain substantial quantities of organic matter, N and P that if managed inappropriately may be lost from animal housing, manure storages or after field application.

Nitrogen and organic matter losses to aquatic systems mainly occur by leaching through the soil profile and through surface run-off when the infiltration capacity of the soil is exceeded. Point-source emissions can also be acutely damaging to local environments, for example, in the case of slurry store leakages. In surface waters, the losses cause problems with eutrophication and algal bloom, and in areas that rely on the use of groundwater, high nitrate concentrations can be a problem for the potable water quality. For drinking water the EU limit has been set at a nitrate (NO_3^-) concentration of 50 mg L^{-1} (EU Drinking Water Directive, 98/83/EC). Once leached to surface waters this N may also become a source of emissions of nitrous oxide, which is a potent greenhouse gas. In addition, significant loss of N resources is also an economic cost for the farmer, and N fertilizer production uses substantial amounts of fossil energy, causing global warming and other environmental emissions. Appropriate management and use of manures is therefore essential for minimizing nutrient leaching and the environmental impact of agriculture.

1.4 Consideration of nitrogen flows

Measures to reduce nitrogen losses from livestock feeding, housing and manure processing need to be seen in relation to other measures described in this guidance document. 'Manure management is a continuum from generation by livestock to storage and treatment and finally to land spreading' (Chadwick et al., 2011). This means that there is the potential for nitrogen, carbon and phosphorus losses at each stage of this continuum. A 'mass flow' approach has been used by Webb and Misselbrook (2004) to estimate NH_3 emissions from the manure management continuum. This approach allows effects of measures to reduce emissions and conserve manure N at one state to be considered as the manure passes to the next stage in the continuum. Similarly, other gaseous N losses, including N_2O, NO_x and N_2, may be assessed using a mass flow approach in a manner similar to that of Dämmgen and Hutchings (2008).

The importance of such a *whole system approach* is that effects of abatement methods at one stage are considered in downstream stages (Sommer et al., 2009, 2013), including losses of nitrogen to water through leaching and run-off.

1.5 Approach used to describe abatement measures

The following sections present the main management practices and abatement/mitigation measures that will influence N utilization and losses from housed livestock, manure storage, manure treatment and manure processing. Some measures will mitigate all forms of N loss, whereas others may mitigate a specific N loss pathway with either little impact or a negative impact on other N loss pathways. Enhanced abatement may be possible through the combined implementation of certain packages of measures. Linked management of nutrient cycles is required for effective environmental protection (Sutton et al., 2013)

The sections are based on information given in 'Options for Ammonia Mitigation: Guidance from the UNECE Task Force on Reactive Nitrogen' (Bittman et al., 2014), 'Best Available Techniques (BAT) Reference Document for the Intensive Rearing of Poultry or Pigs' (Santonja et al., 2017), 'Code for Good Agricultural Practice for Reducing Ammonia Emissions' (UNECE, 2015) and 'Framework Code for Good Agricultural Practice for Reducing Ammonia Emission' (Economic Commission for Europe, 2015). The information shall also be published as part of a Guidance Document on Integrated Sustainable Nitrogen Management developed by members of the Task Force on Reactive Nitrogen and submitted for adoption as UNECE document by the Executive Body of the UNECE.

2 Livestock feeding and housing

2.1 Livestock feeding

Ammonia emissions result from the degradation of urea by the ubiquitary enzyme urease which results in NH_4^+ formation. Urea is mainly excreted in the urine and once it is hydrolysed it is much more prone to ammonia losses than organic nitrogen excreted in the faeces. The crude protein content and composition of the animal diet is the main driver of urine excretion. Excess crude protein (CP) that is not needed by the animal is excreted and can easily be lost in the manure management chain. Adaptation of crude protein in the diet to the animals' needs is therefore the first and most efficient measure to mitigate nitrogen emissions. This measure reduces the loss of all N forms (Fig. 1) because it reduces the amount of excreted nitrogen. Groenestein et al. (2019) compared ammonia emissions related to nitrogen-use efficiency of livestock production in Europe. They come to the conclusion that expressing ammonia losses as a fraction of feed N would be preferable in expressing it per animal place as it give a clearer picture of the actual fraction of N in the feed that is lost to the environment.

Reduction of CP in animal feed is one of the most cost-effective ways of reducing N emissions throughout the entire manure management chain. For each per cent (absolute value) decrease in protein content of the animal feed, NH_3 emissions from animal housing, manure storage and the application of animal manure to land are decreased by 5–15% depending also on the pH of the urine and dung. Low-protein animal feeding also decreases N_2O emissions, and increases the efficiency of N use in animal production. Potential trade-offs with CH_4 emissions from enteric fermentation are not yet fully researched and need to be assessed. However, efficient N use is crucial for environmental-friendly milk production. Moreover, there are no animal health and animal welfare implications as long as the requirements for all amino acids are met. As there is much natural variation in nitrogen use efficiency (NUE) between individual animals, targeted breeding for better NUE can also be an option.

Low-protein animal feeding is most applicable to housed animals. It is less applicable for grassland-based systems with grazing animals, because grass is eaten by the animals in an early physiological growth stage and thus is typically high in degradable protein. It should be noted that grassland with leguminous species (e.g. clover and lucerne) also have a relatively high protein content, and so may be associated with provision of excess dietary N for livestock. Strategies to lower the protein content in herbage include: balanced N fertilization, grazing/harvesting the grassland at later physiological growth stage, and so on, and alteration of the ration of grassland-based systems, such as use of supplementary feeding with low-protein feeds.

2.1.1 Feeding strategies for dairy and beef cattle

Adapt protein intake in diet: Lowering CP of ruminant diets is an effective strategy for decreasing NH_3 loss. The following guidelines hold:

1. The average CP content of diets for dairy cattle should not exceed 15–16% in the dry matter (DM) (Broderick, 2003; Swensson, 2003). For beef cattle older than 6 months this could be further reduced to 12%;
2. Phase feeding can be applied in such a way that the CP content of dairy diets is gradually decreased from 16% of DM just before parturition and in early lactation to below 14% in late lactation and the main part of the dry period; and
3. Phase feeding can also be applied in beef cattle in such a way that the CP content of the diets is gradually decreased from 16% to 12% over time.

More information and associated costs can be found in the TFRN costs assessment (Chapter 3.4 'Low nitrogen feeding strategies in dairy cattle' in Reis et al., 2015).

In general, increasing the energy/protein ratio in the diet by using 'older' grass (higher sward surface height) or swathed forage cereal and/or supplementing grass by high energy feeds (e.g. maize silage) is a well-proven strategy to reduce levels of crude protein. However, for grassland-based ruminant production systems, the feasibility of these strategies may be limited, as older grass may reduce feeding quality, especially when conditions for growing high-energy feeds are poor (e.g. warm climates), and therefore such feeds have to be purchased. Hence, full use of the grass production would no longer be guaranteed. In the absence of other measures, such a strategy may also risk increasing methane emissions.

In many parts of the world, cattle production is grassland-based or partly grassland-based. In such systems, protein-rich grass and grass products form a significant proportion of the diet, and the target values for CP may be difficult to achieve, given the high CP content of grass from managed grasslands. The CP content of fresh grass in the grazing stage (2000–2500 kg DM/ha) is often in the range of 18–20% (or even higher, especially when legumes are present), whereas the CP content of grass silage is often between 16% and 18% and the CP content of hay is between 12% and 15% (e.g. Whitehead, 2000). In contrast, the CP content of maize silage is only in the range of 7–8%. Hence, grass-based diets often contain a surplus of protein and the magnitude of the resulting high N excretion strongly depends on the proportions of grass, grass silage and hay in the ration and the protein content of these feeds. The protein surplus and the resulting N excretion and N losses will be highest for grass-only summer rations (or grass-legume rations) with grazing of young, intensively fertilized grass or grass legume mixtures.

Urine excreted by grazing animals typically infiltrates into the soil. This means that NH_3 emissions per animal are reduced by extending the periods in which animal graze compared with the time spent with animals housed, where the excreta is collected, stored and applied to land. It should be noted that grazing of animals may increase other forms of N emissions (e.g. nitrate-N leaching and N_2O emissions). However, given the clear and well-quantified effect on NH_3 emissions, increasing the period that animals are grazing all day can be considered as a strategy to reduce emissions.

Increase productivity: Overall, increasing the productivity of dairy cattle in terms of milk or meat can decrease emissions per unit of animal production. Optimized productivity will also result in a reduction of enteric methane emissions. However, optimum productivity levels vary according to breed and region and must also consider that ruminants can only cope with a certain amount of concentrates and require sufficient roughage in their diet to stay healthy.

Increase longevity: Productivity can be increased though increasing milk production per year and through increasing the amount of milk production cycles per animal. Optimized diet and housing conditions enable a higher longevity of dairy cattle. Improving the longevity of dairy cattle also decreases the number of young cattle necessary for replacement. Reducing endemic disease and genetic gain through targeted breeding can also offer value.

2.1.2 Feeding strategies for pigs

Adapt protein intake in diet: Feeding measures in pig production include phase feeding, formulating diets based on digestible/available nutrients, using low-protein amino acid-supplemented diets and feed additives/supplements. Further techniques are currently being investigated (e.g. different feeds for males (boars and castrated males) and females) and might be additionally available in the future.

The crude protein (CP) content of the pig ration can be reduced if the amino acid supply is optimized through the addition of synthetic amino acids (e.g. lysine, methionine, threonine, tryptophan, typically limiting amino acids, which are too low in normal grain rations) or special feed components, using the best available information on 'ideal protein' combined with dietary supplementation. Lassaletta et al. (2019) performed a global analysis for pig systems that included the simulation of changes in CP. More information and associated costs can be found in the TFRN Costs Assessment (Chapter 3.2 'Low nitrogen feeding strategies in pigs', in Reis et al., 2015).

A CP reduction of 2–3% in the feed can be achieved, depending on the pig production category and the current starting point (Cahn et al., 1998). It has been shown that a decrease of 1% CP in the diet of finishing pigs results in a 10% lower total ammoniacal nitrogen (TAN) content of the pig slurry and 10% lower NH_3 emissions (Cahn et al., 1998). The inclusion of processed household and industry residues or wastes in the feed rations with a controlled energy/protein ratio is a complementary measure that reduces dependence on imported feedstuff. This measure also represents a reduction of upstream N_r emissions associated with feed production and downstream emissions associated with waste management (Lassaletta et al., 2019; zu Ermgassen et al., 2016).

2.1.3 Feeding strategies for poultry

Adapt protein intake in diet: For poultry, the potential for reducing N excretion through feeding measures is more limited than for pigs because the conversion efficiency currently achieved on average is already high and the variability within a flock of birds is greater. A CP reduction of 1–2% may be achieved depending on the species and the current starting point, but is already a well-proven measure for growers and finishers. Further applied nutrition research is currently being carried out in EU member states and North America and this may support further possible reductions in the future.

2.2 Livestock housing

When using measures to abate emission from livestock houses, it is important to minimize loss of the conserved nitrogen during downstream handling of the manure, in storage and spreading to maximize the benefit from the cost of abatement.

2.2.1 Cattle housing

Housing systems for cattle vary across the ECE region. While loose housing is most common, dairy cattle are still kept in tied stalls in some countries. In loose housing systems all or part of the excreta is collected in the form of slurry. In systems where solid manure is produced (such as straw-based systems), it may be removed from the house daily or it may remain there for up to the whole season, such as in deep litter stables. The system most commonly researched is the 'cubicle house' for dairy cows, where substantial NH_3 emissions arise from fouled slatted and/or solid floors and from manure in pits and channels beneath the slats/floor. There has been much less research to measure NO_x, N_2O and N_2 emissions from cattle housing, so recommendations in some cases have to be based on general principles, and are therefore subject to larger uncertainty than for NH_3 emissions from such systems.

Housed-cattle systems are generally set on stone or concrete bases, so direct nitrate leaching is not expected, unless there are cracked bases associated with poor maintenance. Run-off of N_r compounds from cattle housing systems may occur if ponded excreta is not correctly drained into storage tanks (e.g. associated with flooding events).

While 'hard standings' (typically concrete areas adjacent to dairies) provide a significant source of ammonia emissions outside of animal houses, in some parts of the UNECE region, cattle are kept in confined areas outside (e.g. feed lots), where N_r leaching, run-off and gaseous N losses may be substantial.

Animal welfare considerations tend to lead to an increase of soiled walking area per animal, increased ventilation and an overall increase in emissions. Changes in building design to meet the new animal welfare regulations in some countries (e.g. changing from tied stall to cubicle housing) will therefore increase NH_3 emissions unless abatement measures are introduced at the same time to combat this increase.

Solid versus slurry manure systems: Straw-based systems producing solid manure for cattle are not likely to emit less NH_3 in the animal houses than slurry-based systems. Further, N_2O, NO_x and N_2 losses due to (de)nitrification tend to be larger in litter-based systems than slurry-based systems (Webb et al., 2012; Groenestein and van Faassen, 1996).

While straw-based solid manure can emit less NH_3 than slurry after surface spreading on fields (e.g. Powell et al., 2008), slurry provides a greater opportunity for reduced emission application methods. Verification of any NH_3 emission reductions from using solid-manure versus slurry-based systems and from solid-liquid separation should consider all the stages of emission (housing, storage and land application).

2.2.2 Mitigation measures for cattle housing

Mitigation options for cattle housing can be grouped into the following types:

- Floor-based systems and related management techniques (including scrapers and cleaning robots);
- Litter-based systems (use of alternative organic material);
- Slurry management techniques at pit level;
- Indoor climate control techniques; and
- End-of-pipe techniques (hybrid ventilation + air cleaning techniques) and GHGs abatement/mitigation techniques.

Several pathways can be identified to further optimize existing and develop new abatement techniques. In this respect emission reduction techniques at animal housing level should aim to affect one or more of the following important key factors and/or driving forces of the nitrogen emission process:

- Draining capacity of the floor for direct transportation of urine to the manure storage;
- Residence time of open urine/manure sources;
- Emitting surface area of open urine/manure sources;
- Urease activity in urine puddles;
- Urine/manure pH and temperature (see Housing Measures 6 and 8);
- Indoor air temperature;
- Air velocities at emitting surfaces (urine puddles and manure surface in the pit);
- Air exchange between pit headspace and indoor air; and
- Exhaust of indoor air.

Immediate segregation of urine and faeces: A physical segregation (i.e. keeping separately) of faeces, which contain *urease*, and urine in the housing system reduces hydrolysis of urea, resulting in reduced emissions from both housing and manure spreading (Burton, 2007; Fangueiro et al., 2008a,b; Møller et al., 2007). Both acidification and alkalization of the in-house segregated urine reliably inhibits urea hydrolysis. The duration of the inactivation period can be adjusted by the dosage of acid or alkali addition

Verification of any NH_3 emission reductions from using solid-manure versus slurry-based systems and from solid-liquid separation should consider all the stages of emission (housing, storage and land application). Additional advantages of solid-liquid separation can also be expected during land-application, where urine (containing most of the available ammoniacal N) infiltrates more easily due to its lower dry-matter content than slurry, reducing NH_3 emissions. Although solid manure does not infiltrate, it mainly consists of organic N forms, which are much less liable to NH_3 emissions. Less is known about the consequences of solid-liquid separation on the emissions of N_2O, NO_x, N_2 and nitrate leaching, although substantial adverse effects are not expected.

Regular cleaning of floors in cattle houses by 'toothed scrapers': The 'grooved floor' system for dairy and beef cattle housing, employing 'toothed' scrapers running over a grooved floor, is a reliable technique to abate NH_3 emissions. Grooves should be equipped with perforations to allow drainage of urine. This results in a cleaner, low-emission floor surface with good traction for cattle to prevent slipping. Ammonia emission reduction ranges from 25% to 46% relative to the reference system (Smits, 1998; Swierstra et al., 2001). In the absence of measurement data, it is expected that use of the grooved floor system would have little impact on other N_r and N_2 losses since it is mainly directed to reduce immediate exposure to air of ammonium-rich excreta.

Regular cleaning of floors in cattle houses: Thorough cleaning of walking areas in dairy cattle houses by mechanical scrapers or robots has the potential to substantially reduce NH_3 emissions. The automatic cleaning should be performed at regular intervals (e.g. on an hourly basis) to achieve full benefits of the measure.

Frequent slurry removal: Regular removal of liquid manure from under the slats in the house to an outside store can substantially reduce NH_3 emissions by reducing the emitting surface and the slurry storage temperature. A reduced storage temperature will also result in a reduction of methane emissions.

Increase bedding material: Bedding material in animal housing can affect NH_3, N_2O, NO_x and N_2 emissions. The physical characteristics (urine absorbance capacity, bulk density) of bedding materials are of more importance than their chemical characteristics (pH, cation exchange capacity, carbon to nitrogen ratio) in determining NH_3 emissions from dairy barn floors and pig houses (Misselbrook and Powell, 2005; Powell et al., 2008; Gilhespy et al., 2009; Groenestein et al., 2007). However, further assessment is needed on the effect of bedding on emissions for specific systems while taking into account the whole manure management path. The approach can have a positive interaction with animal welfare measures.

Barn climatization to reduce indoor temperature and air flow: In houses with traditional slats (either non-sloping, 1% sloping, or grooved), optimal barn climatization with roof insulation (RI) and/or automatically controlled natural ventilation (ACNV) can achieve a moderate emission reduction (20%) of NH_3 due to the *decreased temperature* (especially in summer) and *reduced air velocities* (Braam et al., 1997a,b; Smits, 1998; Monteny, 2000). To the extent that such systems cool stored manure, emissions of methane will also be reduced.

Use of acid air scrubbers: Chemical or acid air scrubbers are effective in decreasing NH_3 emissions from force-ventilated pig housing. However, they cannot yet be generally implemented in cattle housing because these are mostly naturally ventilated across the ECE region. Also, there are few data for scrubbers on cattle (Ellen et al., 2008). In any situations where cattle are housed with forced-ventilation, this measure can be considered as Category 1. Recent developments consider combining targeted ventilation of naturally ventilated barns with air scrubbers. More research and development are needed here.

Floor type: Different improved floor types based on slats or solid, profiled concrete elements have been tested. These designs combine emission reduction from the floor (increased run-off of urine) and from the pit (reduction of air exchange by rubber flaps in the floor slots). The emission abatement efficiency depends on the specific technical characteristics of the system.

Increased grazing: Decreasing the amount of animal excrement in animal housing systems through increased grazing is an effective measure to decrease NH_3 emissions. Total annual emissions (including housing, storage and spreading) from dairy systems may decrease by up to 50% with nearly all-day grazing, as compared with animals that are fully confined. While increased grazing is a reliable NH_3 emission reduction measure for dairy cows, the amount of emission reduction depends on the daily grazing time and the cleanliness of the house and holding area. In some cases grazing may also contribute to increased run-off and leaching of NO_3^- and other N_r compounds, as well as N_2O and NO_x emissions. Grazing can also be associated with increased pathogen mobilization.

2.2.3 Pig housing

Designs to reduce NH_3 emissions from pig housing systems have been described in detail in Economic Commission for Europe (2015) and in the IPPC document on Best Available Techniques (BAT) (Santonja et al., 2017). These apply the following main elements:

- Reducing manure surfaces such as soiled floors, using channels for slurry holding surfaces and using sloped walls. Partly slatted floors (~50% area), generally emit less NH_3, particularly if the slats are metal- or plastic-coated rather than concrete, allowing the manure to fall rapidly and completely into the pit below. Emissions from the non-slatted areas are reduced by inclined, smooth surfaces, by locating the feeding and watering facilities to minimize fouling of these areas, and by good climate control in the building;
- Removing the slurry from the pit frequently to an external slurry store with vacuum or gravity removal systems or by flushing systems at least twice a week;
- Additional treatment, such as liquid/solid separation; provided that the storage of the separated fractions maintains low emissions;
- Circulating groundwater or other cooling agents in floating heat exchangers or walls of slurry pits to cool the surface of the manure in the under-floor pit to at least below 12°C. Constraints include costs and need to locate a source of groundwater away from the source of drinking water;
- Changing the chemical/physical properties of the manure, such as decreasing pH;
- Using surfaces which are smooth and easy to clean (see above);
- Treatment of exhaust air by acid scrubbers or biotrickling filters;
- Lowering the indoor temperature and ventilation rate, taking into account animal welfare and production considerations; and
- Reducing airflow over the manure surface.

For a given floor slat width, manure drains from concrete slats less efficiently than from steel- and plastic-covered slats and this is associated with greater emissions of NH_3. Note that steel slats are not allowed in some countries for animal welfare reasons. Cross-media effects have been taken into account in defining BAT for the various housing designs. For example, frequent flushing of slurry (normally once in the morning and once in the evening) causes nuisance odour events. Flushing slurry also consumes energy unless manually operated passive systems are used.

Use of straw litter in pig housing is expected to increase due to concern for the welfare of the pigs. In conjunction with (automatically controlled) naturally ventilated housing systems, straw allows the animals to self-regulate their temperature with less ventilation and heating, reducing energy consumption. In systems with litter, the pen is sometimes divided into solid areas with litter and slatted dunging areas. However, pigs do not always use these areas in the desired way, using the littered area to dung and the slatted area to cool off in warm weather. Generally, pens should be designed to accommodate desired excreting behaviour of pigs to minimize fouling of solid floors. However, this is more difficult in regions with a warm climate. Note that integrated evaluation of straw use should consider (a) the added cost of the straw and mucking out the pens, (b) the possible increased emissions from storage and application of manure with straw and (c) the benefit of adding organic matter from straw to the soil.

2.2.4 Mitigation measures for pig housing

The reference system, used commonly in Europe, is a fully slatted floor with a deep manure pit underneath and mechanical ventilation; emission ranges from 2.4 kg NH_3 to 3.2 kg NH_3 per finisher pig place per year. Since growers/finishers are always housed in a group, most systems used for group housing of sows are applicable to growers. Emissions from different abatement/mitigation approaches are compared with this reference system in terms of the emission reduction amount (Bittman et al., 2014). Most data are available for NH_3, with little data concerning effects on N_2O, NO_x, N_2 and nitrate leaching. The underlying principles are largely similar for these losses as for cattle housing systems, recognizing the different housing needs of pigs and the particular characteristics of pig excreta.

Acidification of slurry: Reductions in NH_3 emissions can be achieved by acidifying slurry to shift the chemical balance from the gaseous NH_3 to the ionic and soluble NH_4^+. The manure (especially the liquid fraction) is collected into a tank with acidified liquid (usually using sulphuric acid, but organic acids can be used as well, though at higher cost) maintaining a pH of less than 6 (Bittman et al., 2014; Fangueiro et al., 2015). In pig housing systems, emission reduction of 60% or more have been observed (Kai et al., 2008). The measure is not anticipated to affect other N_r or N_2 losses. Acidification of slurry is anticipated to be effective for both cattle and pig slurry, though measurements have so far concentrated on investigating pig slurry. One study (Petersen et al., 2012) showed acidification of cattle slurry to pH 5.5 reduced the NH_3 emissions by more than 90% and at the same time reduced emissions of the GHG CH_4 by 67–87%. Attention should be given to monitoring soil pH and metal content if acidified slurry is to be used in agriculture. In-house acidification will reduce NH_3 emissions throughout the manure management chain. Furthermore, slurry acidified with sulphuric acid is not suitable as the sole feedstock for biogas production (but can be used as a smaller proportion).

Reduce emitting surface: Ammonia emissions can be reduced by 25% by decreasing the surface area of the emitting floor through frequent and complete vacuum-assisted drainage of slurry from the floor of the pit. Where this is possible, this technique has no cost. Partly slatted floors covering 50% of floor area generally emit 15–20% less NH_3, particularly if the slats are metal or plastic-coated which is less sticky for manure than concrete. Decreasing the risk of emissions from the solid part of the floor can be achieved by: (a) using an inclined (or convex), smoothly finished surface, (b) by appropriate siting of the feeding and watering facilities to minimize fouling of the solid areas and (c) by good climate control (Aarnink et al., 1996; Guingand and Courboulay, 2007; Ye et al., 2008a,b). Further reduction of the emitting area can be achieved by making both the partly slatted area and the pit underneath smaller. With the smaller slatted area, the risk of greater fouling of the solid area can be mitigated by installing a small second slatted area with a water canal underneath at the other side of the pen where the pigs tend to eat and drink. The canal is filled with about 2 cm of water to dilute any manure that might eventually drop into it. This slatted area will have low emissions because any manure dropped here will be diluted. This combined manure-canal and water-canal system can reduce NH_3 emissions by 40–50% depending on the size of the water canal. This approach is not expected to have a significant effect on emissions of N_2 or other N_r compounds.

Reducing the emitting surface area by having one or two slanted pit walls, in combination with partly slatted floors and frequent manure removal, can reduce emissions by up to 65%. Reducing the emitting surface area with shallow V-shaped gutters (maximum 60 cm wide, 20 cm deep) can reduce emission in pig houses by 40–65%, depending on the pig category and the presence of partly slatted floors. The gutters should be flushed twice a day with the liquid (thin) fraction of the slurry rather than water; flushing with water dilutes the manure and increases the cost of transporting and applying it in the field.

Increase bedding material (solid manure housing): Bedding material in animal housing can affect NH_3, N_2O, NO_x and N_2 emissions. The physical characteristics (urine absorbance capacity, bulk density) of bedding materials are of more importance than their chemical characteristics (pH, cation exchange capacity, carbon to nitrogen ratio) in determining NH_3 emissions from dairy barn floors (Misselbrook and Powell, 2005; Powell et al., 2008; Gilhespy et al., 2009). However, further assessment is needed on the effect of bedding on emissions for specific systems while taking into account the whole manure management path. The approach can have a positive interaction with animal welfare measures. However, approaches benefiting animal welfare can also be operated as slurry-based systems, with only little straw supply.

Regular cleaning of floors: Cleaning of floors in pig houses by mechanical scrapers or robots has the potential to substantially reduce NH_3 emissions. The automatic cleaning should be performed at regular intervals to achieve full benefits of the measure (Amon et al., 2007). It is worth mentioning that in warm countries (e.g. Mediterranean region), for sanitary reasons, floor cleaning is done more frequently with consequences in the slurry composition, which may reach up to 98% water.

Frequent slurry removal: Regular removal of slurry from under the slats in the house to an outside store can substantially reduce NH_3 emissions by reducing the emitting surface and the slurry storage temperature. A reduced storage temperature will also result in a reduction of methane (Amon et al., 2007).

Barn climatization to reduce indoor temperature and air flow: Surface cooling of manure with fans using a closed heat exchange system is a technique with a reduction efficiency of 45–75% depending on animal category and surface of cooling fins. This technique is most economical if the collected heat can be exchanged to warm other facilities such as weaner houses (Huynh et al., 2004). In slurry systems this technique can often be retrofitted into existing buildings. However, this system is not applicable when straw bedding is used or when the feed contains a lot of roughage. This is because a layer of floating residue may develop on top of the slurry.

Use of acid air scrubbers: Treatment of exhaust air by acid scrubbers (mainly using sulphuric acid) or biotrickling filters has proven to be practical and effective for large-scale operations in Denmark, Germany, France and the Netherlands (e.g. Melse and Ogink, 2005; Guingand, 2009). This is most economical when installed in new houses, because retrofitting in existing housing requires costly modification of ventilation systems. Acid scrubbers have demonstrated NH_3 removal efficiencies of more than 90%, depending on their pH-set values. Scrubbers and biotrickling filters also reduce odour and particulate matter by 75% and 70%, respectively (Guingand, 2009). Further information is needed on the suitability of these systems in South and Central Europe. Operational costs of both acid scrubbers and trickling filters are especially dependent on the extra energy use for water recirculation and to overcome increased back pressure on the fans. Optimization methods are available to minimize costs (Melse et al., 2012) and costs will be lower for large operations. The approach may also contribute to reducing N_2O and NO_x emissions, but more research is needed here.

Use of biological air scrubbers: Biological air scrubbers operate with bacteria that remove ammonia and odours from the exhaust air. Ammonia captured in biological air scrubbers typically undergoes nitrification and denitrification associated with increased emissions of N_2O, NO_x and N_2. Recovery of the collected N_r in bio-scrubbers may help offset this increase by reducing the need for fresh N fixation and production of chemical fertilizers.

2.2.5 Poultry housing

Designs to reduce NH_3 emissions from poultry housing systems have been described in detail in Economic Commission for Europe (2015) and in the document on BAT under the EU Industrial Emissions Directive (Santonja et al., 2017), and applying the following principles:

- Reducing the open surface area of emitting manure;
- Removing the manure frequently from the poultry house to an external slurry store (e.g. with belt removal systems);
- Quickly drying the manure to reduce hydrolysis of uric acid to ammonia;
- Using surfaces which are smooth and easy to clean;
- Treatment of exhaust air by acid scrubbers or biotrickling filters (i.e. biological air scrubbers); and
- Lowering the indoor temperature and ventilation as animal welfare and/or production allow, reducing microbial processes that mobilize N_r losses.

Many of the measures listed for cattle and pigs are also applicable for poultry systems, especially reduction of emitting surface, barn climatization to reduce indoor temperature and air flow), and acid air scrubbers. This section therefore focusses on additional considerations for poultry housing. Further information can be found in European Commission (2015), in the IPPC BREF document (Santonja et al., 2017) and in the UNECE Ammonia Guidance Document (Bittman et al., 2014).

Where poultry houses are disconnected from the ground (e.g. concrete base), emission reduction measures for NH_3 are not directly expected to affect nitrate and other N_r leaching and run-off. For smaller farms, which are not required to comply with national legislation (e.g. BAT) for layers, and for free-range poultry, pathways to the soil can also be anticipated. In such cases, NH_3 emission reduction including rapid drying and dry storage of poultry litter may also have benefits to reduce N_r leaching. In addition, expert observations have shown that downward pointing air exhausts onto porous ground surfaces surrounding poultry houses can lead to localized increases of N_r leaching and run-off into ground waters. Reduction of NH_3 emissions (and N_r-containing dusts) can therefore also contribute to reducing such hotspots of N_r leaching and run-off.

2.2.6 Mitigation measures for housing systems for laying hens

A wide range of regulations and minimum standards for protecting laying hens exist across the UNECE region. For example, in the EU, regulations apply under Council Directive 1999/74/EC. This Directive has prohibited the use of conventional cage systems since 2012. Instead, only enriched cages (also called 'furniture cages'), or non-cage systems, such as litter (or deep litter) housing systems or aviary systems, are allowed.

Rapid drying of poultry litter: Ammonia emissions from battery deep-pit or channel systems can be lowered by reducing the moisture content of the manure by ventilating the manure pit. The collection of manure on belts and the subsequent removal of manure to covered storage outside the building can also reduce NH_3 emissions, particularly if the manure has been dried on the belts through forced ventilation. The manure should be dried to 60–70% DM to minimize the subsequent formation of NH_3. Manure collected from the belts into intensively ventilated drying tunnels, inside or outside the building, can reach 60–80% DM content in less than 48 h, but in this case exposure to air is increased, risking an increase in NH_3 emissions. Weekly removal from the manure belts to covered storages reduces emissions by 50% compared with bi-weekly removal. In general, emission from laying hen houses with manure belts will depend on: (a) the length of time that the manure is present on the belts, (b) the drying systems, (c) the poultry breed, (d) the ventilation rate at the belt (low rate = high emissions) and (e) the feed composition. Aviary systems with manure belts for frequent collection and removal of manure to closed storages reduce emission by more than 70% compared with the deep litter housing system. While the primary drying poultry litter has been on reducing NH_3 emissions, keeping excreted N in the form of uric acid can be expected also to reduce N_2O, NO_x and N_2, since this will also reduce nitrification and denitrification. Dried poultry litter will therefore have a higher fertilizer value for farmers, which should be compensated by using reduced doses during land application, as compared with decomposed poultry litter.

Use of acid or biological air scrubbers: Treatment of exhaust air by acid or biological scrubbers (=biotrickling filters) has been successfully employed in several countries (Melse and Ogink, 2005; Ritz et al., 2006; Patterson and Adrizal, 2005; Melse et al., 2012). Hahne et al. (2016) in Germany counted 179 installed air scrubbers in poultry installations and 1012 scrubbers installed in pig houses, respectively. The main difference from pig systems is that poultry houses (especially with dried litter) typically emit a much larger amount of dust. Acid or biological scrubbers remove 70–90% of NH_3, and also remove fine dust and odour. To deal with the high dust loads, multistage air scrubbers with pre-filtering of coarse particles have been developed (Ogink and Bosma, 2007; Melse et al., 2008).

2.2.7 Mitigation measures for housing systems for broilers

To minimize NH_3 emission in broiler housing, it is important to keep the litter dry. Litter moisture and emissions are influenced by:

- Drinking-water design and function (leakage and spills);
- Animal weight and density, and duration of the growing period;
- Ventilation rate, use of in-house air purification and ambient weather;
- Use of floor insulation;
- Type and amount of litter; and
- Feed.

Reducing spillage of water from the drinking system: A simple way to reduce spillage of water from the drinking system is using a 'nipple drinkers' instead of 'bell drinkers'. This approach should be integrated into wider systems designed to keep poultry litter dry.

Air scrubber technologies to remove NH_3 from ventilation air are highly effective, but not currently widely implemented because of high installation and running costs. Packed-bed filters and acid scrubbers currently available in the Netherlands and Germany remove 70–90% of NH_3 from exhaust air. Comprehensive measuring of air scrubbers is done by the German Agricultural Association (DLG, 2020) based on a scientific standard testing frame. As with such systems for laying poultry, questions about long-term reliability due to high dust loads need to be further clarified. Various multi-pollutant scrubbers have been developed to also remove odour and particulate matter (PM_{10} and $PM_{2.5}$) from the exhaust air (Zhao et al., 2011; Ritz et al., 2006; Patterson and Adrizal, 2005). Implementation of both acid scrubbers and biological air scrubbers for broiler housing is largely similar to that for laying hens

3 Manure storage, treatment and processing

3.1 Principles of emissions from manure storage, treatment and processing

For livestock agriculture to become sustainable, an optimal and efficient use of manure nutrients and organic matter is essential. However, manure nitrogen may be easily lost via gaseous emissions (NH_3, N_2O, NO_x and N_2) and leaching of nitrate (NO_3^-) and other N_r compounds. Besides nitrogen losses, animal and manure emissions of methane (CH_4) to the atmosphere must be reduced as far as possible, to limit climate change impacts.

Significant N losses may occur during storage of urine, faeces or mixtures (slurries and farmyard manures/deep litters), and simple treatment (e.g. solid-liquid separation) or more advanced processing (e.g. anaerobic digestion, ultrafiltration) may enable more appropriate manure management with lower N losses.

The *treatment of manures* typically involves a one-step operation to improve the properties of the manure. Expected effects include the improvement of the fluid properties (by adding water or by separating solids), the stabilization of volatile nutrients (by acidification) or a reduction in odour nuisance (e.g. aeration). Single-stage treatment of manures is typically applied on farms in the proximity of livestock buildings. The mass and ingredients of manures are not or only slightly changed by treatment systems.

The *processing of manures* generally describes more complex and multi-step processes, which are used specifically to produce new products, for example,, higher nutrient content, less water content, free of undesirable odours and hygienically safe. In most cases, manure processing is used to produce marketable products that can be used as fertilizers and soil conditioners, as well as secondary raw materials (e.g. fibres). Manure processing technologies may either be located on farms or operated as central/decentral plants.

Manure treatment and processing always come at a cost, both in economic, energy and environmental terms, so the simplest option fulfilling the goal(s) should always be the priority option: (1) direct land application, (2) simple treatment second, (3) advanced processing (with (1) first, according to local limitations, including those related to pollution). Simple treatment and advanced processing are most relevant when conditions (e.g. high regional livestock density, large manure N surplus relative to local crop demand) favour overall environmental benefits from treatment or processing. Such systems should be designed with awareness of the need to avoid pollution swapping (e.g. reducing ammonia loss, but increasing nitrate leaching somewhere else and vice versa).

Animal slurry composition is typically not ideal with regard to low emission handling and crop fertilizing properties. In particular, the high DM and carbon content pose several problems during slurry storage, application and crop utilization (Table 1). This points to the opportunity for increased development of systems to collect and store urine and dung separately, or to apply manure treatment by solid-liquid separation.

High slurry DM tends to result in crust formation on the slurry surface and/or to sedimentation on the bottom of the slurry tank. In order to achieve an even distribution of nutrients in the slurry, slurry must be mixed/homogenized prior to application. Homogenization of slurry with high DM content is energy-consuming and increases NH_3 emissions as a larger volume of the slurry comes in close contact with the atmosphere.

Slurry contains considerable amounts of easily degradable carbon that serves as substrate for microbes. During slurry storage a continuous degradation of organic matter can be observed. Degradation intensity is strongly dependent on the slurry DM content. Amon et al. (1995) investigated changes in slurry composition over a 200-day storage period for stored cattle, beef and pig slurry. Degradation of organic matter was found to be significantly greater with higher slurry DM content. Such slurry degradation will include mineralization to form of ammonium (NH_4^+) from organic matter. This points to the opportunity for increasing the immediate fertilizer value of the slurry, provided that storage is covered, thereby avoiding NH_3 emissions and benefiting from increased slurry NH_4^+ content.

As conditions in slurry are anaerobic, degradation of organic matter is always dominated by anaerobic pathways. This means, that both CH_4 and CO_2 are formed as end-products of the degradation process. It is thus to be assumed that high DM slurry bears a greater risk for CH_4 emissions, contributing

Table 1 Challenges and benefits resulting from slurry high dry matter and carbon content, low nutrient content

	Problems
Storage	• natural crust formation and sedimentation of solids, giving heterogeneous concentration of nutrients • high energy consumption per unit of nutrient for pumping and mixing • potentially higher emissions of NH_3, N_2O, N_2, CH_4 and odour
Field application	• high potential risk of NH_3 losses due to slow infiltration • high technical effort required (at high economic cost) for even and low emission application • suffering of crop plants due to scorching by broadcasted slurry
Crop utilization	• less effective crop uptake of slurry N than from mineral fertilizer • increased temporary N immobilisation in the soil, increasing risk of lower crop N effect • higher risk of denitrification and subsequent N_2O and N_2 emissions • crop N effect less predictable/more variable than from mineral fertilizer
	Benefits
Storage	• natural crust formation may serve as a natural barrier, inhibiting NH_3 transport to the atmosphere; furthermore, the crust may have significant capacity for CH_4 oxidation, due to its partial aerobic conditions and high microbial activity
Field/soil	• high dry matter and carbon content contributes to maintenance of soil organic matter content and biologically active soil

significantly to climate change. This also points to the opportunity for CH_4 and CO_2 recovery, for example, linked to anaerobic digestion for production of biogas.

Environmental-friendly slurry application in the field requires that the slurry is more evenly applied near or below the soil surface. It is much more complicated to fulfil this requirement when the slurry has a higher DM content, causing a higher viscosity and less easy flow through band spreading hoses. Following application of slurry, NH_3 emissions can be substantial and are found to increase with an increase in slurry DM content, due to slower soil infiltration (Sommer et al., 2013; Bittman et al., 2014). This emphasizes the importance of maintaining low DM contents of slurries. By reducing NH_3 and other nitrogen losses, available N resources on farms are increased, decreasing the need for additional N to be bought as manufactured inorganic fertilizer.

The N availability to plants is difficult to calculate with high DM slurry, because a high DM content drives increased microbial immobilization right after application. The more narrow the C/N-ratio, and the higher NH_4-N content, the more slurry N is potentially available to plants, whereas with a wide C/N-ratio,

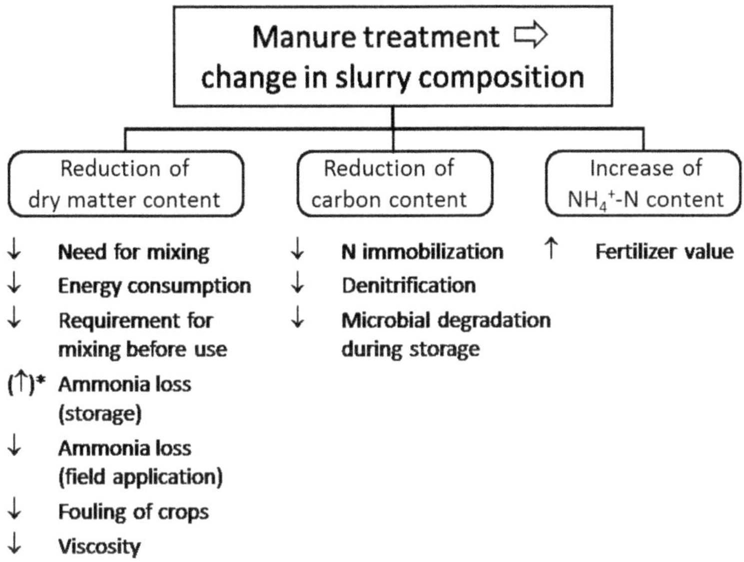

Figure 4 Effect of changes in slurry composition achieved by manure treatment. Arrows indicate decrease (↓) or increase (↑) in the listed property. * If depending on natural crusting of manure to reduce emissions rather than other types of cover.

part of the slurry N is immobilized in the soil N pool and becomes available only at a later and often unpredictable or even too late stage, causing increased risk of nitrate leaching. In addition, an increase in slurry DM and subsequent soil N content has the potential to increase rates of nitrification and denitrification, increasing subsequent N_2O, NO_x and N_2 losses (e.g. Dosch, 1996). It may thus be beneficial to reduce slurry DM and carbon content at an early stage of manure management. This leads to several manure treatment options that can be evaluated in relation to the requirements listed in Fig. 4.

In line with the objectives of the EU Circular Economy Action Plan,[1] there is an opportunity to encourage the use of recycled nutrients that can replace nutrients otherwise obtained from primary raw materials. The main challenge is to use recycled nutrient resources that have an equal or better environmental performance than the primary nutrient resources they replace. Efforts are ongoing across the EU to develop manure-processing technologies that allow manure to be turned into a safe and agronomically valuable resource that could be more widely used.[2]

Techniques for simple manure treatment can be classified as physical, chemical or biological (Fig. 5, Bernal et al., 2015). Furthermore, a number of

1 https://eur-lex.europa.eu/legal-content/EN/TXT/?uri=CELEX:52015DC0614
2 https://ec.europa.eu/jrc/en/research-topic/waste-and-recycling

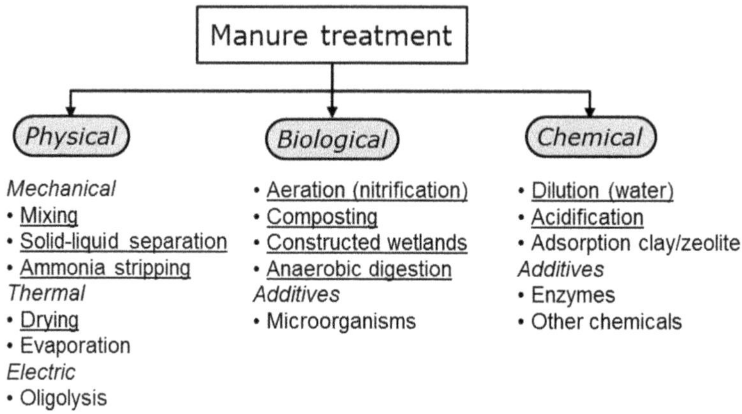

Figure 5 Options for simple manure treatment. Options underlined are in some regions commonly applied in full scale on commercial farms (mainly pig farms); other options are applied either rarely or only in experimental/pilot scale - these are not dealt with further here, pending the availability of proof-of-concept and documentation.

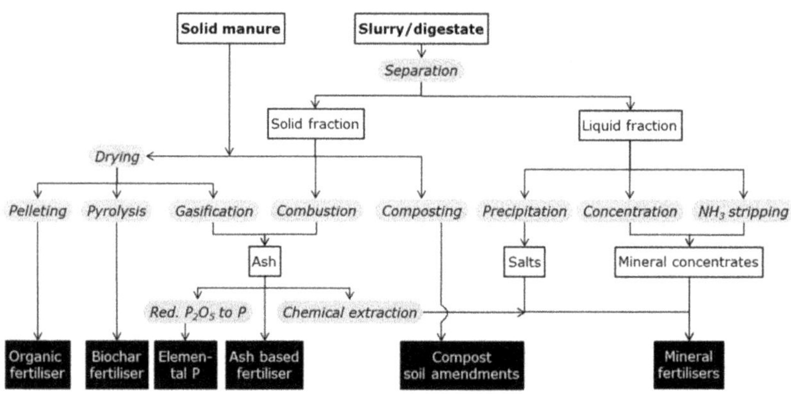

Figure 6 Options for combining simple treatment with more advanced processing of manures to recover and upgrade nutrient and energy, resulting in widely different bio-based fertilizers (modified from Jensen, 2013). Only a few are yet applied in full commercial scale; other are still in experimental/pilot stage (and are therefore not dealt with further here).

different options/technologies are available for further and more advanced processing of raw or treated manures for recovering and upgrading nutrients and organic matter from different manure types (Fig. 6). For slurries or other liquid manures, such as digestate from anaerobic digestion of manure and other bio-waste, all treatment steps start with mechanical separation into (a) a *solid fraction* which is relatively rich in organic N and P, and (b) a *liquid fraction*, with low P, but relatively high mineral N and K contents. Different simple techniques can be combined with each other. This allows a wide variety of by-products to be combined, resulting in highly variable distribution of organic nitrogen, ammoniacal nitrogen, phosphorus, carbon and other nutrients, which must be taken into account when managing the different fractions.

There may be additional possible treatments of the liquid phase. In order to save water without increasing the amount of nitrogen supplied to the soil, and to favour the circular economy of water, it is common to carry out successive treatments of the liquid phase, so that the resulting product can be used in fertigation. For example, in the south of Spain wetlands are being constructed to allow the reuse of water for irrigation, in areas of scarce availability. In addition to nitrogen, many other characteristics have an influence in the decision of choosing a procedure, such as the contribution of organic matter, the formation of methane and other GHGs, the presence of other nutrients, type of agricultural systems, salinity, weather and, very important in the countries of southern Europe, the water footprint.

Each of these processing pathways and resulting products (Fig. 6) has certain advantages and disadvantages, and the net environmental benefits/impacts and economic costs/profits differ greatly. A number of factors must be considered when prioritizing the processing options (Jensen, 2013):

- The primary aim should be nutrient recycling, mainly N and P; N is consumed in the largest quantities, is expensive and has impacts on energy consumption and GHG emissions, while P is a scarce and non-renewable resource, with the highest price.
- Splitting N and P into different fractions is generally beneficial, as this enables more flexible and balanced fertilization in accordance with the needs of many crops.
- The technology or combination of technologies applied should preferably also produce energy or consume relatively little energy, so net energy production should be taken into account for both environmental and economic reasons.
- Local solutions should be preferred, avoiding too high transport cost and impacts; regional or more central solutions are therefore only justified if the economy of scale via higher efficiency outweighs the negative impacts of transporting the manure to a common facility.
- The quality of end-products and byproducts is assessed differently depending on the perspective of the user. For instance, a manure combustion ash, where the majority of the N has been lost, will not be appreciated by an organic farmer, while a compost is highly appreciated for its soil ameliorating effect and slow release of N, even if some N has been lost in the process.
- Biochars and compost may be valued highly by orchard and vineyard producers for its effects on soil-water-holding capacity and nutrient

retention, whereas conventional crop production farmers may value mineral concentrates and salts more highly. Production of recovered, bio-based fertilizer products should not be supply-driven (trying to solve a waste problem), but rather demand-driven (bio-based fertilizers that the farmers want).

3.2 Mitigation measures for manure storage, treatment and processing

3.2.1 Covered storage of slurry (natural crust)

Where slurries have a high DM content, these may form a natural crust during storage, which is associated with substantially reduced ammonia emission (Bittman et al., 2014). There is large agreement that crusting impacts the gas release in many ways: enhanced resistance to mass transfer (Olesen and Sommer, 1993), oxidation of NH_3 (Nielsen et al., 2010) and CH_4 (Petersen et al., 2005) and formation of N_2O related to nitrification and denitrification occurring in liquid-air interfaces near air-filled pores present in crusts (Petersen and Miller, 2006). NH_3 and CH_4 may be consumed due to microbial activity in the crust leading to an emission reduction (Petersen and Ambus, 2006; Nielsen et al., 2010) while N_2O production may be enhanced (VanderZaag et al., 2009). A comprehensive assessment of the current knowledge on the effect of natural crusts can be found in Kupper et al. (2020). The reference is taken as uncovered storage, including on an impermeable surface, which explains the benefit for nitrate leaching.

3.2.2 Covered storage of manure (solid cover)

A wide range of options are available for covered manure storage using solid covers, including use of metal tanks with solid lids, floating covers on lagoons and use of slurry bags, most of which are associated with negligible ammonia emission if well operated. Further details of such systems are provided by Bittman et al. (2014). Less focus has been given to ensuring that solid manure (e.g. farmyard manure and poultry manure) are covered, for example, through use of plastic sheeting. The reference is taken as uncovered storage, including on an impermeable surface, which explains the benefit for nitrate leaching.

3.2.3 Covered storage of manure (dispersed coverings)

Ammonia emissions can be significantly reduced when covering solid organic fertilizers with dispersed coverings such as peat, clay, zeolite and phosphogypsum. The basis of the approach is to prevent contact of NH_3 emitting surfaces with the air, especially when covering them with ammonium absorbing substances. Lukin et al. (2014) found that total NH_3 emissions from poultry manure amounted to 5.9% when it was covered with peat, 4.7% when it was covered with loam, 1.3% when it was covered with zeolites and 16.9% when it was covered with phosphogypsum. These values are relative to NH_3 emissions in the reference variant with no covering. Use of these simple materials to cover piles of organic fertilizers thereby substantially reduces NH_3 emissions into the atmosphere (Lukin et al., 2014). Protocols are needed to specify minimum thickness of each type of covering material. Further testing is needed to assess the effect on N_2O, NO and N_2 emissions. Unless an impermeable base is used, the approach risks significant nitrate leaching. In combination with an impermeable base, it can reduce both N_r emissions to air and leaching losses to water.

3.2.4 Storage of solid manure under dry conditions

Simply storing manure in a dry place, out of the rain, can also reduce nitrogen emissions from a range of N_r compounds and N_2. This is even more important for dried poultry litter, where keeping manure dry and out of the rain helps to avoid hydrolysis of uric acid to form ammonia. However, poultry litter is hydroscopic and will emit some ammonia when in humid atmospheres, even when kept free of rain (e.g. Elliot and Collins, 1982). Keeping solid manure dry during storage minimizes mineralization and denitrification which can give rise to N_2O, NO_x and N_2 emissions, as well as reducing nitrate and other N_r leaching.

3.2.5 Storage of manure on a solid concrete base with walls

Investments in this approach have been motivated out of the need to reduce nitrate leaching and other N_r leaching by avoiding run-off and infiltration into the soil. The approach has the benefit of being low-cost, but risks substantial NH_3 emissions, while also being ineffective at avoiding nitrification and denitrification, which contribute to N_2O, NO_x and N_2 emissions. The approach is preferable to open-field storage of solid manure on a permeable surface. Storage of solid manure on concrete areas is considered good agricultural practice for nitrate pollution, but makes no contribution to reducing NH_3 emissions.

3.2.6 Slurry mixing

Slurry mixing in the storage is one of the most commonly applied manure treatment technologies. Slurry is thereby homogenized, typically shortly prior to field application, in order to achieve a more homogenous distribution of nutrients across the field(s) to which the volume of the slurry storage is applied. Apart from this, mixing does not offer any additional benefits compared to untreated slurry. Neither DM nor carbon content is reduced, and the C/N-ratio is not altered. No significant changes in N_2O or CH_4 emissions are expected, but NH_3 may tend to increase, depending on the extent and timing of mixing (mixing will tend to increases pH by promoting CO_2 loss from slurry), so mixing should only be done shortly before field application.

3.2.7 Adsorption of slurry ammonium

Slurry additives can act on a chemical, physical or biological basis. Clay/zeolite mineral additives have been shown to adsorb NH_4-N and can thus potentially reduce NH_3 losses. However, this can only be achieved effectively with high amount of additives, for example, 25 kg of zeolite per m^3 slurry have been shown necessary to adsorb 55% of NH_4-N (Kocatürk-Schumacher et al., 2017, 2019). On most commercial farms it is neither logistically possible nor economically profitable to add such high amounts of slurry additives. Addition of biochar may also reduce NH_3 emissions from stored manure.

3.2.8 Slurry acidification during storage

An obvious way to minimize ammonia emissions from stored slurry is to decrease pH by addition of strong acids or other acidifying substances. This can also be done in the animal house (Housing Measure 8). Care must be taken that the low pH is maintained to get the full benefit of this measure. Slurry with a sufficiently reduced pH will also emit less methane. This solution has been used commercially since 2010 in countries such as Denmark (by 2018, around 15–20% of all slurry applied in Denmark was acidified; Birkmose, pers. comm.), and its high efficiency for minimizing NH_3 emissions has been documented in many studies (see review by Fangueiro et al., 2015, with emission reductions by >80% possible). It is most typical to acidify slurry using sulphuric acid (cheapest industrial acid; also, the sulphate added serves as a relevant plant nutrient source), though use of other acids is also possible. Acidification also reduces methane formation very effectively, by up to 67–87% (Petersen, 2018). Reduced nitrification and denitrification decrease the potential for N_2O and N_2 emission, though further studies are required to demonstrate efficiency for this. In one novel variant of this method, electricity is used to produce a plasma, which oxidizes N_2 to NO and thence to nitrogen dioxide (NO_2), which converts in slurry to produce nitric acid (HNO_3). In this way, slurry acidification is achieved while augmenting the nutrient value of the manure (Graves et al., 2019). More research is needed to assess this option fully.

Costs for in-house acidification systems can be higher than acidification during field application (Manure Measure 9), but are counteracted by additional benefits including improved in-house air quality benefiting animal and staff, which may influence productivity, retention of more slurry

N throughout the manure management chain, and associated savings in fertilizer costs.

3.2.9 Slurry aeration

Slurry aeration introduces oxygen rapidly into the slurry in order to allow aerobic microbes to develop. Oxidation of organic matter to CO_2 and H_2O increases, and thus CH_4-production and emission is reduced. Odorous compounds are degraded. Slurry DM content decreases. Thus, less mixing is needed and technical properties of slurry are often improved. However, successful aeration requires 200 m^{-3} oxygen per t of slurry (Burton, 1998).

Slurry aeration increases NH_3 emissions and energy consumption. The potential for NO_x emissions is also expected to increase, as increased oxygen availability promotes nitrification, while subsequently higher levels of nitrate availability may increase other oxidized N_r losses and denitrification. The extent of these increases has so far been quantified only in few studies (Amon et al., 2006) and more research is necessary to allow a complete evaluation. In the present context, an increase in denitrification to form N_2 is considered a waste of available N_r resources.

3.2.10 Mechanical solid-liquid separation of slurry fractions

During slurry separation, solids and liquids are mechanically separated from each other. This results in two fractions: a liquid slurry fraction, with relatively low DM content compared with the slurry, and a solid fraction that can be stored in heaps. Energy consumption for slurry separation is relatively low, but depends on the technology used for separation. DM content in the liquid fraction is reduced by 40–45%, and vice versa for the solid. Carbon content in the liquid is typically reduced by 45–50%, with the C/N-ratio of the liquid decreasing from about 10:1 to about 5:1 (Amon et al., 1995; Sommer et al., 2013). As carbon is removed from the slurry, microbial degradation of organic matter during slurry storage is reduced. However, the opposite may be the case for the solid fraction, depending on storage conditions.

The removal of solids reduces crust formation and sedimentation of the liquid fraction compared to raw slurry. Thus, less-intensive mixing is necessary to homogenize the slurry prior to application. Efforts for low-emission application techniques are also reduced as separated slurry has a lower viscosity and flows more easily through band spreading hoses (Owusu-Twuma et al., 2017). Slurries with very low DM content can be spread with simple nozzle-beam-dischargers that can be operated on slopes >10%, which is not possible with other band-spreading techniques. Furthermore, separated slurry liquid fraction has a low viscosity and infiltrates rapidly into the soil. Thus, plants get less dirty, and ammonia emissions after liquid fraction spreading are typically reduced.

A substantial reduction of ammonia emissions by slurry separation is therefore possible for the liquid phase, especially following land application (e.g. Amon et al., 2006).

The liquid fraction of separated slurry has a narrow C/N-ratio which reduces the potential for microbial N immobilization in the soil and the potential for N_2O emissions. Crop N availability of the liquid fraction is therefore more predictable and can be better calculated in order to match nutrient requirements of crops to actual fertilization. Dosch (1996) investigated fertilization with untreated and separated slurries and found significantly higher denitrification rates with untreated slurry. Separated slurry liquid fraction on the other hand resulted in significantly higher crop yield. However, the solid fraction needs to be handled with care during storage to avoid elevated ammonia emissions. Furthermore, the solid fraction may become a source of methane emissions, if not properly treated. Alternatively, if the solid fraction is used as feedstock for biogas production, this methane potential may be recovered and utilized as renewable energy source. After application, the solid fraction serves mainly as soil improvement and slow release N fertilizer.

Slurry separation fulfils most requirements of appropriate manure treatment. Costs could be further reduced if the technology was more widespread and more separators were on the market and available to farmers. As the fertilizer value of the liquid fraction from separated slurry is improved, mineral N fertilizer input can be reduced. The slurry liquid fraction can be applied at the soil surface in a growing crop with very simple low-cost slurry band spreaders with a high uptake efficiency and fertilizer replacement value. The main caveat of the method is the difficulty of appropriate storage, handling and utilization of the solid fraction; this needs to be low emission, in order not to compromise benefits of the liquid fraction. An alternative is to use the solid fraction as a feedstock in nutrient anaerobic digestion with nutrient recovery.

3.2.11 Anaerobic digestion

Anaerobic digestion of animal manures is mainly implemented for bio-energy production reasons. Improvement of manure quality is therefore typically considered to be a 'by-product' of anaerobic digestion. However, when combined with nutrient recovery methods (Fig. 6), nutrient management can be considered as fully integrated as a key goal in implementation of anaerobic digestion.

Biogas production from animal manures through anaerobic digestion aims at maximizing the bio-methane yield. Where no biogas recovery system is available, unintended anaerobic degradation of organic substances into methane during manure storage should be limited as far as possible, to prevent emission to the atmosphere of this strong GHG. This also maximizes the resource availability for subsequent biogas production when facilities are available. Under these circumstances (including heating of the manure to promote digestion) methane production is enhanced, allowing its collection and use (e.g. in combined heat and power production). Anaerobic digestion not only reduces methane emissions from subsequent storage of the manure digestate, but the energy produced typically substitutes consumption of use of fossil energy. Both effects reduce anthropogenic GHGs.

Anaerobic digestion reduces manure carbon and DM content by about 50% (Amon and Boxberger, 2000). NH_4^+ content and pH in digested slurry are higher than in untreated slurry. Thus, potential for ammonia emissions during subsequent slurry storage are increased. Digested slurry therefore has to be stored in covered slurry stores. These should be connected to the gas-bearing system of the biogas plant, because methane is still formed after the main digestion phase has taken place in the heated digester.

Due to the reduced DM content, biogas slurry can infiltrate more rapidly into the soil, which tends to reduce ammonia emissions after slurry application. However, the increased NH_4^+ content and pH give rise to higher potential for ammonia loss especially after surface application. It is therefore strongly recommended to apply biogas slurry with low-emission techniques near or below the soil surface (e.g. band application or injection).

The combined implementation of anaerobic digestion (reducing DM content, increased NH_4^+ and pH) and low-emission land-spreading application (e.g. trailing hose, injection) considerably reduces ammonia emissions. In addition, N immobilization and N_2O losses are likely to be smaller than from untreated slurry, due to the removal of easily degradable organic substances during the anaerobic digestion process. Energy consumption for pumping and mixing is considerably reduced due to the reduced DM content. When combined with appropriate methods for low-emission land-spreading of the digestate, anaerobic digestion therefore has multiple benefits. In addition, it provides the opportunity for further processing for more advanced forms of nutrient recovery, including nutrient precipitation, concentration and ammonia stripping (Fig. 6).

3.2.12 Manure composting

Composting of manure is done in order to create a stable and odourless bio-based fertilizer product, with lower moisture content, while containing most of the initial nutrients, free of pathogens and seeds (Jensen, 2013). Composting significantly reduces mass (as a result of water evaporation and volatile solids decomposition to release CO_2) and hence transport costs. However, it is difficult to avoid some loss of manure N in the form of NH_3 and the process also emits greenhouse gases, with potential for increased N_2O and CH_4 emissions, in addition to NO_x and N_2 (Chowdhury et al., 2014). The N fertilizer value of composts is often significantly lower than the N-rich manure components it is made from, which is largely a result of associated NH_3 and N_2 emissions (Jensen, 2013). Composting on porous soil surfaces may also be associated with significant leachate, including NH_4^+, NO_3^- and other N_r compounds. Composting is typically a low-cost technology, but implies space requirements and energy consumption. Overall, it can therefore not usually be recommended to mitigate nitrogen losses, but may be preferred on other criteria (e.g. volume and weight reduction, compost product stability, reduced odour, improved marketability and soil amelioration).

3.2.13 Nutrient recovery by drying and pelletizing of manure solids

Drying and pelletizing of solid manures, slurry or digestate solids can be done to create a more stable and odourless bio-based fertilizer product. Drying is energy-intensive and thereby relatively expensive, unless excess energy (e.g. from the combined heat and power plant engine on a biogas plant) is freely or cheaply available. Increased ammonia loss is inevitable in the process, unless exhaust filtering or scrubbing and recovery is applied, or the solids are acidified prior to drying. Drying is usually combined with a pelletizing process to facilitate handling. The pelleted material can be marketed as an organic matter and P-rich soil amendment; if acidified prior to drying, the resulting product may also be rich in plant available N (Pantelopoulos et al., 2017).

3.2.14 Nutrient recovery by combustion, gasification or pyrolysis

Combustion, thermal gasification or pyrolysis of manure and digestate solids can be used to generate a net energy output for heat and/or electricity production. However, at present the method leads to a more or less complete loss of the manure N, which is converted into gaseous N_2, as well as NO_x and NH_3. Available advanced technologies (e.g. selective non-catalytic reduction) focus on denitrifying these N_r gases to N_2. Until systems are implemented to minimize N_2 formation and recover the N_r gases, this measure cannot be considered appropriate for abating overall N loss.

At the same time, the approach produces ash or biochar residuals. These ashes contain the non-volatile nutrients, concentrated relative to the solids. They can be used as an ash-based, P- and K-rich soil amendment or bio-based fertilizer. The availability of the remaining nutrients in the ash is generally much lower than for the raw manure, whereas for biochar it is in-between ash and raw manure. Organic compounds in the biochar that are produced are very recalcitrant to biological decay and have a very large specific surface area, being potentially charged. This means that such biochar may be used for soil amendment, ameliorating soil pH and organic matter positively.

3.2.15 Nutrient recovery by precipitation of nitrogen salts

Struvite ($MgNH_4PO_4{\cdot}6H_2O$) can be precipitated from liquid manures, provided that the appropriate conditions are present (pH ~9, a molar ratio 1:1:1 of Mg^{2+}:NH_4^+:PO_4^{3-}, conducive physical settling conditions). As such the precipitation of struvite is a method for removal and recovery of both N and P from liquid manures. The method has been developed for wastewater treatment, where P removal can easily reach more than 70% and it is commercially available for sewage treatment plants, although not yet widely applied. For manures, the struvite technique is particularly relevant for anaerobically digested slurries and the liquid fraction from digestate separation; hence, it has been the subject of massive research in the past decade and quite high removal efficiencies have been achieved (56–93%; see further review in Jensen, 2013). However, it only works for the N already present as NH_4^+ and further development is needed for appropriate application to liquid manures and digestates. So far, only a few commercial-scale plants are in operation worldwide. The main advantage of struvite is its high concentration and similarity in physical-chemical properties to conventional mineral N fertilizer.

3.2.16 Nutrient recovery by concentration of nitrogen salts and solutions

Mineral concentrates are highly nutrient-rich solutions that may be obtained via ultrafiltration, evaporation or reverse osmosis of the liquid fraction from separation of slurry or digestate. These mineral concentrates (the retentate) may be directly applied to agricultural land and the byproduct water low in nutrients (the permeate) may be directly discharged to surface waters or the sewage system. The greatest experience with these technologies in Europe are from the livestock regions of the Netherlands and Belgium, where a number of centralized and large-scale manure processing plants utilizing a range of technologies in combination (e.g. anaerobic digestion, solid-liquid separation, ultrafiltration/reverse osmosis/solids drying). Provided that the losses can be kept to a minimum, the mineral fertilizer replacement value of the mineral concentrates can be relatively high, as they resemble commercial liquid fertilizers, with nearly all the nutrients in a mineral, plant-available form. However, to avoid gaseous NH_3 losses this may require prior acidification or injection of the concentrate into the soil (Jensen, 2013). As these technologies are still under investigation, the UNECE Categories are currently uncertain (i.e. Category 3, pending further assessment).

3.2.17 Nutrient recovery by ammonia stripping

Air stripping of NH_3 is a process whereby the liquid fraction after manure separation is brought into contact with air, upon which NH_3 evaporates and is carried away by the gas. Instead of ambient air, 'steam stripping', can be used where steam is replaces use of air as the ammonia carrier. Since evaporation occurs from the liquid surface, it is advantageous to ensure that the liquid has a large surface area. This can be achieved in a stripping column with structured packing, where it spreads over the packing material in a thin film and therefore has a considerably larger surface. The mass transport also increases with the concentration of $NH_{3(aq)}$ in the liquid phase, hence, if pH and/or temperature is increased, an increasing part of total ammoniacal nitrogen is in $NH_{3(aq)}$ form and the mass transport of NH_3 increases (Sommer et al., 2013). Altogether this makes the technology relatively energy demanding and costly, though cheap/free surplus energy from, for example, a biogas- combined heat and power plant may reduce energy costs. Alternatively, using selectively permeable membrane contact systems at lower temperatures may offer a cheaper solution, if membrane fouling can be avoided. Ammonia released from an NH_3 stripping column or from a manure drying facility can be collected using wet scrubbing with an acid solution, typically sulphuric acid to make ammonium sulphate (most common), but also has been reported with nitric acid to make ammonium nitrate. Both compounds can serve as raw materials for mineral fertilizers, and thus provides opportunity for circular economy development as part of the fertilizer industry's commitment to include recovered and recycled N_r. In general, this is a well-known and generally effective technology. The main barriers are the relatively low N scrubber-liquid concentrations achievable (and thus high logistic costs), and the quality requirements for introduction of the scrubber-liquid into the raw materials market for the fertilizer industry. Actual developments are working on process improvement through applying sulphur or nitrate instead of an acid and on developing a system with CO_2 to generate ammonium carbonate which can also be used as fertilizer and bring C back into the soil. The ammonia-low liquid fraction is then used to flush the manure channel to reduce NH_3 and CH_4 emissions in the house.

4 Best practices and priority measures

Best practices and priorities for the selection of abatement/mitigation measures must be based on the following criteria: (i) implementability; (ii) effectiveness; (iii) impact on environmental emissions; (iv) secondary effects; (v) controllability; and (vi) cost efficiency. Based on these criteria, we suggest the following priority measures:

4.1 Livestock feeding

The following priorities through livestock feeding help to reduce nitrogen losses:

- Avoid N surplus from the very beginning of the manure management continuum;
- Adjust animal diet to animal performance (in line with existing guidance in the UNECE Ammonia Framework Code);
- Adapt animal diet to shift N excretion from urine to faecal excreta;
- Dairy cattle:
    - Reduction of crude protein content in the diet;
    - Adapt diet and dairy production system to site specific conditions;
    - Increase milk yield with moderate level of concentrates;
    - Increase production cycles per cow.
- Pigs:
    - Reduction of crude protein content in the diet;
    - Multiphase feeding;
    - More use of food wastes (inc. from processing and retail) as a way to reduce upstream and downstream emissions.

4.2 Livestock housing

The following priorities help to reduce nitrogen losses from livestock housing:

- Reduction of indoor temperature;
- Reduction of emitting surfaces, reduction of soiled areas;
- Reduction of air flow over soiled surfaces;
- Use of additives (e.g. acidification);
- Frequent removal of slurry to an outside store; and
- In the longer term: smart barns with optimized ventilation (open housing) or ventilation air scrubbing (closed housing), immediate segregation of urine and faeces components, in-house acidification of slurry (pigs and cattle).

4.3 Manure storage, treatment and processing

The following priorities help to reduce nitrogen loses and to mobilize nitrogen recovery and reuse from manure storage, treatment and processing:

- Store solid manures outside the barn on a solid concrete base in a dry/ covered location;
- Ensure tight slurry stores, and cover either by a solid cover, or by ensuring sufficient natural crust formation;

- Use manure treatment where relevant to:
  - Homogenize nutrient content for more even field spreading to ensure that all available nutrient resources are used effectively for crop growth;
  - Reduce slurry DM content, for example, by solid-liquid separation, to enhance soil infiltration and limit NH_3 loss;
  - Increase slurry NH_4^+ content to maximize crop N availability;
  - Lower pH by acidification to reduce NH_3 volatilisation and enhance fertilizer value; and
  - Apply manure treatment methods to enable combined energy and nutrient recovery, that is, anaerobic digestion, where relevant.

The use of manure advanced processing for N recapture and production of value-added nutrient products from recycled manure N resources should be focused on situations where other effective options are not available. Examples of such advanced processing technologies are high-tech separation by filtration, reverse osmosis and NH_3 scrubbing, drying of manure and digestate solids for organic fertilizer production.

Production of recovered, bio-based fertilizer products should not be supply-driven (trying to solve a waste problem), but rather demand-driven (bio-based fertilizers the farmers want).

5 Conclusion and future trends in research

It is clear that manure management impacts quantities of N_r emissions (NH_3, direct and indirect N_2O emissions, NO_x emissions, NO_3^- leaching) and N_2 emissions, as well as emissions of CH_4 and CO_2. This applies at each stage of the manure management continuum (Chadwick et al., 2011). Since production of these gases, as well as of leachable N_r, is of microbial origin, the DM content and temperature of manure and soil are key factors for farm manure management decisions that influence the magnitude of N and GHG losses. There remains a degree of uncertainty in emission rates of N and GHGs from different stages of manure management, and researchers continue to investigate interactions of the management and environmental factors that control emissions. Some specific approaches to reducing N and greenhouse gas emissions from livestock housing and manure storage include optimizing diet formulation, low-emission housing technologies, manure processing and nutrient recovery. The technologies include air scrubbers, covered manure storage, slurry separation and anaerobic digestion, nitrogen concentration and stripping methods.

Existing legislation across the UNECE region offers opportunities to find 'win-win' scenarios, with benefits in reducing multiple forms of pollution. An example is the EU Nitrates Directive (91/676/EEC), which has led to development of Nitrate Vulnerable Zone action plans to prevent application of animal manure, slurry and poultry manure (with high available N content) in autumn, a practice which reduces N losses, as well as direct and indirect N_2O losses. Care is needed to ensure that legislation does not lead to potential 'pollution swapping' (e.g. unadjusted use of slurry injection to reduce NH_3 emissions at the expense of an increase in N_2O emissions, with no modification of N inputs). A core principle is that measures that reduce one form of N loss need to either be accompanied by a reduction of fresh nitrogen inputs, or be accompanied by an increase in harvested products, to maintain mass consistency. In this way, what may first seem a trade-off at the field-scale, can be seen at the landscape and regional scale as an opportunity to move towards a more circular system with lower overall N losses.

The nature of the N cycle and its interaction with the C, P and other nutrient cycles demands a holistic approach to addressing N and GHG emissions and mitigation research at a process level of understanding. Systems-based modelling must play a key role in integrating the complexity of management and environmental controls on emissions. Progress has been made to this end (Sommer et al., 2009), with some studies producing whole farm models encompassing livestock production (del Prado et al., 2010).

Concepts for best practices to reduce adverse environmental impacts depend on the following integrated concepts:

- Relationship between nitrogen and GHG emissions;
- Influence of climate change on nitrogen emissions;
- Interaction between abatement/mitigation and adaptation measures;
- Interaction between nitrogen emissions and animal welfare;
- Integrated assessment of the whole manure management continuum;
- Integrated assessment considering the three pillars of sustainability: economy, environment, society;
- Interaction between consumer demand and nitrogen emissions;
- Development of region-specific concepts for sustainable intensification;
- Modelling of livestock production at regional, national and global scale; and
- Economic impact of both the cost of the techniques and the benefit to the farmer of reducing emissions and retaining nitrogen as a fertilizer.

Concepts to reduce adverse environmental impacts depend on the *understanding at a process level* of the following:

- Assessment of emissions from naturally ventilated barns;
- Assessment of emissions from new, animal friendly housing systems;

- Development of abatement/mitigation measures especially for naturally ventilated dairy barns (e.g. targeted ventilation and air scrubber, manure acidification);
- Interaction between climate change and heat stress/animal behaviour/emissions;
- Interaction between low protein diets and N and GHG emissions;
- Interactions between N and GHG emissions during housing, storage and application to field;
- Life-cycle assessment: for example, grass-based dairy feeding versus low protein dairy feeding;
- Feed and manure additives for improved N use efficiency; and
- Manure treatment for higher N use efficiency (increase of nutrient availability, decrease of emissions) and potential of processing to recover manure N into bio-based fertilizers in a circular economy.

Concepts to reduce adverse environmental impacts depend on the development of *flexible concepts* for environmental improvement:

- Climate and site-specific conditions vary across the UNECE region and globally;
- All three columns of sustainability must be considered: economic, environmental and social sustainability;
- Conflicts of interest must be addressed; and
- Targeted approaches according the needs of different regions.

Concepts to reduce adverse environmental impacts depend on *effective communication and interaction*:

- Establishing networks to exchange manure management information, connect people, and forge partnerships;
- Launching an on-line knowledge hub - on best practices for livestock housing and manure management; and
- Establishing a roster of experts to provide targeted technical assistance and training, analysis and practical implementation and policy support, relying heavily on co-financing and in-kind resources from partners.

The development of best practice concepts is challenging. Climate and site-specific conditions are highly variable. It is essential to consider the three columns of sustainability: economy, environment and society and to address synergies and potential conflicts of interest. This inevitably leads to the conclusion that there will be no 'one size fits all solution'. Best-practice concepts provide the basis to guide on the development of flexible measures targeted for each specific region and context.

6 References

Aarnink, A. J. A., van den Berg, A. J., Keen, A., Hoeksma, P. and Verstegen, M. W. A. (1996). Effect of slatted floor area on ammonia emission and on the excretory and lying behaviour of growing pigs. *Journal of Agricultural Engineering Research* 64(4), 299–310.

Amon, T. and Boxberger, J. (2000). Biogas production from farmyard manure. In: *Management Strategies for Organic Wastes in Agriculture*, FAO European Cooperative Research (Ed.), Network on Recycling of Agricultural. Municipal and industrial residues in. Agriculture (RAMIRAN), 9th International Conference, 6–9 September 2000, Gargnano, Italy.

Amon, T., Boxberger, J., Gronauer, A. and Neser, S. (1995). Einflüsse auf das Entmischungsverhalten, Abbauvorgänge und Stickstoffverluste von Flüssigmist während der Lagerung. In: *der landwirtschaftlichen Nutztierhaltung, Beiträge zur 2. Internationalen Tagung am* 14./15. März 1995 in Potsdam, Institut für Agrartechnik Bornim, MEG, KTBL, AEL (Eds), 91–98.

Amon, B., Kryvoruchko, V., Amon, T. and Zechmeister-Boltenstern, S. (2006). Methane, nitrous oxide and ammonia emissions during storage and after application of dairy cattle slurry and influence of slurry treatment. *Agriculture, Ecosystems and Environment* 112(2–3), 153–162.

Amon, B., Kryvoruchko, V., Frohlich, M., Amon, T., Pollinger, A., Mosenbacher, I. and Hausleitner, A. (2007). Ammonia and greenhouse gas emissions from a straw flow system for fattening pigs: housing and manure storage. *Livestock Science* 112(3), 199–207.

Bernal, M. P., Bescós, B., Bonmatí, A., Burgos, L., Bustamante, M. Á., Clemente, R., Fabbri, C., Flotats, X., García-González, M. C., Herrero, E., Mattachini, G., Moscatelli, G., Noguerol, J., Palatsi, J., Piccinini, S., Proniewicz, M., Provolo, G., Riaño, B., Riau, V., Sáez, J. A., Teresa, M., Tey, L., Torrellas, M., Valli, L., Ward, A. J. and Wiśniewska, H. (2015). Evaluation of manure management systems in Europe. LIFE + MANEV report. LIFE09 ENV/ES/000453, Published by SARGA. Available at: https://core.ac.uk/download/pdf/46606176.pdf.

Bittman, S., Dedina, M., Howard, C. M., Oenema, O. and Sutton, M. A. (Eds.) (2014). *Options for Ammonia Mitigation: Guidance from the UNECE Task Force on Reactive Nitrogen*. Centre for Ecology and Hydrology, Edinburgh, UK.

Braam, C. R., Ketelaars, J. J. M. H. and Smits, M. C. J. (1997a). Effects of floor design and floor cleaning on ammonia emission from cubicle houses for dairy cows. *Netherlands Journal of Agricultural Science* 45(1), 49–64.

Braam, C. R., Smits, M. C. J., Gunnink, H. and Swierstra, D. (1997b). Ammonia emission from a double-sloped solid floor in a cubicle house for dairy cows. *Journal of Agricultural Engineering Research* 68(4), 375–386.

Broderick, G. A. (2003). Effects of varying dietary protein and energy levels on the production of lactating dairy cows. *Journal of Dairy Science* 86(4), 1370–1381.

Burton, C. H. (1998). Processing strategies for organic wastes. In: *Management Strategies for Organic Waste Use in Agriculture* Martinez, J. (Ed). Abstracts of the Papers of 8th International Conference of the FAO Network on Recycling of Agricultural, Municipal and Industrial Residues in Agriculture.

Burton, C. H. (2007). The potential contribution of separation technologies to the management of livestock manure. *Livestock Science* 112(3), 208–216.

Butterbach-Bahl, K., Per Gundersen, P., Ambus, P., Augustin, J., Beier, C., Boeckx, P., Dannenmann, M., Sanchez Gimeno, B., Ibrom, A., Kiese, R., Kitzler, B., Rees, R. M., Smith, K. A., Stevens, C., Vesala, T. and Zechmeister-Boltenstern, S. (2011a). Chapter 6. Nitrogen processes in terrestrial ecosystems. In: *The European Nitrogen Assessment* Sutton, M. A., Howard, C. M., Erisman, J. W., Billen, G., Bleeker, A., Grennfelt, P., van Grinsven, H. and Grizzetti, B. (Eds). Cambridge University Press, New York, pp 99–125.

Butterbach-Bahl, K., Nemitz, E., Zaehle, S., Billen, B., Boeckx, P., Erisman, J. W., Garnier, J., Upstill-Goddard, R., Kreuzer, M., Oenema, O., Reis, S., Schaap, M., Simpson, D., de Vries, W., Winiwarter, W. and Sutton, M. A. (2011b). Chapter 19. Effect of reactive nitrogen on the European greenhouse balance. In: *The European Nitrogen Assessment* Sutton, M. A., Howard, C. M., Erisman, J. W., Billen, G., Bleeker, A., Grennfelt, P., van Grinsven, H. and Grizzetti, B. (Eds). Cambridge University Press, New York, pp 434–462.

Cahn, T. T., Aarnink, A. J. A., Schutte, J. B., Sutton, A., Langhout, D. J. and Verstegen, M. W. A. (1998). Dietary protein affects nitrogen excretion and ammonia emission from slurry of growing-finishing pigs. *Livestock Production Science* 56(3 (December)), 181–191.

Chadwick, D., Sommer, S., Thorman, R., Fangueiro, D., Cardenasa, L., Amon, B. and Misselbrook, T. (2011). Manure management: implications for greenhouse gas emissions. *Animal Feed Science and Technology* 166–167, 514–531.

Chowdhury, M. A., de Neergaard, A. and Jensen, L. S. (2014). Composting of solids separated from anaerobically digested animal manure: effect of different bulking agents and mixing ratios on emissions of greenhouse gases and ammonia. *Biosystems Engineering* 124, 63–77 http://doi.org/10.1016/j.biosystemseng.2014.06.003.

Dämmgen, U. and Hutchings, N. J. (2008). Emissions of gaseous nitrogen species from manure management: a new approach. *Environmental Pollution* 154(3), 488–497.

del Prado, A., Chadwick, D., Cardenas, L., Misselbrook, T., Scholefield, D. and Merino, P. (2010). Exploring systems responses to mitigation of GHG in UK dairy farms. *Agriculture, Ecosystems and Environment* 136(3–4), 318–332.

Denmead, O. T., Freney, L. R. and Simpson, J. R. (1982). Dynamics of ammonia volatilization during furrow irrigation of maize. *Soil Science Society of America Journal* 46, 149–155.

DLG (2020). DLG-Testzentrum Technik und Betriebsmittel: Prüfrahmen Abluftreinigungs-systeme für Tierhaltungsanlagen, Groß-Umstadt. Available at: https://www.dlg.org/de/landwirtschaft/tests/suche-nach-pruefberichten/#!/p/3/u/95/1?locale=de&locale=en.

Dosch, P. (1996). Optimierung der Verwertung von Güllestickstoff durch Separiertechnik und kulturartspezifische Applikationstechniken. *Bayerisches Staatsministerium für ELuF, Gelbe Reihe, Landtechnische Berichte aus Praxis und Forschung* 56.

Economic Commission for Europe (2015). Executive body for the convention on long-range transboundary air pollution. In: *Framework Code for Good Agricultural Practice for Reducing Ammonia Emission, ECE/EB.AIR/129*. Available at: https://undocs.org/ECE/EB.AIR/129.

Ellen, H. H., Hol, J. M. G., Hoofs, A. I. J., Mosquera, J. and Bosma, A. J. J. (2008). Ammoniakemissie en kosten van chemische luchtwasser met bypassventilatoren bij vleesvarkens (Ammonia Emission and Costs of a Chemical Air Scrubber with

Bypass Ventilation at a Pig House). Animal Sciences Group Report 151. Wageningen University and Research Centre, Wageningen, the Netherlands. Available at: http://edepot.wur.nl/35138.

Elliott, H. A. and Collins, N. E. (1982). Factors affecting ammonia release in broiler houses. *Transactions of the ASCE* 25, 413-418.

Fangueiro, D., Pereira, J., Chadwick, D., Coutinho, J., Moreira, N. and Trindade, H. (2008a). Effect of cattle slurry separation on greenhouse gas and ammonia emissions during storage. *Journal of Environmental Quality* 37(6) (November), 2322-2331.

Fangueiro, D., Pereira, J., Chadwick, D., Coutinho, J., Moreira, N. and Trindade, H. (2008b). Laboratory assessment of the effect of cattle slurry pre-treatment on organic N degradation after soil application and N_2O and N_2 emissions. *Nutrient Cycling in Agroecosystems* 80(2), 107-120.

Fangueiro, D., Hjorth, M. and Gioelli, F. (2015). Acidification of animal slurryea review. *Journal of Environmental Management* 149, 46-56.

Firestone, M. K. and Davidson, E. A. (1989). Microbial basis of NO and N2O production and consumption in soil. In: *Exchange of Trace Gases between Terrestrial Ecosystems and the Atmosphere* Andreae, M. O. and Schimel, D. S. (Eds). Wiley, New York, NY, 7-21.

Gilhespy, S. L., Webb, J., Chadwick, D. R., Misselbrook, T. H., Kay, R., Camp, V., Retter, A. L. and Bason, A. (2009). Will additional straw bedding in buildings housing cattle and pigs reduce ammonia emissions? *Biosystems Engineering* 102(2), 180-189.

Graves, D. B., Bakken, L. B., Jensen, M. B. and Ingels, R. (2019). Plasma activated organic fertilizer. *Plasma Chemistry and Plasma Processing* 39(1), 1-19. https://doi.org/10.1007/s11090-018-9944-9.

Groenestein, C. M. and Van Faassen, H. G. (1996). Volatilization of ammonia, nitrous oxide and nitric oxide in deep-litter systems for fattening pigs. *Journal of Agricultural Engineering Research* 65(4, December), 269-274. https://doi.org/10.1006/jaer.1996.0100.

Groenestein, C. M., Hutchings, N. J., Haenel, H. D., Amon, B., Menzi, H., Mikkelsen, M. H., Misselbrook, T. H., van Bruggen, C., Kupper, T. and Webb, J. (2019). Comparison of ammonia emissions related to nitrogen use efficiency of livestock production in Europe. *Journal of Cleaner Production* 211, 1162-1170. https://doi.org/10.1016/j.jclepro.2018.11.143.

Groenestein, C. M., Monteny, G. J., Aarnink, A. J. A. and Metz, J. H. M. (2007). Effect of urinations on the ammonia emission from group-housing systems for sows with straw bedding: model assessment. *Biosystems Engineering* 97(1), 89-98. https://doi.org/10.1016/j.biosystemseng.2007.02.004.

Guingand, N. (2009). Wet scrubber: one way to reduce ammonia and odours emitted by pig units. In: Paper presented at the sixtieth meeting of the European Association for Animal Production, Barcelona, Spain, 24-27 August 2009.

Guingand, N. and Courboulay, V. (2007). Reduction of the number of slots for concrete slatted floor in fattening buildings: consequences for pigs and environment. In: *Proceedings of the International Conference on Ammonia in Agriculture: Policy, Science, Control and Implementation, 19-21 March 2007* Monteny, G. J. and Hartung, E. (Eds). Ede. Netherlands. Wageningen Academic Publishers, Wageningen, the Netherlands, pp 147-148.

Hahne, J., Arends, F., Beverborg, R., Niehoff, A.-L., Bönsch, S. and Hortmann-Scholten, A. (2016). Aktuelle Entwicklung Kosten-Nutzenanalyse und Vollzugsempfehlungen für den Einsatz von Abluftreinigungsanlagen in der Tierhaltung. *UBA Texte* 61, 41-65.

Huynh, T. T. T., Aarnink, A. J. A., Spoolder, H. A. M., Verstegen, M. W. A. and Kemp, B. (2004). Effects of floor cooling during high ambient temperatures on the lying behavior and productivity of growing finishing pigs. *Transactions of the ASAE*, 47(5), 1773–1782.

Hutchinson, G. L. and Davidson, E. A. (1993). Processes for production and consumption of gaseous nitrogen oxides in soil. *ASA Special Publications* 55, 79–93.

Jensen, L. S. (2013). Chapter 14. Animal manure residue upgrading and nutrient recovery in biofertilizers. In: *Animal Manure Recycling–Treatment and Management* Sommer, S. G., Christensen, M. L., Schmidt, T. and Jensen, L. S. (Eds). John Wiley & Sons Ltd., Hoboken, NJ, pp 271–294.

Kai, P., Pedersen, P., Jensen, J. E., Hansen, M. N. and Sommer, S. G. (2008). A whole-farm assessment of the efficacy of slurry acidification in reducing ammonia emissions. *European Journal of Agronomy* 28(2), 148–154.

Kocatürk-Schumacher, N. P., Bruun, S., Zwart, K. and Jensen, L. S. (2017). Nutrient recovery from the liquid fraction of biogas digestate by adsorption to clinoptilolite. *CLEAN–Soil, Air, Water* 45(6) http://doi.org/10.1002/clen.201500153 1500153.

Kocatürk-Schumacher, N. P., Zwart, K., Bruun, S., Stoumann Jensen, L., Sørensen, H. and Brussaard, L. (2019). Recovery of nutrients from the liquid fraction of digestate: use of enriched zeolite and biochar as nitrogen fertilizers. *Journal of Plant Nutrition and Soil Science* 182(2), 187–195. https://doi.org/10.1002/jpln.201800271.

Kupper, T., Häni, C., Neftel, A., Kincaid, C., Bühler, M., Amon, B. and VanderZaag, A. (2020). Ammonia and greenhouse gas emissions from slurry storage–a review. *Agriculture, Ecosystems and Environment* 300, 106963. ISSN 0167-8809, https://doi.org/10.1016/j.agee.2020.106963.

Lassaletta, L., Estellés, F., Beusen, A. H. W., Bouwman, L., Calvet, S., van Grinsven, H. J. M., Doelman, J. C., Stehfest, E., Uwizeye, A. and Westhoek, H. (2019). Future global pig production systems according to the shared socioeconomic pathways. *Science of the Total Environment* 665, 739–751.

Lukin, S. M., Nikolskiy, K. S., Ryabkov, V. V. and Rysakova, I. V. (2014). Chapter 20. Methods to reduce ammonia nitrogen losses during production and application of organic fertilizers. In: *Ammonia Workshop 2012 Saint Petersburg. Abating Ammonia Emissions in the UNECE and EECCA Region* Van der Hoek, K. W. and Kozlova, N. P. (Eds). RIVM Report 680181001/SZNIIMESH Report. Bilthoven, The Netherlands, ISBN: 978-90-6960-271-4, pp 169–175.

Melse, R. W., Ogink, N. W. M. and Bosma, B. J. J. (2008). Multi-pollutant scrubbers for removal of ammonia, odor, and particulate matter from animal house exhaust air. In: Proceedings of the Mitigating Air Emissions from Animal Feeding Operations Conference, 19-21 May 2008, Des Moines, Iowa, United States of America.

Melse, R. W., Hofschreuder, P. and Ogink, N. W. M. (2012). Removal of particulate matter (PM10) by air scrubbers at livestock facilities: results of an on-farm monitoring program. *Transactions of the ASABE* 55, 689–698.

Melse, R. W. and Ogink, N. W. M. (2005). Air scrubbing techniques for ammonia and odor reduction at livestock operations: review of on-farm research in the Netherlands. *Transactions of the ASAE* 48, 2303-2313.

Misselbrook, T. H. and Powell, J. M. (2005). Influence of bedding material on ammonia emissions from cattle excreta. *Journal of Dairy Science* 88(12), 4304–4312.

Møller, H. B., Hansen J. D. and Sørensen C. A. G. (2007). Nutrient recovery by solid-liquid separation and methane productivity of solids. *Transactions of the ASABE* 50(1), 193-200.

Monteny, G. J. (2000). Modelling of ammonia emissions from dairy cow houses. PhD thesis. Wageningen University, Wageningen, the Netherlands (with summaries in English and Dutch).

Nielsen, D. A., Nielsen, L. P., Schramm, A. and Revsbech, N. P. (2010). Oxygen distribution and potential ammonia oxidation in floating, liquid manure crusts. *Journal of Environmental Quality* 39(5), 1813-1820.

Ogink, N. W. M. and Bosma, B. J. J. (2007). Multi-phase air scrubbers for the combined abatement of ammonia, odor and particulate matter emissions. In: Proceedings of the International Symposium on Air Quality and Waste Management for Agriculture, Broomfield, Colorado, 16-19 September 2007. ASABE. Available at: http://elibrary.asabe.org/conference.asp?confid=aqwm2007.

Olesen, J. E. and Sommer, S. G. (1993). Modelling effects of wind speed and surface cover on ammonia volatilization from stored pig slurry. *Atmospheric Environment. (Part A: General Topics)* 27, 2567-2574.

Owusu-Twuma, M. Y., Polastre, A., Subedi, R., Santos, A. S., Ferreira, L. M. M., Coutinho, J. and Trindade, H. (2017). Gaseous emissions and modification of slurry composition during storage and after field application: effect of slurry additives and mechanical separation. *Journal of Environment Management* 200, 416-422.

Pantelopoulos, A., Magid, J., Jensen, L. S. and Fangueiro, D. (2017). Nutrient uptake efficiency in ryegrass fertilized with dried digestate solids as affected by acidification and drying temperature. *Plant and Soil* 421(1-2), 401-416.

Patterson, P. H. and Adrizal (2005). Management strategies to reduce air emissions: emphasis–dust and ammonia. *Journal of Applied Poultry Research* 14(3 (Fall)), 638-650.

Petersen, S. O. (2018). Greenhouse gas emissions from liquid dairy manure: prediction and mitigation. *Journal of Dairy Science* 101(7), 6642-6654, ISSN 0022-0302, https://doi.org/10.3168/jds.2017-13301.

Petersen, S. O. and Ambus, P. (2006). Methane oxidation in pig and cattle slurry storages, and effects of surface crust moisture and methane availability. *Nutrient Cycling in Agroecosystems* 74(1), 1-11.

Petersen, S. O., Amon, B. and Gattinger, A. (2005). Methane oxidation in slurry storage surface crusts. *Journal of Environmental Quality* 34(2), 455-461.

Petersen, S. O., Andersen, A. J. and Eriksen, J. (2012). Eff ects of cattle slurry acidification on ammonia and methane evolution during storage. *Journal of Environmental Quality* 41(1), 88-94 https://doi.org/10.2134/jeq2011.0184.

Petersen, S. O. and Miller, D. N. (2006). Greenhouse gas mitigation by covers on livestock slurry tanks and lagoons? *Journal of the Science of Food and Agriculture* 86(10), 1407-1411.

Powell, J. M., Misselbrook, T. H. and Casler, M. D. (2008). Season and bedding impacts on ammonia emissions from tie-stall dairy barns. *Journal of Environmental Quality* 37(1), 7-15.

Reis, S., Howard, C. and Sutton, M. A. (Eds.) (2015). *Costs of Ammonia Abatement and the Climate Co-Benefits*. Springer, Netherlands, 284.

Ritz, C. W., Mitchell, B. W., Fairchild, B. D., Czarick, M. and Worley, J. W. (2006). Improving in-house air quality in broiler production facilities using an electrostatic space charge system. *Journal of Applied Poultry Research* 15(2 (summer)), 333-340.

Santonja, G. G., Georgitzikis, K., Scalet, B. M., Montobbio, P., Roudier, S. and Delgado Sancho, L. (2017). Best Available Techniques (BAT) Reference document for the intensive rearing of poultry or pigs. EUR 28674 EN. https://doi.org/10.2760/020485.

Smits, M. C. J. (1998). Groeven maken in een dichte V-vormige vloer: enkele observaties naar loopgedrag en ammoniakemissies (Grooving a solid V-shaped floor: some observations on walking behaviour and ammonia emission). DLO18-IMAG19 Report P 98-60. Wageningen, the Netherlands.

Sommer, S. G., Olesen, J. E., Petersen, S. O., Weisbjerg, M. R., Valli, L., Rohde, L. and Béline, F. (2009). Region-specific assessment of greenhouse gas mitigation with different manure management strategies in four agroecological zones. *Global Change Biology* 15(12), 2825-2837.

Sommer, S. G., Christensen, M. L., Schmidt, T. and Jensen, L. S. (Eds.) (2013). *Animal Manure Recycling–Treatment and Management*. John Wiley & Sons Ltd., Hoboken, NJ, 384 pp.

Sutton, M. A., Bleeker, A., Howard, C. M., Bekunda, M., Grizzetti, M., Vries, W. de, van Grinsven, H. J. M., Abrol, Y. P., Adhya, T. K., Billen, G., Davidson, E. A., Datta, A., Diaz, R., Erisman, J. W., Liu, X. J., Oenema, O., Palm, C., Raghuram, N., Reis, S., Scholz, R. W., Sims, T., Westhoek, H., Zhang, F. S., with contributions from Ayyappan, S., Bouwman, A. F., Bustamante, M., Fowler, D., Galloway, J. N., Gavito, M. E., Garnier, J., Greenwood, S., Hellums, D. T., Holland, M., Hoysall, C., Jaramillo, V. J., Klimont, Z., Ometto, J. P., Pathak, H., Plocq Fichelet, V., Powlson, D., Ramakrishna, K., Roy, A., Sanders, K., Sharma, C., Singh, B., Singh, U., Yan, X. Y. and Zhang, Y. (2013). Our nutrient world: the challenge to produce more food and energy with less pollution. Global Overview of Nutrient Management. Centre of Ecology and Hydrology, Edinburgh on behalf of the Global Partnership on Nutrient Management and the International Nitrogen Initiative.

Swensson, C. (2003). Relationship between content of crude protein in rations for dairy cows, N in urine and ammonia release. *Livestock Production Science* 84(2 (December)), 125-133.

Swierstra, D., Braam, C. R. and Smits, M. C. (2001). Grooved floor systems for cattle housing: ammonia emission reduction and good slip resistance. *Applied Engineering in Agriculture* 17(1), 85-90.

United Nations Economic Commission for Europe Framework (2015). *Code for Good Agricultural Practice for Reducing Ammonia Emissions*. Available at: https://www.unece.org/fileadmin/DAM/env/lrtap/Publications/Ammonia_SR136_28-4_HR.pdf.

VanderZaag, A. C., Gordon, R. J., Jamieson, R. C., Burton, D. L. and Stratton, G. W. (2009). Gas emissions from straw covered liquid dairy manure during summer storage and autumn agitation. *Transactions of the ASABE* 52(2), 599-608.

Webb, J. and Misselbrook, T. H. (2004). A mass-flow model of ammonia emissions from UK livestock production. *Atmospheric Environment* 38(14), 2163-2176.

Webb, J., Sommer, S., Kupper, T., Groenestein, K., Hutchings, N. J., Eurich-Menden, B., Rodhe, L., Misselbrook, T. H. and Amon, B. (2012). Emissions of ammonia, nitrous oxide and methane during the management of solid manures. In: *Agroecology and Strategies for Climate Change. Sustainable Agriculture Reviews* Lichtfouse, E. (Ed.) (vol. 8). Springer, Dordrecht.

Whitehead, D. C. (2000). *Nutrient Elements in Grassland: Soil-Plant-Animal Relationships*. CABI Publishing, Wallingford, UK.

Ye, Z. Y., Zhang, G., Li, B., Søberg Strøm, J., Tong, G. and Dahl, P. J. (2008a). Influence of airflow and liquid properties on the mass transfer coefficient of ammonia in aqueous solutions. *Biosystems Engineering* 100(3), 422–434.

Ye, Z. Y., Zhang, G., Li, B., Strøm, J. S. and Dahl, P. J. (2008b). Ammonia emissions affected by airflow in a model pig house: effects of ventilation rate, floor slat opening and headspace height in a manure storage pit. *Transactions of the ASABE* 51(6), 2113–2122.

Zhao, Y., Aarnink, A. J. A., de Jong, M. C. M., Ogink, N. W. M. and Groot Koerkamp, P. W. G. (2011). Effectiveness of multi-stage scrubbers in reducing emissions of air pollutants from pig houses. *Transactions of the ASABE* 54(1), 285–293.

zu Ermgassen, E. K. H. J., Phalan, B., Green, R. E. and Balmford, A. (2016). Reducing the land use of EU pork production: where there's swill, there's a way. *Food Policy* 58, 35–48.

Chapter 6

Developments in anaerobic digestion to optimize the use of livestock manure

Mingxue Gao, Danmeng Wang, Chunlan Mao, Yongzhong Feng, Zhiyuan Zhu, Xiaojiao Wang, Guangxin Ren and Gaihe Yang, Northwest A&F University, China

1 Introduction

For thousands of years, agricultural production has provided humans with the food needed for survival. However, the fact that agriculture can provide energy has received less attention, starting with burning firewood and crop straw to produce heat and energy. Non-renewable energy sources such as oil and coal are declining, which means alternative energy sources such as biomass energy are urgently needed. As a renewable resource, biomass energy is also environmental-friendly.

Anaerobic digestion (AD) is an important method for converting biomass into bioenergy. AD can use a range of substrates including kitchen waste, crop straw, energy crops, municipal solid waste, sewage sludge and livestock manure. Although biogas production from livestock and poultry manure cannot achieve zero carbon emissions, it does not produce the level of carbon

http://dx.doi.org/10.19103/AS.2020.0077.09

emissions produced by fossil fuels. Livestock manure is still a widely used biomass material for production of bioenergy.

The production of biogas through AD of livestock manure is a complex process. It involves a variety of complex physiological and biochemical metabolic pathways, the essence of which is the material and energy metabolism of microorganisms under anaerobic conditions. AD is typically into three stages according to the utilization and transformation of organic matter (Fig. 1):

- hydrolysis;
- acidogenesis; and
- methanogenesis.

In the hydrolysis step, macromolecular organic matters (fat, carbohydrate, protein, etc.) are hydrolyzed into small molecules such as monosaccharides, amino acids, fatty acids and so on by the action of extracellular enzymes. Then the small molecule organic compounds are converted to a volatile organic acid, ethanol and so on by the acidified bacteria. H_2, CO_2 and acetic acid are then formed under the action of hydrogenic bacteria and acetogenic bacteria during this acidogenesis step. Finally, the methanogenic bacteria synthesize methane using acetic acid, H_2, CO_2 and so on in the methanogenesis stage.

It can be seen from the above that the AD process can produce a variety of substances, including H_2, CH_4, alcohol, lower fatty acids and so on. These substances can also be reused as biofuels, biodiesel, biogas, power generation and so on. The production of bioenergy by AD is thus very clean and efficient. There are many kinds of substrates in AD. Livestock and poultry manure is one of the commonly used raw materials, but a large amount can cause pollution, a key issue to address in its effective use.

The purpose of this chapter is to discuss developments in AD to optimize the use of livestock manure, particularly the use of livestock in the production of biogas. It also discusses the use of biogas slurry and residues. The chapter shows how AD can play an important role in promoting circular agriculture. It includes a case study on the use of AD in practice in Henan Province, China.

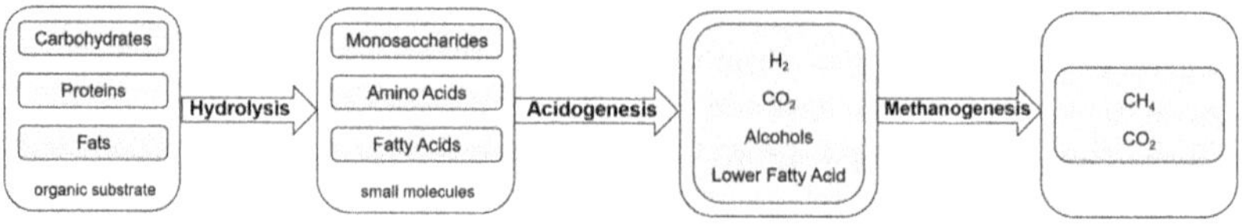

Figure 1 Steps of anaerobic digestion process.

2 Livestock manure: quantities and risks

Since ancient times, pigs, cattle, sheep and poultry have been kept by humans. However, it has not been possible to measure the amount of that feces produced by these livestock species exactly. Fortunately, the numbers of livestock and poultry raised in various countries and regions are now more accurately counted, so that the amount of manure can be estimated. Numbers of livestock in countries and regions are counted by the Food and Agriculture Organization of the United Nations (FAO), and can be found through its database (http://www.fao.org/home/en/; accessed November 10, 2019). A method for estimating the amount of manure by the quantity of livestock and poultry has been proposed using the following formula that is as follows (Gao and Li, 2015):

$$M_i = \sum_{i}^{n} Q_i \cdot d_i \cdot m_i \quad (1)$$

where Q_i is the total number of livestock, d_i is the excretion coefficient (the daily excretion of each animal) and m_i is the animal slaughter period.

This formula has been used to estimate the quantity of livestock manure produced in China (Lin et al., 2012; Li et al., 2010; Liu et al., 2005) (Fig. 2). The estimate for 2017 is up to 4 trillion tons of livestock manure that is produced

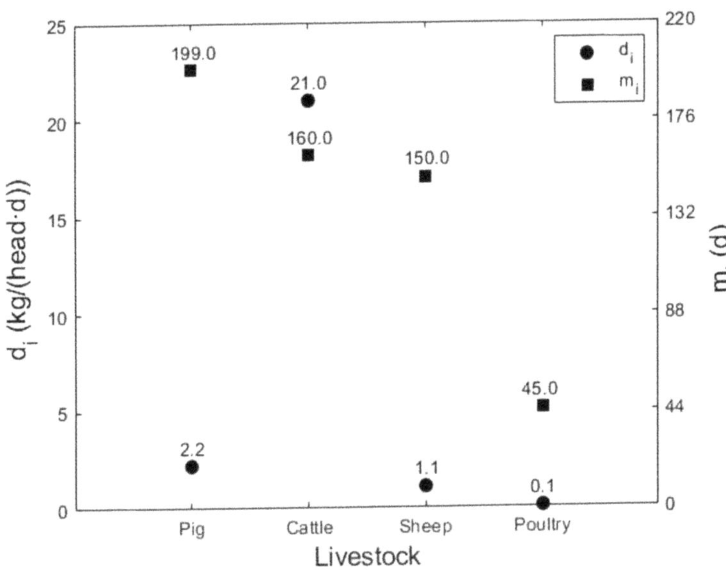

Figure 2 The estimated parameters of livestock manure quantity (Note: d_i is the excretion coefficient; m_i is the animal slaughter period).

in China. If such large amounts of manure were directly discharged without proper treatment, they would cause a huge pollution risk.

The first pollution consequence is air pollution. Livestock and poultry manure contain harmful gases such as hydrogen sulfide and amine, which can produce methyl mercaptan, dimethyl sulfide, methyl sulfide, dimethylamine and a variety of low-grade fatty acids or other odorous gases. If not treated in time, they cause a relatively low oxygen content in the air and increase the degree of turbidity (Liu et al., 2010). Moreover, greenhouse gases from livestock manure are also an important factor in global climate change. In addition to emitting CH_4, it is estimated that N_2O produced by livestock manure accounts for 30–50% of global agricultural N_2O emissions (Oenema et al., 2005). These harmful gases are caused by the unique environment of the feces and the large amount of microbial activity. It is striking that since the pre-industrial era, the concentration of CH_4 in the atmosphere has increased by 2.5 times and the concentration of N_2O has increased by about 20% (IPCC, 2006). Although this situation is caused by many factors, the treatment of livestock manure is a significant issue in controlling greenhouse gas (GHG) emissions (https://www.ipcc.ch/; accessed November 6, 2019).

The second consequence is water pollution. The effects of pollution of livestock and poultry manure to water bodies can be divided into three types:

- the release of acid gas into the air and then precipitation in the form of acid rain;
- direct discharge of pollution into surface water which flows into rivers and lakes; and
- entry into groundwater by osmosis.

Manure contains a lot of nitrogen and phosphorus. Once manure enters a water body and the quantity exceeds the water's purification ability, it causes algae to multiply in the water, and aquatic animals die of oxygen deficiency. Physical and chemical water properties will change followed by microbial composition, which finally makes the water toxic (Hooda et al., 2000).

The third consequence is soil pollution. Heavy metals are ingested through feed during livestock and poultry farming. Most are not used but excreted. When these manures are used as fertilizers, heavy metals including copper, zinc, cadmium, chromium, lead, mercury and arsenic will accumulate in the soil and crops, posing a potential risk to soil and human health (Feng et al., 2018b). In addition, when excessive sodium and potassium are in the feces, it will decrease soil micropores, destroy soil structure and reduce plant growth if applied directly to the soil. In addition, animal antibiotics are commonly used to treat and prevent animal diseases. Excessive antibiotics cannot be completely absorbed in animals but are excreted, accumulating in the soil and posing a significant environmental risk (Pu et al., 2018).

Finally, pollution from livestock also causes harm to human health. Air quality is reduced and the odor causes potentially fatal diseases of the respiratory system. Unclean drinking water systems are associated with waterborne disease outbreaks (Oun et al., 2014). Moreover, there are many pathogens in feces and, if humans eat plants grown in contaminated land, they are also likely to get sick. Thus, livestock manure can directly or indirectly affect human health. In summary, the large amount of livestock manure produced and unsuitable handling creates enormous environmental and health risks. Rational use of livestock and poultry manure to reduce these hazards and turn them into useful products is an important area of research.

3 The biogas potential of livestock manure

The biogas potential of a region refers to the amount of biogas produced by the waste in the region under appropriate conditions. Since not all waste is used for AD, and production conditions are not always optimal, biogas potential is an estimate under ideal conditions. Nevertheless, it is of great significance as a guide to the development of biogas production in a particular area.

Different types of animal manure have different capacities to produce biogas. The difference between them is called the biogas conversion coefficient of manure. Many studies have shown that livestock manure has moderate biogas potential (Nasir et al., 2012). Many factors involved in production processes also affect the production of biogas from livestock manure. Previous studies have suggested the following livestock manure biogas conversion parameter and formula for estimating manure biogas potential (MBP) (Yao, 1988; Zhang et al., 2012) (Fig. 3). Eq. (2) is one of the methods to calculate the biogas potential.

$$MBP = M_i \cdot b_i \tag{2}$$

where M_i is the amount of livestock manure produced (kg) from livestock i, and b_i is the livestock manure biogas conversion parameter.

This method is a simple calculation but subjected to a high potential level of error. In order to make up for the shortcomings of this method, some scholars have proposed a new method to improve biogas potential estimation. Biogas production is affected by the total solids (TS) of livestock manure because the presence of manure and urine in the excrement means it cannot be used entirely to produce methane. The total solid coefficients (TSC) of the feces can be further calculated by Eq. (3) (Zhang et al., 2019):

$$TSC(\%) = \frac{F_i \cdot F_s + U_i \cdot U_s}{F_i + U_i} \tag{3}$$

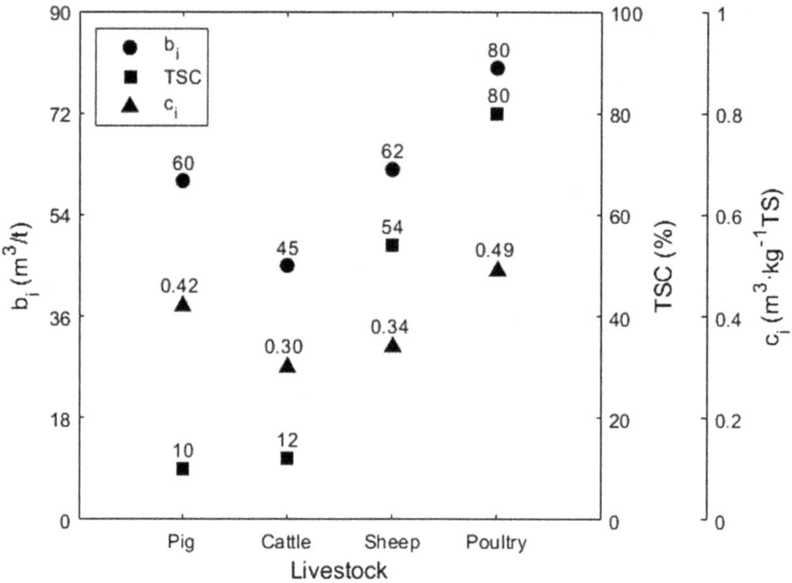

Figure 3 The estimated parameters of livestock manure biogas potential (Note: b_i is the livestock manure biogas conversion parameter. TSC is the total solid coefficients of livestock manure. c_i is raw biogas production rate per kg of dry matter).

where F_i represents the daily output of feces, F_s represents the total solid of the feces, U_i is the daily yield of urine and U_s is the urine's total solid.

Eq. (4) is used to calculate the biogas potential:

$$MBP = TSC \cdot M_i \cdot c_i \tag{4}$$

where M_i is the amount of livestock manure, c_i is the raw biogas production rate per kg of dry matter ($m^3 \cdot kg^{-1}TS$).

For reference, the TSC and c_i of some livestock have been listed in Fig. 3. It should be noted that, if the statistical data of animals is used to directly estimate the potential (where the amount of manure used for gasification, returning to the field and other industrial production has not been removed), the estimated value is relatively large.

4 Anaerobic mono-digestion and co-digestion

Depending on the different feeding modes of the substrate, the AD of biogas is divided into two types:

- anaerobic mono-digestion; and
- anaerobic co-digestion.

These are discussed in the following sections.

4.1 Anaerobic mono-digestion

At present a diverse range of livestock manure can be used in anaerobic mono-digestion raw material, in which feces from swine, cattle and chicken are richer. In addition, there are some AD that uses the droppings of ducks, sheep and rabbits (Song et al., 2010). The growth characteristics of cattle mean a large amount of excretion. Its abundance makes cattle manure one of the most commonly used raw materials for AD. Fodder is excreted after being digested through the rumen, and so cattle manure is rich in microorganisms, making cow dung a good raw material for AD. However, the content of cellulose and hemicellulose in cattle dung is high, and microorganisms are difficult to decompose and utilize, so there is still much space for improvement of biogas production from this source.

Swine manure is also a good AD substrate, with rich organic matter and high biogas yield. But the rapid hydrolysis caused by the large amount of carbohydrates can also have a negative effect. The accumulation of volatile fatty acids (VFA) in the system leads to acidification, imbalance of microbial growth and even digestion failure. Swine manure also contains ammonia, which is prone to ammonia inhibition. Chicken manure contains ammonia nitrogen which interferes with digestion because ammonia poisoning affects the performance of reactor. However, chicken feed is mainly based on grains and protein, and chicken manure has high nutrient content and high utilization value.

As this suggests, the performance of different manures in anaerobic mono-digestion varies because of their different characteristics. Gao and Li (2015) found methane yields for swine, cattle and chicken manure of 410, 270, 377 mL CH_4/g VS respectively, at initial volatile solid loading (VSL) of 8 g VS/L. Due to the variable presence of nutrients, anaerobic manure mono-digestion often faces problems such as poor stability, long lag time and low yield. However, anaerobic co-digestion can balance the carbon-nitrogen ratio (C/N) of the system to make microorganisms grow. It can also supplement the trace elements needed for microorganism growth and metabolism which further improves the fermentation effect.

4.2 Anaerobic co-digestion

Single manure substrate fermentation often suffers from C/N imbalance. A high C/N ratio results in a low protein dissolution rate. The total ammonia nitrogen (TAN) and free ammonia nitrogen (FA) concentrations are both low (Xue et al., 2019). However, an excessively high C/N ratio is not sufficient to maintain cell biomass, resulting in a reduction of biogas production. Substrates with a low

C/N ratio increase ammonia inhibition, which is noxious to methanogens and decreases underutilization of carbon sources (Mao et al., 2015). Optimizing the substrate C/N ratio by mixing suitable raw materials thus plays an important role in the AD process. The optimal C/N ratio for AD is between 20 and 35, with the most commonly used ratio at 25 (Yen and Brune, 2007; Zhang et al., 2013; Punal et al., 2000). It is worth mentioning that, whether anaerobic mono- or co-digestion is used, the optimal pH in systems is in the range of 6.8–7.4, because pH can directly affect microbial activity.

Mixing raw materials is an effective way to reduce the negative impact of anaerobic mono-digestion and expand the production of biogas. The mixing of raw materials for AD is not limited to feces, which can also extend to other substances like municipal solid waste, food waste, sewage sludge and crop straw. Through the mixing of substrates, nutrients such as C and N in the digestion system can be balanced, the microbial community in reactor can be enriched and the methane conversion efficiency of the organic substance can be improved. Studies have shown that the synergistic effect of mixed fermentation has a positive effect on the production of biogas.

Co-digestion of dairy manure, chicken manure and wheat straw performed better in methane production than digestion using a single substrate, with a maximum potential of CH_4 production predicted to be 394 mL/g VS (Wang et al., 2013). A synergistic effect was found when corn straw and chicken manure was mixed in proportions of 3:1, regardless of pretreatment (Feng et al., 2018a). The co-digestion of pig manure and dry maize straw is superior to the mono-digestion of pig manure, improving the C/N ration and enhancing the buffer capacity of the system (Song et al., 2016). When dairy manure was co-digested with food wastes, the highest cumulative biogas yield reached 459.4 mL/g VS (Batool et al., 2020). Livestock manure can also be co-digested with a variety of organic matter, such as the *Pennisetum* hybrid (Lianhua et al., 2020), mango leaves (Abudi et al., 2020) and animal carcasses (Tapparo et al., 2020). As these examples show, in general, mixed fermentation is better than single fermentation.

5 Factors affecting the efficiency of anaerobic digestion

There are many factors affecting AD which can be divided into the following categories:

- Substrate regulation;
- Raw material pretreatment;
- Process control; and
- Accelerators.

These are discussed in the following sections.

5.1 Substrate regulation

Each kind of animal excrement has unique physicochemical characteristics. Substrate regulation includes factors such as substrate concentration, C/N ratio and inoculum substrate ratio. The digestion system can be affected by changing these conditions. The fact that different livestock and poultry manures produce different biogas yields is one example of the way substrate properties and their combinations influence AD. The effect of inoculum on AD is also multifaceted. It can increase the number of microorganisms and ensure a suitable pH, but also accelerates the start of fermentation and shorten the lag phase. Selecting the proportion of inoculation to match the properties of the raw materials is important for optimizing AD.

5.2 Raw material pretreatment

The importance of pretreatment in substrate regulation needs to be emphasized. Low biodegradability usually occurs because the substrate contains something which is difficult to degrade, for example, the undigested plant fiber in cow dung (Sawatdeenarunat et al., 2016). Pretreatment of the substrate before fermentation is a good solution to this problem (Mosier et al., 2005). Different pretreatments have different effects on substrates.

Physical pretreatment mainly consists of mechanical pulverization and grinding to change the external morphology or internal structure of the raw material in order to improve biodegradability and biogas production (Elliott and Mahmood, 2012; Carrere et al., 2010). Larger granules of substrate reduce the degradation rate of chemical oxygen demand (COD) in the fermentation process, thereby reducing biogas production (Esposito et al., 2011). Chemical pretreatment mainly uses chemical reagents such as acid, alkali or oxidizing agent. Alkali reagents like ammonium hydroxide and sodium hydroxide are often used as alkaline pretreatment. Wahid et al. (2020) have shown that mixing wheat straw, solid cattle manure and solid slaughterhouse carcass material can improve anaerobic biodegradability after alkali and ultrasonic pretreatment. Acetic acid and organic acids provide the desired organic acid components to some microorganisms. Hydrogen peroxide and especially ozone are common oxidants, which avoid excessive accumulation of intermediate products in the digestion process and eliminate pathogens (Kianmehr et al., 2010). However, oxidant pretreatment is relatively costly. Other pretreatments include thermal pretreatment (Kaar and Holtzapple, 1998) and biological pretreatment (Zhong et al., 2011). All pretreatment methods have advantages and disadvantages. Various factors such as likely costs and benefits need to be considered, after which a single pretreatment can be chosen or combined to achieve the desired effect.

5.3 Process control

Digestion temperature, pH, hydraulic retention time (HRT) and organic loading rate (OLR) are all factors that can be regulated during the fermentation process. Some of these factors are discussed in more detail below. In addition, factors such as the type of reactor (Fig. 4), management, construction time and power generation equipment will impact the generation and utilization of biogas.

The usual temperature for thermophilic AD is 55°C and 35°C for mesophilic digestion. Compared to mesophilic AD, thermophilic AD has a fast response speed and high carrying capacity, resulting in higher yield. But it also has shortcomings, including lower system stability, acidification, more polluting discharge, higher net energy input and more investment, given the temperature requirements (Bowen et al., 2014). In actual production, many biogas digesters on farms depend on seasonal temperature without additional heating, reducing process stability and gas production due to ambient temperature changes.

pH directly affects the digestion process. The growth rate of microorganisms is very sensitive to changes in pH. The relative abundance of microorganism species is 6 when pH is 4.0 and increases to 14 when pH rises to 7.0 (Fang and Liu, 2002). Suitable pH ranges for methanogenic and acidogenic microorganisms are different. At pH 6.5–8.2, the efficiency of methanogenesis is optimal with the best pH at 7.0 (Lee et al., 2009). Methanogen growth slowed down visibly when the pH is less than 6.6

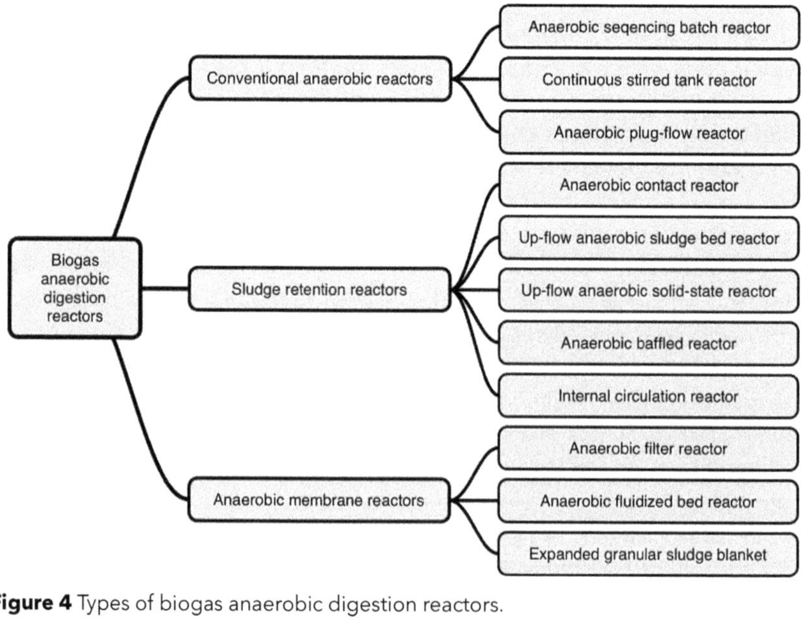

Figure 4 Types of biogas anaerobic digestion reactors.

(Zhang et al., 2009b). For acidogenesis, 5.5–6.5 is the most suitable pH range (Kim et al., 2003). Separating AD into a two-stage process is the preferred mode of operation with separate hydrolysis/acidification and acetogenesis/methanogenesis stages.

The time required for the decomposition of organic matter is defined as the retention time. Decomposition is affected by microbial growth rate, the process temperature, OLR and the matrix substance (Aboudi et al., 2015). The quantity of volatile solids in the digester is represented by OLR. Biogas production will improve to some extent with higher OLR (Mao et al., 2015).

5.4 Accelerators

AD additives can be divided into inhibitors and accelerators. Artificial inhibitors are not generally used in most biogas production processes, and so this section focuses on the use of accelerators. Accelerators can be further subdivided into biological accelerators and inorganic chemical accelerators.

Biological accelerators include several kinds of fungal microbial agents, which are mainly used for the treatment process of lignocellulosic biomass (Zheng et al., 2014). These include rumen fungi and protozoa (Zheng et al., 2019; Yildirim et al., 2017). Other biological accelerators include functional enzymes, for example, extracted from 3-day cultures of *T. harzianum* (Enzyme T) and *Aspergillus* spp. to improve methane yield (Zhao et al., 2018). But the cost of enzyme preparation is high, limiting its use.

Inorganic chemical accelerators are widely used, including acids, bases, inorganic salts and elements such as N, P, K and S. These substances are essential nutrients for the growth and metabolism of microorganisms that can be directly utilized in the AD process with significant effects (Weiland, 2010). However, inappropriate doses of these elements may result in reduced system stability. Trace elements such as Fe, Ni, Mg and Ca can also promote biochemical reactions (Demirel and Scherer, 2011). However, if too acidic or alkaline, and not utilized by microorganismss, these accelerators may affect the stability of the process.

6 Products from biogas digestate

During the AD process, livestock and poultry manure are decomposed through the action of microorganisms into biogas slurry containing proteins, amino acids and other water-soluble substances. Substances that cannot be completely decomposed and other impurities are deposited at the bottom of the reactor by gravity and become biogas residue after drying. The effluent of many biogas digesters is a solid-liquid mixture referred to as biogas digestate.

6.1 Digestate value-added products

The components of biogas slurry are complex, including nutrients such as N, P and K, as well as trace elements like Fe, Zn, Mn and Mo, which can be utilized by animals and plants. Microorganisms, antibiotics, enzymes and other unknown substances may also be present in the slurry. This means that the reuse value of biogas slurry is very high. In addition, the use of digestate to produce value-added products can reduce odor by up to 80% and make positive changes in odor composition (Weiland, 2010). The AD process is also able to partially inactivate weed seeds, bacteria, viruses, fungi and parasites in feedstock from livestock manure, improving safety (Mata-Alvarez et al., 2000).

Biogas slurry and biogas residue can be used directly or processed to produce value-added products such as biogas fertilizers and soil improvers (Nkoa, 2014; Edwards and Daniel, 1992) and animal feed (Zhang et al., 2008). Because biogas digestate can contain plant hormones including heteroauxin, gibberellins (GA4, GA19, GA53), plant growth agents can be produced (Huo et al., 2011). Biogas residue can also produce bedding materials for animal housing.

Biogas digestate is widely used as a base fertilizer, foliar fertilizer, top dressing and in soilless cultivation. As an example, the recommended application rate of biogas slurry on coastal-reclaimed farmland is 480 and 9.00–11.25 $m^3\ ha^{-1}$ for rice and wheat, respectively (Tang et al., 2019). The optimal proportion of chemical fertilizer nitrogen substitution by biogas slurry was 70% with a biogas slurry application rate of 278.56×10^3 kg hm^{-2} (Wang et al., 2018). Digestate as a fertilizer has a positive function on plant biomass (Barbosa et al., 2014). Digestate promotes the activities of various enzymes such as nitrate reductase and glutamine synthetase (Jabeen and Ahmad, 2017). It has also been shown to improve yield (Holm and Heinsoo, 2013, 2014). The use of digestate also promotes recycling (Tampio et al., 2016). Indirect effects include promotion of soil microbial activity (Hupfauf et al., 2016). Biogas fertilizer also changes soil composition through processes such as ammonia and nitrogen oxide emissions (Nkoa, 2014), nitrate leaching potential (Riva et al., 2016), transformation of organic carbon and GHG emissions (Knudsen et al., 2014), as well as leaching and precipitation of phosphorus. These cause changes in elements such as Mg, Fe, Mn and so on (Zirkler et al., 2014). Biogas digestate also increases pH, electrical conductivity, air porosity and bulk density (Meng et al., 2018). In addition to biogas fertilizer, biogas slurry can also promote germination and seedling by soaking of seeds (Ni et al., 2015).

Biogas slurry can also be used for aquaculture and livestock feed. Biogas residue as an alternative feed saves feed costs and improves digestion without affecting pork quality (Zhang et al., 2009a). In aquaculture, biogas digestate nutrients can be eaten by fish or promote the growth of plankton as a source of nutrition.

6.2 Disadvantages of biogas digestate

There are some shortcomings in using biogas digestate. Harmful substances such as heavy metals may still be present in biogas fertilizer and will still contaminate the soil (Duan et al., 2012). The bioavailability of heavy metals in AD is an important criterion to predict the ecological risk of heavy metals, which is determined by the form of each metal (Thanh et al., 2016). AD operational parameters (pH, redox potential and VFAs concentration) determine bioavailability (Knoop et al., 2017). Heavy metals are not biodegradable and can accumulate to potentially toxic concentrations as digestion proceeds (Chen et al., 2014). In the process of AD, S^{2-}, CO_3^{2-}, PO_4^{3-} and so on will cause reactions and precipitation will change the solubility and availability of heavy metals (Insam et al., 2015). The toxic effects of heavy metals depend on their solubility and availability (Choong et al., 2016). Vegetables and grains produced after irrigation with heavy metals such as zinc, lead and cadmium are harmful to human health (Bian et al., 2015). Although the maximum allowable concentrations of heavy metals in the digestate are currently set in many countries, they still present an environmental risk (Gusiatin and Kulikowska, 2014). However, proper management can reduce this risk.

7 Ecological agriculture models for biogas utilization

Agriculture needs to reduce its environmental impact by adopting ecological principles such as reducing waste and creating circular, more self-sufficient systems which recycle waste into useful by-products such as fertilizer and biogas which can then be fed back into the system. This section introduces four typical ecological agriculture models: ecological farms, ecological orchards, ecological greenhouses and 'five supporting' mode.

7.1 Ecological farms

Ecological farms are a widely used model. They can be divided into two types:

- Small-scale household farms; and
- Large-scale 'engineering' farms.

The household ecological farm consists of a single household or several neighboring households on a small scale. Processes are simple: human feces and livestock manure are added to a fermentation tank to produce biogas for human. The 'engineering' type is mainly for farms that cultivate animals and crops on a large scale. Biogas digesters are constructed to eliminate agricultural waste such as straw and manure. Equipment and capacity of biogas digester are usually well developed. Biogas is not only used for heating but also for generating electricity which can be used on the farm with any surplus supplied into the local power grid to generate additional value too.

7.2 The ecological orchard and other three-in-one models

The ecological orchard model is one example of a circular agricultural system. The three-in-one model (Fig. 5) combines animals, microorganisms and plants. First, animals can eat plant stalks, weeds, fruits and so on to grow. Secondly, biogas is produced by AD when animal manure is fed to biogas digesters to produce biogas for heat and power. Thirdly, biogas slurry can be applied to the plants as organic fertilizer to stimulate plant growth, and biogas residue can also be used as animal feed. This creates a circular, more self-sufficient system with fewer inputs and outputs. Plants can be orchard trees, arable crops, vegetables, tea and so on. Animals can be livestock, poultry and aquatic species. Typical examples of this three-in-one model are the 'biogas-rice-duck', 'chicken-biogas-fruit' and 'pig-biogas-fish' models (Chen, 1997; Yang et al., 2018; Yang et al., 2012). Successful models of this type save feed and energy costs, increase the variety and quantity of products produced, and can also improve product quality while reducing environmental impact.

7.3 Ecological greenhouse: the four-in-one model

Temperature is a key limiting factor for the production of biogas. Both ecological farms and ecological orchard models can use their own production of biogas as an energy source for heating. However, in some areas, the winter is cold and low temperatures prevent the fermentation tank from operating properly (Pham et al., 2014). Adding thermal insulation can be expensive. For the circular agriculture model to operate properly and economically, a four-in-one model has emerged. This model has a greenhouse in addition to biogas digesters and provides an effective ecological cycle from livestock feeding to

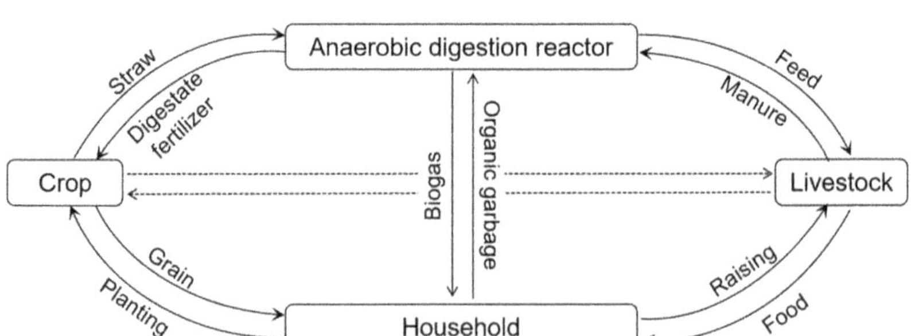

Figure 5 Eco-agriculture model of biogas anaerobic digestion.

vegetable production, resulting in a higher conversion efficiency in nutrient cycles and energy flows (Qi et al., 2005). An example is combined pig and vegetable greenhouse production, by combining a plastic solar greenhouse, an underground 6–10 m^3 biogas digester, a pigsty of about 10–20 m^2 and toilet facilities, parallel to the vegetable greenhouse. This forms an energy ecosystem (Fig. 6).

This model has the following advantages:

1 Due to the presence of the greenhouse, the biogas digesters are warmed up by the solar energy, solving the problem of gas production in cold winter temperatures.
2 The temperature of the pigsty can also be increased by 3–5°C in winter, providing suitable growth conditions for pigs.
3 The toilet (for human personnel) and pigsty provide sufficient fermentation feed for the biogas tank.
4 The pigs exhale a large amount of CO_2, and increase CO_2 concentration in the greenhouse by a factor of 4–5 which can greatly improve vegetable growth conditions in the greenhouse, and increase yield and quality.

7.4 Five-in-one model

The four-in-one model has improved ecological agricultural production in cold areas, but what kind of ecological agricultural model should be developed in areas of water shortage? As an example, the northwestern region of China has sufficient sunshine and a large temperature variation between day and night

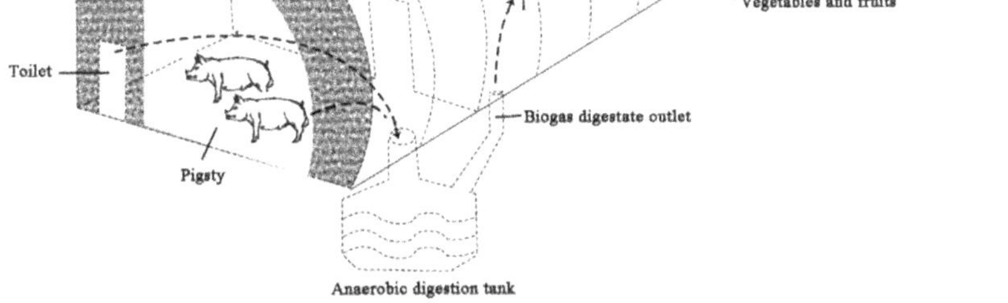

Figure 6 Structure of 'four-in-one' eco-agriculture model.

which could favor planting high-quality fruit and cash crops. However, the regional climate is arid and semi-arid climates with problems such as drought and soil erosion due to lack of water. The five-in-one model incorporates the components of the four-in-one system but adds water storage capacity (to capture the available natural precipitation) to enable drip irrigation (Qiu et al., 2001; Wu et al., 2015). The biogas slurry can, for example, be applied to the roots of fruit trees using drip irrigation. As with the four-in-one model, each component supports the others.

8 Case study: biogas production in Henan Province, China

China is a country with abundant resources and well-developed agriculture, including the production of large volumes of agricultural, forestry and domestic waste as well as sewage sludge. Bio-gasification is a promising way of recycling this large amount of waste. Henan is the country's major agricultural province. In 2014, the output of major cereal and meat products in Henan Province reached 55.227 million tons and 69.91 million tons, respectively, as recorded by the Henan Statistical Yearbook (http://www.ha.stats.gov.cn/hntj/lib/tjnj/2015/indexch.htm; accessed November 8th 2019). Henan Province also has a large number of biogas projects combining the development of biogas and agriculture.

8.1 Biogas potential of Henan Province

The 2014 China Statistical Yearbook shows the number of pigs, cattle, sheep, chickens and ducks raised in Henan Province during the year. The total number of these animals reached 959.62 million. The amount of cattle manure and pig manure produced is 88.77 million ton and 61.68million ton, respectively, sheep manure is 11.12 million ton and poultry is 4.05 million ton, respectively. Figure 7 shows the numbers of sheep, cattle, pigs and poultry, the volume of manure and MBP. It can be seen that biogas potential and fecal volume do not increase with the number of animals in a linear way. Although poultry is the largest group, accounting for 84% of the total, the biogas potential is the smallest, accounting for only 4% at 324 million m^3. Biogas potential also corresponds to the kinds of manure produced: the MBP of cattle, pig, sheep and poultry reached 3.99 billion m^3, 3.70 billion m^3, 0.68 billion m^3 and 0.32 billion m^3, respectively. This shows that agricultural waste resources have great potential to supply energy via biogas. However, the biogas production of current biogas projects is far below the potential value (Gao et al., 2019). This may be due to geographical conditions, technical equipment, management methods, and policy and economic constraints. However, this estimate shows the biogas potential of agricultural waste.

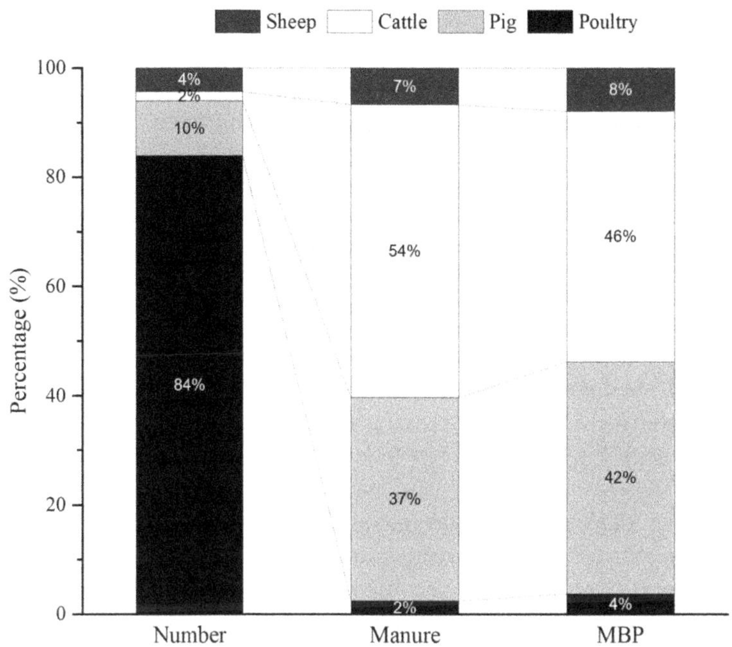

Figure 7 Livestock number, manure and MBP volume of Henan Province in 2014.

8.2 Eco-agriculture model

Henan Province has pioneered an eco-agriculture model called '100 mu-1000 heads ecological quadrant' (mu is a unit of area of about 0.0667 ha). It consists of a 3 mu livestock production line and 100 mu of farmland as a unit with 1000 pigs produced. Pig manure is digested and turned into organic fertilizer to use on the farmland, and the feed grain is processed into organic grain in situ. Such a 'pig-fertilizer-grain' recycling production line has the characteristics of low investment, rapid results and environmental benefits. It not only solves the problem of pollution in pig production, but also helps farmers out of poverty.

There are a number of requirements needed to develop this model. First, the government must improve policies related to biogas. Good policy guidelines can lead, supervise, guarantee, promote and support high-quality and sustainable production of biogas locally. High-quality biogas projects are economically and ecologically beneficial while supplying large amounts of bioenergy. Policies are an effective guarantee to achieve these benefits. The types of policies should include guidance, incentives, research, measure to promote markets, as well as standards governing constructive and operation (Feng et al., 2012).

Second, biogas companies must be responsive to policies and regional conditions, particularly in developing technologies that match local conditions. A key aspect is the intelligent control technology to improve the gas production rate of AD. By continuously monitoring and adjusting the actual fermentation environment, it is possible to find the best substrate and conditions for AD. Selecting a fermentation device to suit local conditions can improve resource utilization and biogas production efficiency. There is also a need for supporting infrastructure and maintenance capabilities to support biogas power generation, including skills training (Gao et al., 2019b).

Finally, each region needs to adapt biogas production to local resources and conditions. As an example, it might select crops which benefit most from local conditions while selecting livestock with the greatest regional market potential. Establishing successful, long-running biogas engineering enterprises are the path to promote large-scale, professional development of biogas. Circular agriculture and ecological agriculture models provide the basis for fully utilizing biogas resources, for example, by integration with solar greenhouses and water-saving irrigation systems, and combining biogas production with crop and aquaculture production to achieve sustainable biogas production. These models provide a path toward a sustainable future for agriculture.

9 Summary and future trends

Livestock and poultry manure is a huge potential problem which exposes the environment to pollution and other risks. However, as an organic substance, livestock manure has huge biogas potential in the production of clean energy in the face of energy shortages and environmental change. Developments in biogas production technology have improved the utilization of animal manure through AD, increasing the economic and environmental benefits. However, at present biogas production from animal manure is below its potential. Areas for further study include:

1. Establishing a more accurate biogas estimation method, for combining various wastes with actual research data.
2. Innovating high-efficiency, low-cost biogas production technology and equipment.
3. Developing a variety of eco-agricultural models suitable for different regions to find more efficient ways of using animal manure.
4. Developing evaluation methods to determine the most scientific and effective way to use manure in different areas.

Advances in these and other areas offer the opportunity for a bright future for this sector.

10 Where to look for further information

Further reading:

- A website that can provide information on the size of livestock and poultry farming in various countries and regions around the world is the Food and Agriculture Organization of the United Nations (FAO) database available at: (http://www.fao.org/home/en/).
- A good way to assess the energy distribution and utilization of different countries and regions is the BP Statistical Review of World Energy available at: (https://www.bp.com/en/global/corporate.html).
- *A Comprehensive Review of Anaerobic Digestion Is Review on Research Achievements of Biogas from Anaerobic Digestion by Mao Chunlan* (2015).
- *The Study of Biogas Potential Estimation Is Biogas Potential, Utilization and Countermeasures in Agricultural Provinces by Gao Mingxue* (2019).

Key journals/conferences:

- *The World Energy Congress Organized by the World Energy Council (WEC),* Is the Most Important Meeting for the Development of the Energy Industry.
- *Renewable and Sustainable Energy Review* is a journal that contains reviews of renewable and sustainable energy.
- *Bioresource Technology* publishes articles about bioresource fundamentals, applications and management, including biogas.

11 Acknowledgements

This work was supported by the Shaanxi Innovation Capability Support Program (2019PT-13), and Youth Talent Cultivation Program Funding of Northwest A&F University, Special Research for Rural Vitalization Strategy (Z1090219062), National Natural Science Foundation of China (51508467, 41871205), Shaanxi Engineering Research Center of Circular Agriculture (2019HBGC-13).

12 References

Aboudi, K., Álvarez-Gallego, C. J. and Romero-García, L. I. (2015). Semi-continuous anaerobic co-digestion of sugar beet byproduct and pig manure: effect of the organic loading rate (OLR) on process performance. *Bioresource Technology* 194: 283–290.

Abudi, Z. N., Hu, Z. and Abood, A. R. (2020). Anaerobic co-digestion of mango leaves and pig manure: performance assessment and kinetic analysis. *Biomass Conversion and Biorefinery*. Online: MAR 2020. DOI: 10.1007/s13399-020-00665-6.

Barbosa, D. B. P., Nabel, M. and Jablonowski, N. D. (2014). Biogas-digestate as nutrient source for biomass production of Sida hermaphrodita, Zea mays L. and Medicago sativa L. In: Bruckman, V. J., Hangx, S. and Ask, M. (Eds) *European Geosciences Union General Assembly 2014, Egu Division Energy, Resources & the Environment. Energy Procedia* 59: 120-126.

Batool, N., Qazi, J. I., Aziz, N., Hussain, A. and Shah, S. Z. H. (2020). Bio-methane production potential assays of organic waste by anaerobic digestion and co-digestion. *Pakistan Journal of Zoology* 52(3): 971-976.

Bian, B., Zhou, L. J., Li, L., Lv, L. and Fan, Y. M. (2015). Risk assessment of heavy metals in air, water, vegetables, grains, and related soils irrigated with biogas slurry in Taihu Basin, China. *Environmental Science and Pollution Research International* 22(10): 7794-7807.

Bowen, E. J., Dolfing, J., Davenport, R. J., Read, F. L. and Curtis, T. P. (2014). Low-temperature limitation of bioreactor sludge in anaerobic treatment of domestic wastewater. *Water Science and Technology: A Journal of the International Association on Water Pollution Research* 69(5): 1004-1013.

Carrere, H., Dumas, C., Battimelli, A., Batstone, D. J., Delgenes, J. P., Steyer, J. P. and Ferrer, I. (2010). Pretreatment methods to improve sludge anaerobic degradability: a review. *Journal of Hazardous Materials* 183(1-3): 1-15.

Chen, J. L., Ortiz, R., Steele, T. W. J. and Stuckey, D. C. (2014). Toxicants inhibiting anaerobic digestion: a review. *Biotechnology Advances* 32(8): 1523-1534.

Chen, R. J. (1997). Livestock-biogas-fruit systems in South China. *Ecological Engineering* 8(1): 19-29.

Choong, Y. Y., Norli, I., Abdullah, A. Z. and Yhaya, M. F. (2016). Impacts of trace element supplementation on the performance of anaerobic digestion process: a critical review. *Bioresource Technology* 209: 369-379.

Demirel, B. and Scherer, P. (2011). Trace element requirements of agricultural biogas digesters during biological conversion of renewable biomass to methane. *Biomass and Bioenergy* 35(3): 992-998.

Duan, G., Zhang, H., Liu, Y., Jia, Y., Hu, Y. and Cheng, W. (2012). Long-term fertilization with pig-biogas residues results in heavy metal accumulation in paddy field and rice grains in Jiaxing of China. *Soil Science and Plant Nutrition* 58(5): 637-646.

Edwards, D. R. and Daniel, T. C. (1992). Environmental impacts of on-farm poultry waste-disposal- a review. *Bioresource Technology* 41(1): 9-33.

Elliott, A. and Mahmood, T. (2012). Comparison of mechanical pretreatment methods for the enhancement of anaerobic digestion of pulp and paper waste activated sludge. *Water Environment Research: A Research Publication of the Water Environment Federation* 84(6): 497-505.

Esposito, G., Frunzo, L., Panico, A. and Pirozzi, F. (2011). Modelling the effect of the OLR and OFMSW particle size on the performances of an anaerobic co-digestion reactor. *Process Biochemistry* 46(2): 557-565.

Fang, H. H. P. and Liu, H. (2002). Effect of pH on hydrogen production from glucose by a mixed culture. *Bioresource Technology* 82(1): 87-93.

Feng, J., Li, Y., Zhang, E., Zhang, J., Wang, W., He, Y., Liu, G. and Chen, C. (2018a). Solid-state co-digestion of NaOH-pretreated corn straw and chicken manure Under mesophilic condition. *Waste and Biomass Valorization* 9(6): 1027-1035.

Feng, Y., Guo, Y., Yang, G., Qin, X. and Song, Z. (2012). Household biogas development in rural China: on policy support and other macro sustainable conditions. *Renewable and Sustainable Energy Reviews* 16(8): 5617-5624.

Feng, Z., Zhu, H., Deng, Q., He, Y., Li, J., Yin, J., Gao, F., Huang, R. and Li, T. (2018b). Environmental pollution induced by heavy metal(loid)s from pig farming. *Environmental Earth Sciences* 77(3): 10.

Gao, M., Wang, D., Wang, H., Wang, X. and Feng, Y. (2019). Biogas potential, utilization and countermeasures in agricultural provinces: a case study of biogas development in Henan Province, China. *Renewable and Sustainable Energy Reviews* 99: 191–200.

Gao, M., Wang, D., Wang, Y., Wang, X. and Feng, Y. (2019b). Opportunities and challenges for biogas development: a review in 2013–2018. *Current Pollution Reports* 5(2): 25–35.

Gao, W. Y. and Li, J. M. (2015). Present situation of agricultural bioenergy industry and the effects evaluation. *China Biogas* 33(01): 46–52.

Gusiatin, Z. M. and Kulikowska, D. (2014). The usability of the I-R, RAC and MRI indices of heavy metal distribution to assess the environmental quality of sewage sludge composts. *Waste Management* 34(7): 1227–1236.

Holm, B. and Heinsoo, K. (2013). Influence of composted sewage sludge on the wood yield of willow short rotation coppice: an Estonian case study. *Environment Protection Engineering* 39: 17–32.

Holm, B. and Heinsoo, K. (2014). Biogas digestate suitability for the fertilisation of young Salix Plants. *Baltic Forestry* 20: 263-271.

Hooda, P. S., Edwards, A. C., Anderson, H. A. and Miller, A. (2000). A review of water quality concerns in livestock farming areas. *Science of the Total Environment* 250(1–3): 143–167.

Huo, C. Y., Wu, S. B., Guo, J. B., Cheng, H. C., Ren, H., Dong, R. J. and Pang, C. L. (2011). Analysis of phytohormone and Qunolin-ketone components in anaerobic digestion effluent. *China Biogas* 29: 7–10.

Hupfauf, S., Bachmann, S., Juarez, M. F.-D., Insam, H. and Eichler-Loebermann, B. (2016). Biogas digestates affect crop P uptake and soil microbial community composition. *Science of the Total Environment* 542(B): 1144–1154.

Insam, H., Gomez-Brandon, M. and Ascher, J. (2015). Manure-based biogas fermentation residues–friend or foe of soil fertility? *Soil Biology and Biochemistry* 84: 1–14.

IPCC (2006). Guidelines for national greenhouse gas inventories. Agriculture. Forestry and Other Land Use Vol. 4. Institute for Global Environmental Strategies, Kanagawa.

Jabeen, N. and Ahmad, R. (2017). Growth response and nitrogen metabolism of sunflower (Helianthus annuus L.) to vermicompost and biogas slurry under salinity stress. *Journal of Plant Nutrition* 40(1): 104–114.

Kaar, W. E. and Holtzapple, M. T. (1998). Benefits from Tween during enzymic hydrolysis of corn stover. *Biotechnology and Bioengineering* 59(4): 419-427.

Kianmehr, P., Parker, W. and Seto, P. (2010). An evaluation of protocols for characterization of ozone impacts on WAS properties and digestibility. *Bioresource Technology* 101(22): 8565–8572.

Kim, J., Park, C., Kim, T. H., Lee, M., Kim, S., Kim, S. W. and Lee, J. (2003). Effects of various pretreatments for enhanced anaerobic digestion with waste activated sludge. *Journal of Bioscience and Bioengineering* 95(3): 271–275.

Knoop, C., Dornack, C. and Raab, T. (2017). Nutrient and heavy metal accumulation in municipal organic waste from separate collection during anaerobic digestion in a two-stage laboratory biogas plant. *Bioresource Technology* 239: 437–446.

Knudsen, M. T., Meyer-Aurich, A., Olesen, J. E., Chirinda, N. and Hermansen, J. E. (2014). Carbon footprints of crops from organic and conventional arable crop rotations - using a life cycle assessment approach. *Journal of Cleaner Production* 64: 609–618.

Lee, D. H., Behera, S. K., Kim, J. W. and Park, H. S. (2009). Methane production potential of leachate generated from Korean food waste recycling facilities: a lab-scale study. *Waste Management* 29(2): 876–882.

Li, S., Deng, L. W., Yong, Y., Pu, X. D. and Wang, Z. Y. (2010). Biogas production potential and characteristics of manure of sheep, duck and rabbit under anaerobic digestion. *Transactions of the Chinese Society of Agricultural Engineering* 26: 277–282.

Lianhua, L., Shuibin, H., Yongming, S., Xihui, K., Junfeng, J., Zhenhong, Y. and Dingfa, L. (2020). Anaerobic co-digestion of pennisetum hybrid and pig manure: a comparative study of performance and microbial community at different mixture ratio and organic loading rate. *Chemosphere* 247: 125871.

Lin, Y., Ma, J. and Qin, F. (2012). The structure distribution and prospect of China manure resource. *Chinese Agricultural Science Bulletin* 28: 1–5.

Liu, D. J., Gao, G. L., Zhu, Y. M., Li, Y. and Pei, Z. Q. (2005). The comparison experiment of biogas fermentation with pig dung, cow dung, sheep dung. *Journal of Tarim University of Agricultural Reclamation* 2005(02): 10–12.

Liu, Y. F., Mayilaoaini, T. S. Z., BAYINBATE and Wang, W. Q. (2010) Pollution of livestock and poultry feces and counter measures. *Grass-Feeding Livestock* 24(6): 1164–1170.

Mao, C., Feng, Y., Wang, X. and Ren, G. (2015). Review on research achievements of biogas from anaerobic digestion. *Renewable and Sustainable Energy Reviews* 45: 540–555.

Mata-Alvarez, J., Mace, S. and Llabres, P. (2000). Anaerobic digestion of organic solid wastes. An overview of research achievements and perspectives. *Bioresource Technology* 74(1): 3–16.

Meng, X., Dai, J., Zhang, Y., Wang, X., Zhu, W., Yuan, X., Yuan, H. and Cui, Z. (2018). Composted biogas residue and spent mushroom substrate as a growth medium for tomato and pepper seedlings. *Journal of Environmental Management* 216: 62–69.

Mosier, N., Wyman, C., Dale, B., Elander, R., Lee, Y. Y., Holtzapple, M. and Ladisch, M. (2005). Features of promising technologies for pretreatment of lignocellulosic biomass. *Bioresource Technology* 96(6): 673–686.

Nasir, I. M., Ghazi, T. I. M. and Omar, R. (2012). Anaerobic digestion technology in livestock manure treatment for biogas production: a review. *Engineering in Life Sciences* 12(3): 258–269.

Ni, T.-C., Zhou, C.-F., Zhu, H.-G., Zhou, F., Li, Y.-H. and Hu, C. Q. (2015). Effects of soaking of seeds with biogas slurry on germination and seedling growth of rice. *Journal of Ecology and Rural Environment* 31: 594–599.

Nkoa, R. (2014). Agricultural benefits and environmental risks of soil fertilization with anaerobic digestates: a review. *Agronomy for Sustainable Development* 34(2): 473–492.

Oenema, O., Wrage, N., Velthof, G. L., Groenigen, J. W., Dolfing, J. and Kuikman, P. J. (2005). Trends in global nitrous oxide emissions from animal production systems. *Nutrient Cycling in Agroecosystems* 72(1): 51–65.

Oun, A., Kumar, A., Harrigan, T., Angelakis, A. and Xagoraraki, I. (2014). Effects of biosolids and manure application on microbial water quality in rural areas in the US. *Water* 6(12): 3701–3723.

Pham, C. H., Vu, C. C., Sommer, S. G. and Bruun, S. (2014). Factors affecting process temperature and biogas production in small-scale rural biogas digesters in winter in Northern Vietnam. *Asian-Australasian Journal of Animal Sciences* 27(7): 1050–1056.

Pu, C., Liu, H., Ding, G., Sun, Y., Yu, X., Chen, J., Ren, J. and Gong, X. (2018). Impact of direct application of biogas slurry and residue in fields: in situ analysis of antibiotic resistance genes from pig manure to fields. *Journal of Hazardous Materials* 344: 441–449.

Punal, A., Trevisan, M., Rozzi, A. and Lema, J. M. (2000). Influence of C:N ratio on the start-up of upflow anaerobic filter reactors. *Water Research* 34(9): 2614–2619.

Qi, X. S., Zhang, S. P., Wang, Y. Z. and Wang, R. Q. (2005). Advantages of the integrated pig-biogas-vegetable greenhouse system in North China. *Ecological Engineering* 24(3): 175–183.

Qiu, L., Yang, G. H. and Yang, S. Q. (2001). A study into the designs of the optimal ecological orchard project model in the Loess Plateau. *Journal of Northwest Sci-Tech University of Agriculture and Forestry* 2001(05):65–69.

Riva, C., Orzi, V., Carozzi, M., Acutis, M., Boccasile, G., Lonati, S., Tambone, F., D'Imporzano, G. and Adani, F. (2016). Short-term experiments in using digestate products as substitutes for mineral (N) fertilizer: agronomic performance, odours, and ammonia emission impacts. *Science of the Total Environment* 547: 206–214.

Sawatdeenarunat, C., Nguyen, D., Surendra, K. C., Shrestha, S., Rajendran, K., Oechsner, H., Xie, L. and Khanal, S. K. (2016). Anaerobic biorefinery: current status, challenges, and opportunities. *Bioresource Technology* 215: 304–313.

Song, L., Deng, L. W., Yi, Y., Pu, X. D. and Wang, Z. Y. (2010). Biogas production potential and characteristics of manure of sheep, duck and rabbit under anaerobic digestion. *Transactions of the CSAE* 26(10): 277–282.

Song, X., Zhang, K., Han, B., Liang, J., Zhai, Z. and Du, L. (2016). Anaerobic co-digestion of pig manure with dried maize straw. *BioResources* 11(4): 8914–8928.

Tampio, E., Salo, T. and Rintala, J. (2016). Agronomic characteristics of five different urban waste digestates. *Journal of Environmental Management* 169: 293–302.

Tang, Y., Wen, G., Li, P., Dai, C. and Han, J. (2019). Effects of biogas slurry application on crop production and soil properties in a rice-wheat rotation on coastal reclaimed farmland. *Water, Air, and Soil Pollution* 230(3): 1–13.

Tapparo, D. C., Viancelli, A., Amaral, A. C. D., Fongaro, G., Steinmetz, R. L. R., Magri, M. E., Barardi, C. R. M. and Kunz, A. (2020). Sanitary effectiveness and biogas yield by anaerobic co-digestion of swine carcasses and manure. *Environmental Technology* 41(6): 682–690.

Thanh, P. M., Ketheesan, B., Yan, Z. and Stuckey, D. (2016). Trace metal speciation and bioavailability in anaerobic digestion: a review. *Biotechnology Advances* 34(2): 122–136.

Wahid, R., Romero-Guiza, M., Moset, V., Møller, H. B. and Fernández, B. (2020). Improved anaerobic biodegradability of wheat straw, solid cattle manure and solid slaughterhouse by alkali, ultrasonic and alkali-ultrasonic pre-treatment. *Environmental Technology* 41(8): 997–1006.

Wang, G. L., Kou, K. M., Zhang, J. H., Wang, S. W. and Wu, L. M. (2018). Effect of chemical fertilizer nitrogen substitution by biogas slurry on the growth and quality of rice. *Shengtaixue Zazhi* 37: 2672–2679.

Wang, X., Yang, G., Li, F., Feng, Y. and Ren, G. (2013). Response surface optimization of methane potentials in anaerobic co-digestion of multiple substrates: dairy, chicken manure and wheat straw. *Waste Management and Research: The Journal of the International Solid Wastes and Public Cleansing Association, ISWA* 31(1): 60–66.

Weiland, P. (2010). Biogas production: current state and perspectives. *Applied Microbiology and Biotechnology* 85(4): 849–860.

Wu, X., Wu, F., Wu, J. and Sun, L. (2015). Emergy-based sustainability assessment for a five-in-one integrated production system of apple, grass, pig, biogas, and rainwater on the Loess Plateau, Northwest China. *Agroecology and Sustainable Food Systems* 39(6): 666–690.

Xue, S., Zhao, N., Song, J. and Wang, X. (2019). Interactive effects of chemical composition of food waste during anaerobic co-digestion under thermophilic temperature. *Sustainability* 11(10).

Yang, H., Yu, D., Zhou, J., Zhai, S., Bian, X. and Weih, M. (2018). Rice-duck co-culture for reducing negative impacts of biogas slurry application in rice production systems. *Journal of Environmental Management* 213: 142–150.

Yang, Q., Wu, X., Yang, H., Zhang, S. and Chen, H. (2012). Nonrenewable energy cost and greenhouse gas emissions of a "pig-biogas-fish" systemin China. *Scientific World Journal* 2012: 1–7.

Yao, A. L. (1988). Gas potential and characteristics of biogas fermentation raw materials in rural areas in China. *China Biogas* 1988(1): 23–27.

Yen, H. W. and Brune, D. E. (2007). Anaerobic co-digestion of algal sludge and waste paper to produce methane. *Bioresource Technology* 98(1): 130–134.

Yildirim, E., Ince, O., Aydin, S. and Ince, B. (2017). Improvement of biogas potential of anaerobic digesters using rumen fungi. *Renewable Energy* 109: 346–353.

Zhang, H., Lei, Z. M., Dou, X. C. (2008). Analysis of nutritive value of biogas residues and heavy metal content in feedstuff with biogas residues and the pork fed on it. *Chinese Journal of Eco-Agriculture* 16(5): 1298–1301.

Zhang, H. B., Li, J. R. and Saho, K. W. (2009a). Safety and effect of biogas slurry in swine feed. *Journal of Domestic Animal Ecology* 30(03): 68–72.

Zhang, P., Chen, Y. and Zhou, Q. (2009b). Waste activated sludge hydrolysis and short-chain fatty acids accumulation under mesophilic and thermophilic conditions: effect of pH. *Water Research* 43(15): 3735–3742.

Zhang, T., Bu, M. D. and Geng, W. (2012) Pollution status and biogas-producing potential of livestock and poultry excrements in China. *Chinese Journal of Ecology* 31(05): 1241–1249.

Zhang, T., Liu, L., Song, Z., Ren, G., Feng, Y., Han, X. and Yang, G. (2013). Biogas production by co-digestion of goat manure with three crop residues. *PLoS ONE* 8(6): e66845.

Zhang, Y. F., Shao, L., Leng, S. N., Pang, L. H., Jiang, K. Y. and Wang, L. X. (2019). Assessment of livestock and poultry excrement resources in Shandong and its fertilizer and energy utilization. *China Biogas* 37(03): 93–99.

Zhao, X., Luo, K., Zhang, Y., Zheng, Z., Cai, Y., Wen, B., Cui, Z. and Wang, X. (2018). Improving the methane yield of maize straw: focus on the effects of pretreatment with fungi and their secreted enzymes combined with sodium hydroxide. *Bioresource Technology* 250: 204–213.

Zheng, Y., Wang, X. and Yang, F. (2019). Improving the anaerobic digestion of switchgrass via cofermentation of rumen microorganisms (rumen bacteria, protozoa, and fungi) and a biogas slurry. *Energy and Fuels* 33(2): 1185–1195.

Zheng, Y., Zhao, J., Xu, F. and Li, Y. (2014). Pretreatment of lignocellulosic biomass for enhanced biogas production. *Progress in Energy and Combustion Science* 42: 35–53.

Zhong, W., Zhang, Z., Luo, Y., Sun, S., Qiao, W. and Xiao, M. (2011). Effect of biological pretreatments in enhancing corn straw biogas production. *Bioresource Technology* 102(24): 11177–11182.

Zirkler, D., Peters, A. and Kaupenjohann, M. (2014). Elemental composition of biogas residues: variability and alteration during anaerobic digestion. *Biomass and Bioenergy* 67: 89–98.

Part 3

Nutrition

Chapter 7

The impact of improving feed efficiency on the environmental impact of livestock production

James K. Drackley, University of Illinois at Urbana-Champaign, USA; and Christopher K. Reynolds, University of Reading, UK

1 Introduction

As ruminants, cattle are marvellous bioreactors that, through symbiotic rumen fermentation, convert cellulosic plant biomass and other organic materials inedible to humans into high-quality animal proteins for human nutrition. Nevertheless, the conversion is of course not 100% efficient, and so varying quantities of waste products such as carbon dioxide (CO_2), methane (CH_4), and reactive nitrogenous compounds are emitted. As such, producers of ruminant livestock must strive to maximise output for each unit of input, both to enhance enterprise profitability and to minimise the environmental impacts of dairy and meat production. A key metric of this system efficiency is feed conversion efficiency (FCE), which for milk production is usually defined as energy-corrected

http://dx.doi.org/10.19103/AS.2020.0077.10

milk divided by feed dry matter intake (DMI) and for meat production is live weight gain divided by feed DMI. FCE per se also has a genetic component, which can be measured by residual feed intake (RFI). RFI is defined as the actual intake minus the feed intake expected to meet requirements for milk production, growth, reproduction and maintenance (Koch et al., 1963).

FCE has been widely used in beef production, as well as in pork and poultry production, to monitor the efficiency of feed utilization for growth. The dairy industry also recognizes the importance of the metric in management systems, but in addition to milk yield, there is also a need to account for body tissue loss and gain in calculating the efficiency of a lactating dairy cow (VandeHaar, 1998). Maximising the output of saleable product per unit of resource input is a standard principle of all manufacturing industries that relate directly to profitability. Another way of stating this relationship is that producers must minimise their unit cost of product and optimise their total unit output (Colman et al., 2011).

Relative to the reduction of greenhouse gases and contaminants of water, the simple concept is that the more carbon and nitrogen (N) in feedstuffs captured in the product, the less carbon and N are available for conversion into waste products (e.g. CO_2, CH_4 or urea N). By this principle, increasing milk or meat output from the same feed input requires changes in digestibility or postabsorptive nutrient metabolism with the result that less greenhouse gases and other waste products are produced per unit of milk or meat. The same principles apply to phosphorus and other nutrients that may become pollutants when they escape the animal through feaces or urine. This chapter will focus on the efficiency of milk production by dairy cattle related to nutrition and genetics focussing on how improving FCE can decrease the greenhouse gas burden of milk production and how FCE can be improved.

2 Greenhouse gases and dairy production

In the United States during 2008, the dairy sector's contribution to greenhouse gas emissions was estimated to be 134 Tg CO_2 equivalents, which equated to 1.9% of the total U.S. output (Thoma et al., 2013). Of this, CH_4, nitrous oxide (N_2O), and CO_2 contributed 44%, 13%, and 41%, respectively, of total emissions of the sector (Thoma et al., 2013). Emissions of CH_4 attributed to the dairy sector are primarily enteric emissions arising from digestive tract fermentation and stored manure, whereas N_2O emissions are largely attributable to N fertiliser application for the production of feedstuffs and manure application to farm land, including direct deposition by grazing livestock (Uweze et al., 2020; Rotz, 2018). In contrast, direct emissions of CO_2 arising from rumen fermentation and animal metabolism are not considered, as these emissions are a consequence of the digestion and metabolism of plant material incorporating atmospheric

CO_2 captured by photosynthesis. These estimates of the contribution of dairy production systems to global greenhouse gas emissions have been revisited recently because of the shorter half-life of CH_4 relative to CO_2. Methane from ruminants is also derived from the digestion of plant material, and over time CH_4 in the atmosphere is converted to CO_2, which reduces the long-term global warming effect of enteric CH_4 (Cain et al., 2019). In light of these considerations, dairy's contributions to CO_2 and N_2O emissions through fossil fuel consumption, fertilizer production and use, as well as manure management, are a greater concern for global warming in the longer term.

While the dairy sector is a small contributor relative to other industries such as oil and gas, the industry faces pressure to decrease greenhouse gas output. From the viewpoint of dairy producers, this pressure should not be viewed necessarily as burdensome because methane amounts to a 3.8–7.4% (5.6% on average) loss of gross energy intake from feeds (Kebreab et al., 2008). Mitigating greenhouse gas emissions should help not only the environment but also the financial bottom line of dairy production, unless the costs of mitigation become excessive. This is true especially if the feed energy not emitted as methane is captured as additional milk or body tissue (Reynolds et al., 2011). However, in most studies to date decreases in methane emission resulting from feeding methane inhibitory compounds such as 3-nitrooxypropanol (Reynolds et al., 2014; Hristov et al., 2015) or nitrate (Olijhoek et al., 2016) have not been associated with increases in milk energy yield or body energy balance. In this case, a part of the energy not emitted as methane is emitted as hydrogen, but the fate of the remainder of the methane energy not emitted is not certain (Olijhoek et al., 2016). In contrast, dietary strategies such as supplemental fat consistently decrease methane per kg feed DMI and often increase milk yield (Beauchemin and Grainger, 2011), but in this case, the effects on milk yield are through mechanisms independent from the effects on methane emission.

An issue with striving to decrease methane emissions is a decrease relative to what – in other words, what is the denominator of the equation? A reduction in total methane produced by the dairy sector will be difficult to achieve without reductions in animal numbers or effective mitigation strategies (e.g., methane-inhibiting feed supplements). Reductions in methane formation per unit of feed dry matter consumed (methane yield) may require shifts in the microbial community's overall metabolism or methanogenic enzyme inhibition through feed additives such as nitrate (Olijhoek et al., 2016) or 3-nitrooxypropanol (Hristov et al., 2015) as discussed above. Reductions in methane yield can also be achieved by feeding supplemental fat, which provides dietary energy that is not fermented in the rumen to yield hydrogen for methane synthesis (Grainger and Beauchemin, 2011). In contrast, decreases in methane production per unit of milk produced (methane intensity) are possible by boosting production efficiency (milk per unit of feed DMI). Much of this benefit arises from a dilution of the greenhouse gas production associated with the maintenance intake of the animal (Knapp et al., 2014). Cows require a certain amount of feed nutrients to maintain critical life functions even in the absence of milk production; think of it as the 'overhead' digestion and metabolism needed to sustain the basic functions of cows. Increasing the amount of milk produced per unit of additional feed intake thus serves to dilute the amount of maintenance greenhouse gas production over a larger number of milk production units. Other ways that improving feed efficiency may decrease methane output per unit of milk produced relate to increasing rates of nutrient passage through the rumen, shifting site of digestion to the intestine, feeding supplemental fat, or reducing heat production by the animal by alterations in metabolism that result in greater milk yield (discussed in a later section).

3 Origin of methane and reactive nitrogen excretions

In ruminants, enteric methane is mostly produced by the reduction of CO_2 in the rumen and hindgut. Greenhouse gas (primarily methane) output by dairy cattle represents 24.2% of enteric emissions from livestock (USEPA, 2021). The majority of methane production (ca. 85%) occurs in the reticulorumen of cattle, with normally only 13% being produced in the lower gut and rectal emissions constituting only 2–3% of total animal emissions (Murray et al., 1976; Munoz et al., 2012). Thus, factors relating to feed quality, feed consumption, feed degradation, and ruminal metabolism are paramount for the determination of methane output by dairy cattle.

The rich and extensive microbiota in the rumen constantly degrades complex carbohydrates in the diet (cellulose, hemicellulose, pectin, and starch) and metabolizes the constituent monosaccharides to the principal short-chain or volatile fatty acids (VFA), which are acetate, propionate, and butyrate. Conversion of monosaccharides to VFA generates ATP that can be used to drive nucleic acid and protein biosynthesis, that is, the principal components of new microbial cells. Degradation of feed protein and nonprotein N, including urea N recycled to the rumen, provides substrates (NH_3, amino acids, peptides) for microbial amino acid and protein synthesis. Ammonia N not captured as microbial protein is absorbed, converted to urea N in the liver, and either recycled to the gastrointestinal tract or excreted in the urine. Urea recycled to the rumen and hindgut is degraded by microbial urease to CO_2 and NH_3, and the NH_3 is either used for microbial protein synthesis or is reabsorbed. Any microbial protein synthesized in the hindgut is excreted as faeces, and microbial protein can account for more than 50% of faecal non-ammonia N of lactating dairy cattle (Larsen et al., 2001). Urea N in urine is very reactive and can be quickly volatilised as NH_3, especially when urine is mixed with faeces containing microbial urease. Faecal protein N is less reactive than urinary urea N, but as noted above, can contribute to N_2O and nitrate losses to the environment, depending on manure management practice.

Methane production serves a critical role within the microbial community. Production of acetate from fibre fermentation is associated with the co-production of CO_2 and H_2. The partial pressure of H_2 must be kept very low in the rumen to avoid poisoning the fermentation, and the primary means of doing so is the conversion of H_2 and CO_2 to CH_4 and H_2O (Hungate, 1966). The organisms responsible belong to the Archaea, but changes in Archaeal populations correlate only weakly with methane production (Tapio et al., 2017). Rather, methane production is related more to the populations of hydrogen-producing bacteria (Wallace et al., 2017). Production of propionate is not associated with the production of CO_2 or H_2; in fact, propionate uses hydrogen reducing equivalents in its synthesis. Although butyrate production also produces CO_2, it uses hydrogen reducing equivalents, so methane formation is not necessary (Hungate, 1966).

A fundamental relationship exists then in the fermentative breakdown of dietary components within the rumen, which can be appreciated by writing a simplified (non-chemically balanced) equation for rumen fermentation:

$$\text{dietary C and N} \rightarrow \text{microbial biomass} + \text{VFA} + CO_2 + CH_4 + H_2O + NH_3 + \text{heat} \quad (1)$$

From the standpoint of microbial growth efficiency, microbes strive to maximise biomass growth at a minimum VFA production, thus capturing as much of the starting feed substrate in new microbial cells as possible. The mixed rumen microbial population is able to generate 3–4 moles of ATP per mole of glucose fermented, which contrasts with simple monoculture fermentations that may generate only 2 moles of ATP (Hungate, 1966). Therefore, as rumen nutritionists, we understand that maximising microbial efficiency in the rumen ultimately leads to improved efficiency of milk production. Another way to think of this relationship is that the more dietary C and N we can keep in VFA and microbial protein, the less CH_4 and NO_2 are released into the environment. Similarly, the more efficiently dietary protein can be converted to milk protein through microbial protein synthesis and the metabolism of absorbed amino acids, the less excess N is excreted in urine and faeces. This idea forms the basis for why improving FCE decreases methane loss and the total greenhouse gas emissions attributable to dairy production.

4 Feed conversion efficiency

Improvements in milk production over the last several decades provide a clear demonstration of the benefits of improved FCE. Increases in FCE have arisen from increases in milk production, which dilutes the proportion of feed used for maintenance (VandeHaar and St-Pierre, 2006). Milk production has increased from intense genetic selection, improved nutrition and cow health, and other advances in modern management techniques. The improved FCE has benefited dairy producers' profitability, decreased the environmental impact of milk production, reduced the amount of land required for milk production, and decreased the greenhouse gas emission per unit of milk produced (Knapp et al., 2014). Continued gains in FCE by these methods will diminish, however, because the effect diminishes with each increment of dilution of maintenance (VandeHaar, 1998; Reynolds et al., 2011; VandeHaar et al., 2016).

Many environmental factors also affect FCE, including nutrition. Diet digestibility, or more specifically forage digestibility, is a major component of FCE. Dietary protein and carbohydrate balance, supplemental fats and various additives may also impact FCE through effects on digestive function and milk energy yield. Other factors include the stage of lactation, body reserve changes, physical activity, mastitis or other diseases, acidosis and heat and cold stress. From a dairy system standpoint, the total FCE of the farm unit is impacted by numerous factors, including calving interval, days dry, age of first calving, reproductive efficiency, periparturient health disorders, and death loss in calves and heifers. These latter factors centre on the effects of decreased productive days of life for those animals that are not producing milk, or not producing large amounts of milk, but continue to consume the feed.

Herd and Arthur (2009) reviewed the biological basis for differences in RFI for beef cattle. Possible components included feed intake, digestion, metabolism associated with fat and protein deposition, physical activity, thermoregulation, and other components of basal metabolism. As reviewed by Reynolds et al. (2011), digestible energy is the most variable proportion of gross energy intake in ruminants. It accounted for 86% of the variation in dietary net energy for lactation in dairy cattle in the calorimetry studies on which the initial NEmilk system for energy requirements of lactating dairy cows in the USA was based (Moe et al., 1972). This largely reflects differences in forage quality and the digestibility of fibre, which is affected by forage maturity at harvest and processing during conservation and feeding, as well as associative effects of other diet components that impact the rate and extent of fibrolytic activity in the rumen.

Beever and Drackley (2013) calculated methane production per kilogram of milk production as a function of increasing milk production efficiency and found a negative relationship between the two (Fig. 1). That this theoretical calculation was borne out in practice was demonstrated by an analysis of 323 measurements of methane emission by lactating dairy cows at the University of Reading (Fig. 2). When expressed on the basis of MJ of methane emitted per MJ of milk energy production, there was a reduction in methane emission

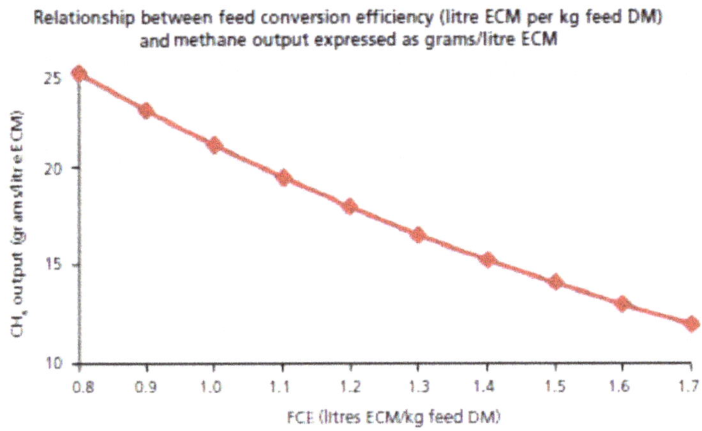

Figure 1 Theoretical relationship between methane output (g/L ECM) and FCE (L ECM/kg DM). Reprinted from Beever and Drackley (2013) with permission.

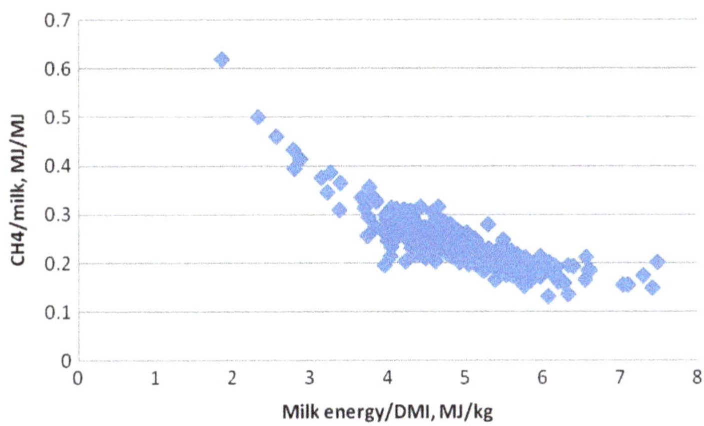

Figure 2 Methane energy/milk energy (MJ/day) as a function of milk energy (MJ/day) per unit feed dry matter intake (kg/day) where $y = 0.833*\exp^{(-0.259*x)}$. Individual observations (*n* = 323) for cows fed various dietary treatments in experiments conducted at the University of Reading with correction for the random effect of experiment on the intercept.

per unit milk energy yield with increasing FCE, expressed as milk energy per kg feed DMI (MJ/kg). An exponential decay curve provided the best fit of the data after correction for the effect of experiment, clearly showing the higher methane 'overhead' of lower-yielding cows and the dilution of maintenance greenhouse gas 'costs' with higher milk yield and FCE. Similar relationships between increased milk yield per se and reduced methane emission per unit milk yield have been noted previously (Reynolds et al., 2011).

5 Nutritional practices to enhance feed conversion efficiency and decrease CH_4 excretion

Because methane formation occurs from microbial fermentation of dietary constituents, it follows that nutrition of the ruminant impacts FCE and methane production (Table 1). The most important aspect of nutrition is feed intake, represented in ruminants as DMI. A meta-analysis of methane measurements from growing and lactating cattle (Mills et al., 2009) revealed that feed DMI was the primary predictor of methane emission (Reynolds et al., 2011). The effect of DMI is explained by the increasing provision of fermentable substrates, although the use of digestible energy only marginally improved the prediction (Mills et al., 2009).

A commonly used equation to estimate methane production (Moe and Tyrrell, 1979) is as follows:

$$\begin{aligned}\text{methane production, MJ/d} = 1.837 &+ 1.142\,(\text{digested soluble residue, kg/d}) \\ &+ 2.142\,(\text{digested hemicellulose, kg/d}) \\ &+ 5.828\,(\text{digested cellulose, kg/d}).\end{aligned} \quad (2)$$

Since this early work of Moe and Tyrrell (1979), there have been numerous subsequent meta-analyses of measurements of methane emission relative to

Table 1 Feeding management variables and effects on methane (CH_4/ECM), dry matter intake (DMI), energy-corrected milk (ECM), and feed conversion efficiency (FCE). Modified from Knapp et al. (2014)

Feeding alteration	CH_4/ECM	DMI	ECM	FCE
Increased DMI	Decrease	Increase	Increase	Increase or no change
Increased forage quality	Decrease	Increase	Increase	Increase
Decreased forage particle size	No change	Increase	Increase or no change	No change
Grain processing	Decrease	No change	Increase or no change	Increase or no change
Increased concentrate feeding	Decrease	Increase	Increase	Increase or no change
Rumen pH <5.5	Decrease	No change or decrease	Decrease	Decrease
Brown midrib corn	Decrease	Increase	Increase	Increase or no change
Fat feeding	Decrease	Decrease or no change	Increase	Increase

diet intake and composition, which have largely confirmed the overriding role of intake of digestible structural and nonstructural carbohydrates, as well as lipids (see below), in determining methane emission and yield. For example, a recent global analysis of 3183 individual observations concluded that prediction models should include diet intake, digestibility, and concentrations of structural and nonstructural carbohydrates and ether extract for greatest accuracy (Benaouda et al., 2019).

According to Equation 2, increments of digested soluble residue, largely comprising starch and sugars, will increase methane to a lesser extent than digested hemicellulose or cellulose from forages. In turn, digested hemicellulose has a much smaller effect than digested cellulose. This is consistent with the fact that starch fermentation yields a greater proportion of propionate compared to fibre fermentation (Hungate, 1966). Increasing concentrate supplementation in the diet increases digestible energy, usually increases FCE, and decreases methane output per unit of milk produced, as predicted by the analysis in Fig. 2 and by the diluting effect of greater productivity on methane production per unit of milk yield. Concentrates containing cereal grains rich in starch will increase the metabolisable energy available to the animal, usually increasing milk yield unless excessive amounts are fed that cause acidosis (Van Soest, 1994).

In contrast, increasing the digestibility of forages, which will usually increase FCE by boosting milk yield, might not reliably change methane production per unit of milk produced because of the increase of fermented cellulose and hemicellulose. However, in practice, improving forage digestibility usually increases FCE and, by increasing milk production, the yield of methane per unit of ECM decreases due to the dilution of the maintenance concept. Variations in concentrations of cellulose, hemicellulose, and starch, as well as their digestibility, interact with differences in DMI that can make it difficult to predict the resulting methane production from differences in FCE (Reynolds et al., 2010; Livingstone et al., 2015). For example, diets higher in maize silage contained more starch and less NDF than diets higher in grass silage, and resulted in greater DMI and milk yields, but FCE was greater for the diets high in grass silage. However, methane intensity was greater for the high grass silage diets (Hammond et al., 2016).

Forage digestibility can be increased in several ways. Harvesting at younger maturity increases digestibility because of greater concentrations of nonstructural carbohydrates, such as sugars and fructans, along with lower lignification of plant cell walls (Van Soest, 1994). Some genetic variants have greater digestibility. For example, the stover from the *bm-3* or brown midrib genetic variant of maize and sorghum is more highly digested than typical varieties (Oba and Allen, 2000). Cellulose, hemicellulose, and NDF were less digestible in diets containing 40% (DM) maize silage and 10% alfalfa silage than in diets containing 10% maize silage and 40% alfalfa silage (Ruppert et al., 2003). Finally, grass forage is typically more digestible than legumes, such as alfalfa, although the rate of fermentation is faster for legumes (Van Soest, 1994).

Processing forages can have contradictory effects on methane yield. Fine grinding or pelleting of forage increases the surface area available for fibrolytic microbiota to attach for fermentation, but the small particle size increases its passage from the rumen (Russell and Hespell, 1981), thus limiting the digested dry matter and so limiting methane formation (Moss et al., 2000). Such effects could decrease methane by 20–40% per unit of DM at high intakes, although at restricted intakes the effects are less (Johnson and Johnson, 1995). However, this also decreases potential energy delivery to the animal, which might decrease milk production, although compensatory hindgut digestion can recover some of the lost energy (Van Soest, 1994). With fine grinding of the forage and the resulting faster passage, however, the animal will likely eat more (Van Soest, 1994), which would tend to increase methane production.

Grains can be processed by grinding, rolling, and steam flaking, all of which increase surface area for digestion and decrease particle size. Grain processing can decrease the amount of starch that passes to the small intestine. Digestion in the small intestine should be more efficient energetically, but results have not borne this out in practice (Huntington et al., 2006). The effect of processing on ruminal degradation of starch varies among cereals and depends on the processing methods used (Svihus et al., 2005). Extensive rolling of barley maximised ruminal and postruminal digestion of starch (Yang et al., 2001). More extreme processing tends to increase total VFA concentration in the rumen, with a greater proportion of propionate that would decrease methane formation. Grinding maize grain increased starch digestibility in the total tract compared with rolling (Knowlton et al., 1998), but ground maize resulted in greater methane production (Wilkerson et al., 1997). Steam flaking maize grain resulted in greater FCE than grinding, regardless of particle size of grind (Ahmadi et al., 2020). However, finely ground maize can produce results similar to steam-flaked maize (Mathew et al., 2011). Heat treatment through pelleting, flaking, extruding, and toasting can change the ruminal degradation rates of protein and carbohydrates and decrease the acetate:propionate ratio in the rumen (Van Nevel and Demeyer, 1996). However, the relationship between such treatments and CH_4 emissions depends on the feed, composition of the total diet, and intake (Knapp et al., 2014).

Supplemental fats and oils are often added to diets for dairy cattle to increase the energy density of diets for high milk production (Palmquist and Jenkins, 2017). According to meta-analyses (Rabiee et al., 2012; Hu et al., 2017), fats usually maintain or decrease DMI while increasing milk yield and milk energy output. Consequently, supplemental fat sources often increase FCE and, as expected, decrease methane output per unit of milk energy (Ruppert et al., 2003; Kliem et al., 2019). The effect of lipids on methane production is dependent on the source, fatty acid profile, inclusion rate, form of lipid, and diet composition (Beauchemin et al., 2008). Lipid sources replace fermentable substrates, and the unsaturated fatty acids provide an alternate sink for hydrogen disposal. Unsaturated fatty acids also directly impact fibre-fermenting microbiota and methanogens in the rumen (Czerkawski et al., 1966; Blaxter and Czerkawski, 1966). In practice, however, changes in methane production are not always observed. Nevertheless, a meta-analysis conducted by Knapp et al. (2014) found that increasing fat content of the diet decreased methane per unit of ECM, and that the type of lipid source affected the response. Free vegetable oils and endogenous lipids resulted in a greater reduction of methane per unit of ECM than did inert fats or seed lipids. Some seeds, vegetable oils, and medium-chain fatty acids (such as those found in coconut oil; Hristov et al., 2009) further decrease methane, but often at the expense of DMI, which could be detrimental to productivity and reproduction over the long term (Reynolds et al., 2011; Knapp et al., 2014).

In addition to 3-nitrooxypropanol and nitrate mentioned previously, a variety of potential nutritional additives have been tested for their ability to improve FCE and decrease methane losses, including ionophores, yeasts, certain seaweeds, and plant bioactive compounds with antimicrobial or digestive effects, such as tannins and essential oils (Beauchemin et al., 2008; McAllister and Newbold, 2008). These compounds have had varying efficacies, especially in the longer term, perhaps due to differences in dose rate and adaptation of the rumen microbes (Beauchemin et al., 2008). Although some have shown potential to decrease methane excretion, for these approaches to be widely adopted their effects will need to be consistent, sustained, without deleterious effects on feed intake, production, FCE or product quality, economic to adopt, and sustainable (Reynolds et al., 2011).

6 Nutritional practices to increase milk protein efficiency and decrease N_2O excretion

In ruminant production systems, enteric CH_4 production is the greatest contributor to greenhouse gas emissions, followed by CH_4 from manure and in beef feedlot systems, N_2O from pen surfaces, and N_2O emissions from soils (Montes et al., 2013). Improving the efficiency of dietary N use by increasing N incorporation into milk protein and decreasing N losses in faeces and urine will have positive effects on N_2O excretion and environmental contamination with N (Uwizeye et al., 2020). Many of the same principles discussed to this point relative to increasing FCE will have beneficial effects on increasing efficiency of dietary N use. From an environmental standpoint, animal practices to decrease urinary N loss have the biggest impact because urinary N is more susceptible to leaching and volatile losses than faecal N and contributes directly to N_2O loss from livestock facility surfaces (Dijkstra et al., 2013a; Montes et al., 2013).

The main driver of N losses from cattle is N consumed in feed. Dairy cows secrete in milk an average of 21–33% of consumed N (Calsamiglia et al., 2010), with almost all the remaining N excreted in faeces and urine. In agreement, Reed et al. (2015) calculated an average total manure N excretion of 69% of N intake from a large database. Dijkstra et al. (2013b) calculated the theoretical upper limit of dietary N incorporation into milk protein to be 43% at maximal milk secretion for a cow weighing 650 kg and producing 40 kg/day of fat and protein corrected milk.

Huhtanen and Hristov (2009) concluded from a meta-analysis that dietary crude protein (CP) concentration is the most important dietary factor influencing milk N efficiency, with ruminal degradation of CP being of lesser importance. Differences in amount and, to a smaller extent, digestibility of N in feed affect not only the total amount excreted but also the partitioning of N among milk, urine, and faeces (Castillo et al., 2001; Kebreab et al., 2002). Partitioning of manure N excretion into faecal and urinary N excretion is important because differences in N intake largely affect urinary N output, which is of greater importance to reduce environmental impact (Dijkstra et al., 2013a).

The carbohydrate portion of the diet may impact the efficiency of dietary N use through differences in the amount or efficiency of N capture in the rumen. Ruminants must have sufficient ruminally degradable protein to maximise fermentation with the amount of fermentable carbohydrates supplied. Wilkinson and Garnsworthy (2017) showed that replacing dietary grazed grass or grass silage with corn silage at similar milk yield increased N efficiency and reduced the C footprint. Grazed grass is particularly problematic in terms of urinary N loss because of the high soluble CP content of the grass and concurrent lower content of rapidly fermentable carbohydrates (Hristov et al., 2013).

Dairy producers have often overfed dietary protein in the mistaken assumption that it would boost milk production (Broderick, 2018). Broderick (2003) compared diets with CP increasing from 15.1% to 16.7% and 18.4% by adding soya meal to the diet. Milk and milk protein yields increased with the first increment of CP but not the second. The only result from increasing dietary CP to 18.4% was the increased excretion of urinary N, which accounted for nearly all of the increment of dietary N. Recently, a trend for lower dietary CP has been observed in high-producing herds. Diets with as little as 14.9% CP can support milk production in excess of 45 kg/day when properly balanced for N and carbohydrate fractions and supplemented with key ruminally protected amino acids (Fessenden et al., 2020). In 2010, the five highest producing herds in the state of Wisconsin were being fed diets containing an average of 16.9% dietary CP, with the lowest at 16.3% (Broderick, 2018). Over the period from 2004 to 2010, CP content of dairy diets in Wisconsin decreased 1.1 percentage units, but milk and protein production increased by 1700 and 51 kg per lactation. Clearly, minimising dietary CP needed to support high milk production pays dividends in the way of decreasing ration cost and environmental excretion of N, without sacrificing productivity.

7 Genetics and feed conversion efficiency

Using RFI as a measure of FCE has the advantage that it is independent of maintenance requirements and is not an efficiency determined by the level of production per se. A disadvantage of the measurement is that animals that are more efficient have negative RFI, which is non-intuitive and difficult for producers to accept. Reported heritability estimates of RFI in dairy cattle generally are low to moderate, with estimates ranging from 0.01 to 0.40 among lactating cows (Connor, 2015). Tempelman et al. (2015) determined that RFI had a heritability of 0.15–0.18. This means that the trait could be improved through genetic selection but that much of the variation in phenotype must be attributable to other environmental factors. Genetic correlations suggest no undesirable relationships detected between RFI and fat-corrected milk yield, productive life or feeding behaviours, and desirable relationships between RFI and predicted methane production in lactating cows (Connor, 2015).

There is currently interest in the potential selection of dairy cattle for improved feed efficiency (e.g. Lin et al., 2013; VandeHaar et al., 2020) and lower methane emission or methane yield (e.g. Breider et al., 2019; Lassen and Løvendahl, 2016). However, these traits must not be considered in isolation; for example, reductions in methane yield in sheep were shown to be driven by higher rates of feed passage through the rumen attributable to smaller rumen volume and thus reduced total feed intake capacity (Goopy et al., 2013).

The benefits of selection for RFI make the prospect of its application promising. For example, Holstein-Friesian heifers in New Zealand and Australia fed a forage-based diet that were in the bottom 10% of the sample population for RFI (i.e. most efficient) consumed 15%–20% less feed relative to heifers in the top 10% for RFI (least efficient; Williams et al., 2011; Waghorn et al., 2012). Differences are similar to differences in DMI of 12–13% reported in low versus high RFI groups of growing Angus-Hereford steers (Cruz et al., 2010). The dairy heifers cited showed no differences in feed intake, yields of milk or milk components, change in BW or body condition score when evaluated during days 75–195 of their first lactation (Macdonald et al., 2014). These results indicate that considerable savings in feed costs can be achieved by maintaining only the most feed-efficient growing heifers in the herd, with no negative consequences on future lactation performance.

In addition to selection for RFI, it might be possible to use genetic selection to improve N efficiency. Marshall et al. (2020) used breeding values for milk urea N (MUN) to create high and low lines of Holstein-Friesian dairy cows. There was a positive relationship between MUN breeding value and MUN, with MUN decreasing 1.61 mg/dL per unit decrease in MUN breeding value. Urinary N concentration decreased 0.67 g/L per unit decrease of MUN breeding value, with no difference in urine volume or urinary frequency, which resulted in a 165 g/day decrease in urinary urea N excretion between animals with the highest and lowest MUN breeding value. At the same milk yield, milk protein percentage increased by 0.09 per unit of MUN breeding value. Such preliminary results are positive both for decreasing environmental excretion of N and for improving N partitioning into milk for greater producer profitability.

Relative to greenhouse gas emissions, cattle with lower RFI have a lower DMI than less efficient animals at similar production levels (Connor, 2015), thus decreasing one major contributor to CH_4 production. By definition, animals with lower RFI have less manure output for a given level of production, which should decrease the release of CH_4 and N_2O from stored manure (Montes et al., 2013). Recent research demonstrated that cows with greater FCE (lower RFI) also used protein more efficiently, which would help reduce N excretion in urine (Liu and VandeHaar, 2020).

8 Postabsorptive metabolism and feed conversion efficiency

Heat energy lost in the conversion of ME to net energy (NE) typically accounts for 20–30% of gross energy intake in dairy cattle. But, in contrast to digestibility, the variation and opportunities to decrease these losses are more limited. The effects of forage type, processing, and forage to concentrate ratio on the efficiency of ME utilisation for milk and meat production (and FCE) have been extensively researched (Reynolds, 2011). Such effects are a greater concern in extensive systems relying heavily on poorer quality forages. Because the digestive system and liver account for as much as 50% of body heat production in ruminants, their metabolism has an impact on the partial efficiency of ME use for production that is disproportionate to their mass. Forage amount and digestibility impact ME utilisation for NE, which is largely attributable to the tissues of the gut and may reflect differences in the work of digestion and gut mass (McLeod and Baldwin, 2000). In contrast, nutrient re-partitioning agents such as growth hormone impact energy utilisation by altering metabolism in the muscle, adipose tissue, and mammary gland to increase milk production, which improves FCE by dilution of maintenance.

Reducing losses of dietary energy as heat may be one of the main factors affected by selection for improved RFI. Only 19% of the variation in RFI among animals may be attributable to differences in diet digestion and heat of fermentation, with the remainder likely due to differences in physical activity, body composition, protein turnover, and other metabolic processes associated with maintenance requirements and partitioning of nutrients between protein and lipid deposition (Herd and Arthur, 2009).

The effects of excess protein on energy utilisation in cattle have generated much discussion. Tyrrell et al. (1970) reviewed results of calorimetry studies and reported a reduction in ME or NE of diets when digestible protein was fed in excess of requirements. This effect was included in the calculation of NE for lactation (Moe et al., 1972). Some have attributed this effect of excess protein to the energy costs of urea synthesis in the liver, but studies in sheep and cattle catheterised for measurements of liver metabolism have consistently failed to show an effect of increased ammonia absorption and subsequent urea production by the liver on hepatic oxygen consumption (Reynolds, 2006). In contrast, supplemental protein fed in excess of requirement does increase liver oxygen use (i.e. heat production), as well as oxygen use by the gut, suggesting that the effect of excess protein on the efficiency of energy utilisation is associated with amino acid catabolism, rather than urea synthesis per se (Reynolds, 2006). Over 100 years ago, Rubner showed that the postprandial rise in heat production in dogs, termed 'specific dynamic action', associated with consumption of protein was greater than for carbohydrates and fats (Brody, 1939), which also reflects the catabolism of amino acids in excess of requirements. These effects might be attributable to stimulation of protein turnover and greater activation of the sympathetic nervous system. Regardless of the mechanism, the potential benefits of feeding protein and ruminally protected amino acids more precisely to cattle extend beyond just the environment and economics to improving FCE.

9 Conclusion

Methane intensity of dairy production is related inversely to FCE. By diluting methane associated with cow maintenance, increasing milk production and increased FCE serve to decrease methane production per unit of milk energy produced. Nutritional practices that boost productivity usually boost FCE and will help to decrease methane intensity. Some of these include improving forage quality, increasing concentrate supplementation, and using supplemental fats and oils. Genetic selection for improved FCE (measured by RFI) will decrease methane intensity. Minimising dietary CP will help decrease urinary N losses, which contribute to increased N_2O and decreased efficiency of N capture into milk protein.

10 Future trends in research

Increasing FCE is directly related to profitability and will be a continued goal of research for improvement in the dairy industry. Decreased methane intensity should be a collateral benefit of such improvement. Determining the genetic basis for FCE, as measured by RFI, may eventually lead to improvements in the overall efficiency of milk production. Improving genetic aspects of forage digestibility and methods to improve use by cows will help decrease methane intensity. The research will continue to find compounds or strategies that will decrease methane yield without compromising DMI or productive efficiency. Novel mitigation strategies might include inhibitors or early life microbial modulation to change the balance between hydrogen producers and methanogens. In addition, the possibility that synergistic effects between strategies might occur, such as combining 3-nitrooxypropanol with lipids, should be investigated. Research will continue to identify genetic opportunities to minimise the environmental burden of livestock production as well as urinary N excretion.

11 Where to look for further information

A comprehensive discussion of strategies to mitigate methane emissions in dairy cattle can be found in Knapp, J. R., Laur, G. L., Vadas, P. A., Weiss, W. P., and Tricarico, J. M. (2014), 'Invited review: Enteric methane in dairy cattle production: Quantifying the opportunities and impact of reducing emissions', *J. Dairy Sci.*, 97, 3231–3261.

12 References

Ahmadi, F., Ghorbani, G. R., Sadeghi-Sefidmazgi, A., Heydari, M., Rafiee, H. and Beauchemin, K. A. (2020). Performance and feeding behavior of dairy cows fed high-concentrate diets containing steam-flaked or ground corn varying in particle size, *J. Dairy Sci.* 103(4), 3191–3203.

Beauchemin, K. A., Kreuzer, M., O'Mara, F. and McAllister, T. A. (2008). Nutritional management for enteric methane abatement: a review, *Aust. J. Exp. Agric.* 48(2), 21–27.

Beever, D. E. and Drackley, J. K. (2013). Feeding for optimal rumen and animal health and optimal feed conversion efficiency: the importance of physical nutrition. In: Makkar, H. P. S. and Beever, D. (Eds), *Optimization of Feed Use Efficiency in Ruminant Production Systems*. Proceedings of the FAO Symposium, 27 November 2012, Bangkok, Thailand, FAO Animal Production and Health Proceedings, No. 16. Rome: FAO and Asian-Australasian Association of Animal Production Societies, pp. 75–92.

Benaouda, M., Martin, C., Li, X., Kebreab, E., Hristov, A. N., Yu, Z., Yáñez-Ruiz, D. R., Reynolds, C. K., Crompton, L. A., Dijkstra, J., Bannink, A., Schwarm, A., Kreuzer, M., McGee, M., Lund, P., Hellwing, A. L. F., Weisbjerg, M. R., Moate, P. J., Bayat, A. R., Shingfield, K. J., Peiren, N. and Eugène, M. (2019). Evaluation of the performance of existing mathematical models predicting enteric methane emissions from ruminants: animal categories and dietary mitigation strategies, *Anim. Feed Sci. Technol.* 255, 114207.

Blaxter, K. L. and Czerkawski, J. (1966). Modifications of methane production of the sheep by supplementation of its diet, *J. Sci. Food Agric.* 17(9), 417–421.
Breider, I. S., Wall, E. and Garnsworthy, P. C. (2019). Short communication: heritability of methane production and genetic correlations with milk yield and body weight in Holstein-Friesian dairy cows, *J. Dairy Sci.* 102(8), 7277–7281.
Broderick, G. A. (2003). Effects of varying dietary protein and energy levels on the production of lactating dairy cows, *J. Dairy Sci.* 86(4), 1370–1381.
Broderick, G. A. (2018). Review: optimizing ruminant conversion of feed protein to human food protein, *Animal* 12(8), 1722–1734.
Brody, S. (1939). Factors influencing the apparent energetic efficiency of productive processes in farm animals, *J. Nutr.* 17(3), 235–251.
Cain, M., Lynch, J., Allen, M. R., Fuglestvedt, J. S., Frame, D. J. and Macey, A. H. (2019). Improved calculation of warming-equivalent emissions for short-lived climate pollutants, *NPJ Clim. Atmos. Sci.* 2, 29.
Calsamiglia, S., Ferret, A., Reynolds, C. K., Kristensen, N. B. and van Vuuren, A. M. (2010). Strategies for optimizing nitrogen use by ruminants, *Animal* 4(7), 1184–1196.
Castillo, A. R., Kebreab, E., Beever, D. E., Barbi, J. H., Sutton, J. D., Kirby, H. C. and France, J. (2001). The effect of protein supplementation on nitrogen utilization in lactating dairy cows fed grass silage diets, *J. Anim. Sci.* 79(1), 247–253.
Colman, D. R., Beever, D. E., Jolly, R. W. and Drackley, J. K. (2011). Commentary: Gaining from technology for improved dairy cow nutrition: economic, environmental, and animal health benefits, *Prof. Anim. Sci.* 27(6), 505–517.
Connor, E. E. (2015). Invited review: improving feed efficiency in dairy production: challenges and possibilities, *Animal* 9(3), 395–408.
Cruz, G. D., Rodríguez-Sánchez, J. A., Oltjen, J. W. and Sainz, R. D. (2010). Performance, residual feed intake, digestibility, carcass traits, and profitability of Angus-Hereford steers housed in individual or group pens, *J. Anim. Sci.* 88(1), 324–329.
Czerkawski, J. W., Blaxter, K. L. and Wainman, F. W. (1966). The effect of linseed oil and of linseed oil fatty acids incorporated in the diet on the metabolism of sheep, *Br. J. Nutr.* 20(3), 485–494.
Dijkstra, J., Oenema, O., van Groenigen, J. W., Spek, J. W., van Vuuren, A. M. and Bannink, A. (2013a). Diet effects on urine composition of cattle and N2O emissions, *Animal* 7 (Suppl. 2), 292–302.
Dijkstra, J., Reynolds, C. K., Kebreab, E., Bannink, A., Ellis, J. L., France, J. and van Vuuren, A. M. (2013b). Challenges in ruminant nutrition: towards minimal nitrogen losses in cattle. In: Oltjen, J. W., Kebreab, E. and Lapierre, H. (Eds) *Energy and Protein Metabolism and Nutrition in Sustainable Animal Production*. Wageningen, The Netherlands: Wageningen Academic Publishers, pp. 47–58.
Fessenden, S. W., Ross, D. A., Block, E. and Van Amburgh, M. E. (2020). Comparison of milk production, intake, and total-tract nutrient digestion in lactating dairy cattle fed diets containing either wheat middlings and urea, commercial fermentation by-product, or rumen-protected soybean meal, *J. Dairy Sci.* 103(6), 5090–5101.
Goopy, J. P., Donaldson, A., Hegarty, R., Vercoe, P. E., Haynes, F., Barnett, M. and Oddy, V. H. (2013). Low-methane yield sheep have smaller rumens and shorter rumen retention time, *Br. J. Nutr.* 111(4), 578–585.
Grainger, C. and Beauchemin, K. A. (2011). Can enteric methane emissions from ruminants be lowered without lowering their production?, *Anim. Feed Sci. Technol.* 166–167, 308–320.

Hammond, K. J., Jones, A. K., Humphries, D. J., Crompton, L. A. and Reynolds, C. K. (2016). Effects of diet forage source and neutral detergent fiber content on milk production of dairy cattle and methane emissions determined using GreenFeed and respiration chamber techniques, *J. Dairy Sci.* 99(10), 7904–7917.

Herd, R. M. and Arthur, P. F. (2009). Physiological basis for residual feed intake, *J. Anim. Sci.* 87(14) (Suppl.), E64–E71.

Hristov, A. N., Oh, J., Giallongo, F., Frederick, T. W., Harper, M. T., Weeks, H. L., Branco, A. F., Moate, P. J., Deighton, M. H., Williams, S. R. O., Kindermann, M. and Duval, S. (2015). An inhibitor persistently decreased enteric methane emission from dairy cows with no negative effect on milk production, *Proc. Natl. Acad. Sci. U. S. A.* 112(34), 10663–10668.

Hristov, A. N., Ott, T., Tricarico, J., Rotz, A., Waghorn, G., Adesogan, A., Dijkstra, J., Montes, F., Oh, J., Kebreab, E., Oosting, S. J., Gerber, P. J., Henderson, B., Makkar, H. P. and Firkins, J. L. (2013). Special topics - Mitigation of methane and nitrous oxide emissions from animal operations: III. A review of animal management mitigation options, *J. Anim. Sci.* 91(11), 5095–5113.

Hristov, A. N., Vander Pol, M., Agle, M., Zaman, S., Schneider, C., Ndegwa, P., Vaddella, V. K., Johnson, K., Shingfield, K. J. and Karnati, S. K. (2009). Effect of lauric acid and coconut oil on ruminal fermentation, digestion, ammonia losses from manure, and milk fatty acid composition in lactating cows, *J. Dairy Sci.* 92(11), 5561–5582.

Hu, W., Boerman, J. P. and Aldrich, J. M. (2017). Production responses of Holstein dairy cows when fed supplemental fat containing saturated free fatty acids: a meta-analysis, *Asian-Australas. J. Anim. Sci.* 30(8), 1105–1116.

Huhtanen, P. and Hristov, A. N. (2009). A meta-analysis of the effects of dietary protein concentration and degradability on milk protein yield and milk N efficiency in dairy cows, *J. Dairy Sci.* 92(7), 3222–3232.

Hungate, R. E. (1966). *The Rumen and Its Microbes*. New York: Academic Press.

Huntington, G. B., Harmon, D. L. and Richards, C. J. (2006). Sites, rates, and limits of starch digestion and glucose metabolism in growing cattle, *J. Anim. Sci.* 84 (Suppl.), E14–E24.

Johnson, K. A. and Johnson, D. E. (1995). Methane emissions from cattle, *J. Anim. Sci.* 73(8), 2483–2492.

Kebreab, E., France, J., Mills, J. A. N., Allison, R. and Dijkstra, J. (2002). A dynamic model of N metabolism in the lactating dairy cow and an assessment of impact of N excretion on the environment, *J. Anim. Sci.* 80(1), 248–259.

Kebreab, E., Johnson, K. A., Archibeque, S. L., Pape, D. and Wirth, T. (2008). Model for estimating enteric emissions from United States dairy and feedlot cattle, *J. Anim. Sci.* 86(10), 2738–2748.

Kliem, K. E., Humphries, D. J., Kirton, P., Givens, D. I. and Reynolds, C. K. (2019). Differential effects of oilseed supplements on methane production and milk fatty acid concentrations in dairy cows, *Animal* 13(2), 309–317.

Knapp, J. R., Laur, G. L., Vadas, P. A., Weiss, W. P. and Tricarico, J. M. (2014). Invited review: enteric methane in dairy cattle production: quantifying the opportunities and impact of reducing emissions, *J. Dairy Sci.* 97(6), 3231–3261.

Knowlton, K. F., Glenn, B. P. and Erdman, R. A. (1998). Performance, ruminal fermentation, and site of starch digestion in early lactation cows fed corn grain harvested and processed differently, *J. Dairy Sci.* 81(7), 1972–1984.

Koch, R. M., Swiger, L. A., Chambers, D. and Gregory, K. E. (1963). Efficiency of feed use in beef cattle, *J. Anim. Sci.* 22(2), 486–494.

Larsen, M., Madsen, T. G., Weisbjerg, M. R., Hvelplund, T. and Madsen, J. (2001). Small intestinal digestibility of microbial and endogenous amino acids in dairy cows, *J. Anim. Physiol. Anim. Nutr. (Berl.)* 85(1-2), 9-21.

Lassen, J. and Løvendahl, P. (2016). Heritability estimates for enteric methane emissions from Holstein cattle measured using noninvasive methods, *J. Dairy Sci.* 99(3), 1959-1967.

Lin, Z., Macleod, I. and Pryce, J. E. (2013). Short communication: estimation of genetic parameters for residual feed intake and feeding behavior traits in dairy heifers', *J. Dairy Sci.* 96(4), 2654-2656.

Liu, E. and VandeHaar, M. J. (2020). Relationship of residual feed intake and protein efficiency in lactating cows fed high- or low-protein diets, *J. Dairy Sci.* 103(4), 3177-3190.

Livingstone, K. M., Humphries, D. J., Kirton, P., Kliem, K. E., Givens, D. I. and Reynolds, C. K. (2015). Effects of forage type and extruded linseed supplementation on methane production and milk fatty acid composition of lactating dairy cows, *J. Dairy Sci.* 98(6), 4000-4011.

Macdonald, K. A., Pryce, J. E., Spelman, R. J., Davis, S. R., Wales, W. J., Waghorn, G. C., Williams, Y. J., Marett, L. C. and Hayes, B. J. (2014). Holstein-Friesian calves selected for divergence in residual feed intake during growth exhibited significant but reduced residual feed intake divergence in their first lactation, *J. Dairy Sci.* 97(3), 1427-1435.

Marshall, C. J., Beck, M. R., Garrett, K., Barrell, G. K., Al-Marashdeh, O. and Gregorini, P. (2020). Grazing dairy cows with low milk urea nitrogen breeding values excrete less urinary urea nitrogen, *Sci. Total Environ.* 739, 139994.

Mathew, B., Eastridge, M. L., Oelker, E. R., Firkins, J. L. and Karnati, S. K. R. (2011). Interactions of monensin with dietary fat and carbohydrate components on ruminal fermentation and production responses by dairy cows, *J. Dairy Sci.* 94(1), 396-409.

McAllister, T. A. and Newbold, C. J. (2008). Redirecting rumen fermentation to reduce methanogenesis, *Aust. J. Exp. Agric.* 48(2), 7-13.

McLeod, K. R. and Baldwin, R. L. 6th. (2000). Effects of diet forage:concentrate ratio and metabolizable energy intake on visceral organ growth and in vitro oxidative capacity of gut tissues in sheep, *J. Anim. Sci.* 78(3), 760-770.

Mills, J. A. N., Crompton, L. A., Bannink, A., Tamminga, S., Moorby, J. and Reynolds, C. K. (2009). Predicting methane emissions and nitrogen excretion from cattle, *J. Agric. Sci.* 147, 741.

Moe, P. W., Flatt, W. P. and Tyrrell, H. F. (1972). Net energy value of feeds for lactation, *J. Dairy Sci.* 55(7), 945-958.

Moe, P. W. and Tyrrell, H. F. (1979). Methane production in dairy cows, *J. Dairy Sci.* 62(10), 1583-1586.

Montes, F., Meinen, R., Dell, C., Rotz, A., Hristov, A. N., Oh, J., Waghorn, G., Gerber, P. J., Henderson, B., Makkar, H. P. S. and Dijkstra, J. (2013). Mitigation of methane and nitrous oxide emissions from animal operations: II. A review of manure management options, *J. Anim. Sci.* 91(11), 5070-5094.

Moss, A. R., Jouany, J. P. and Newbold, J. (2000). Methane production by ruminants: its contribution to global warming, *Ann. Zootech.* 49(3), 231-253.

Munoz, C., Yan ,T., Wills, D. A., Murray, S. and Gordon, A. W. (2012). Comparison of the sulfur hexafluoride tracer and respiration chamber techniques for estimating methane emissions and correction for rectum methane output from dairy cows, *J. Dairy Sci.* 95(6), 3139-3148.

Murray, R. M., Bryant, A. M. and Leng, R. A. (1976). Rates of production of methane in rumen and large-intestine of sheep, *Br. J. Nutr.* 36(1), 1–14.

Oba, M. and Allen, M. S. (2000). Effects of brown midrib 3 mutation in corn silage on productivity of dairy cows fed two concentrations of dietary neutral detergent fiber: 3. Digestibility and microbial efficiency, *J. Dairy Sci.* 83(6), 1350–1358.

Olijhoek, D. W., Hellwing, A. L. F., Brask, M., Weisbjerg, M. R., Højberg, O., Larsen, M. K., Dijkstra, J., Erlandsen, E. J. and Lund, P. (2016). Effect of dietary nitrate level on enteric methane production, hydrogen emission, rumen fermentation, and nutrient digestibility in dairy cows, *J. Dairy Sci.* 99(8), 6191–6205.

Palmquist, D. L. and Jenkins, T. C. (2017). A 100-year Review: fat feeding of dairy cows, *J. Dairy Sci.* 100(12), 10061–10077.

Rabiee, A. R., Breinhild, K., Scott, W., Golder, H. M., Block, E. and Lean, I. J. (2012). Effect of fat additions to diets of dairy cattle on milk production and components: a meta-analysis and meta-regression, *J. Dairy Sci.* 95(6), 3225–3247.

Reed, K. F., Moraes, L. E., Casper, D. P. and Kebreab, E. (2015). Predicting nitrogen excretion from cattle, *J. Dairy Sci.* 98(5), 3025–3035.

Reynolds, C. K. (2006). Splanchnic metabolism of amino acids. In: Sejrsen, K., Hvelplund, T. and Nielsen, M. O. (Eds) *Ruminant Physiology. Digestion, Metabolism and Impact of Nutrition on Gene Expression, Immunology and Stress'*. The Netherlands: Wageningen Academic Publishers, pp. 225-248.

Reynolds, C. K., Crompton, L. A. and Mills, J. A. N. (2011). Improving the efficiency of energy utilisation in cattle, *Anim. Prod. Sci.* 51(1), 6–12.

Reynolds, C. K., Crompton, L. A., Mills, J. A. N., Humphries, D. J., Kirton, P., Relling, A. E., Misselbrook, T. H., Chadwick, D. R. and Givens, D. I. (2010). Effects of diet protein level and forage source on energy and nitrogen balance and methane and nitrogen excretion in lactating dairy cows. In: Corvetto, G. M. (Ed) Proceedings of the 3rd International Symposium on Energy and Protein Metabolism. EAAP Publ. No. 127. Wageningen, the Netherlands: Wageningen Academic Publishers.

Reynolds, C. K., Humphries, D. J., Kirton, P., Kindermann, M., Duval, S. and Steinberg, W. (2014). Effects of 3-nitrooxypropanol on methane emission, digestion, and energy and nitrogen balance of lactating dairy cows, *J. Dairy Sci.* 97(6), 3777–3789.

Rotz, C. A. (2018). Modeling greenhouse gas emissions from dairy farms', *J. Dairy Sci.* 101(7), 6675–6690.

Ruppert, L. D., Drackley, J. K., Bremmer, D. R. and Clark, J. H. (2003). Effects of tallow in diets based on corn silage or alfalfa silage on digestion and nutrient use by lactating dairy cows, *J. Dairy Sci.* 86(2), 593–609.

Russell, J. B. and Hespell, R. B. (1981). Microbial rumen fermentation, *J. Dairy Sci.* 64(6), 1153–1169.

Svihus, B., Uhlen, A. K. and Harstad, O. M. (2005). Effect of starch granule structure, associated components and processing on nutritive value of cereal starch: a review, *Anim. Feed Sci. Technol.* 122(3–4), 303–320.

Tapio, I., Snelling, T. J., Strozzi, F. and Wallace, R. J. (2017). The ruminal microbiome associated with methane emissions from ruminant livestock, *J. Anim. Sci. Biotechnol.* 8, 7.

Tempelman, R. J., Spurlock, D. M., Coffey, M., Veerkamp, R. F., Armentano, L. E., Weigel, K. A., de Haas, Y., Staples, C. R., Connor, E. E., Lu, Y. Y. and VandeHaar, M. J. (2015). Heterogeneity in genetic and nongenetic variation and energy sink relationships for residual feed intake across research stations and countries, *J. Dairy Sci.* 98(3), 2013–2026.

Thoma, G., Popp, J., Nutter, D., Shonnard, D., Ulrich, R., Matlock, M., Kim, D. S., Neiderman, Z., Kemper, N., East, C. and Adom, F. (2013). Greenhouse gas emissions from milk production and consumption in the United States: a cradle-to-grave life cycle assessment circa 2008, *Int. Dairy J.* 31, S3–S14.

Tyrrell, H. F., Moe, P. W. and Flatt, W. P. (1970). Influence of excess protein intake on energy metabolism of the dairy cow. In: Schurch, A. and Wenks, C. (Eds) *Energy Metabolism of Farm Animals*,. EAAP Publ. No. 13. Zurich: Juris Verlag pp.69–72.

USEPA. (2021). *Inventory of U. S. Greenhouse Gas Emissions and Sinks: 1990–2019*. Washington, DC: United States Environmental Protection Agency.

Uwizeye, A., de Boer, I. J. M., Opio, C. I., Schulte, R. P. O., Falcucci, A., Tempio, G., Teillard, F., Casu, F., Rulli, M., Galloway, J. O., Leip, A., Erisman, J. W., Robinson, T. P., Stenfeld, H. and Gerber, P. J. (2020). Nitrogen emissions along global livestock supply chains, *Nat. Food* 1(7), 437–446, https://doi.org/10.1038/s43016-020-0113-y.

Van Nevel, C. J. and Demeyer, D. I. (1996). Control of rumen methanogenesis, *Environ. Monit. Assess.* 42(1–2), 73–97.

Van Soest, P. J. (1994). *Nutritional Ecology of the Ruminant* (2nd edn.). Ithaca, NY: Cornell University Press.

VandeHaar, M. J. (1998). Efficiency of nutrient use and relationship to profitability on dairy farms, *J. Dairy Sci.* 81(1), 272–282.

VandeHaar, M. J., Armentano, L. E., Weigel, K., Spurlock, D. M., Tempelman, R. J. and Veerkamp, R. (2016). Harnessing the genetics of the modern dairy cow to continue improvements in feed efficiency, *J. Dairy Sci.* 99(6), 4941–4954.

VandeHaar, M. J. and St-Pierre, N. (2006). Major advances in nutrition: relevance to the sustainability of the dairy industry, *J. Dairy Sci.* 89(4), 1280–1291.

Waghorn, G. C., Macdonald, K. A., Williams, Y., Davis, S. R. and Spelman, R. J. (2012). Measuring residual feed intake in dairy heifers fed an alfalfa (Medicago sativa) cube diet, *J. Dairy Sci.* 95(3), 1462–1471.

Wallace, R. J., Snelling, T. J., McCartney, C. A., Tapio, I. and Strozzi, F. (2017). Application of metaomics techniques to understand greenhouse gas emissions originating from ruminal metabolism, *Genet. Sel. Evol.* 49(1), 9.

Wilkerson, V. A., Glenn, B. P. and McLeod, K. R. (1997). Energy and nitrogen balance in lactating cows fed diets containing dry or high moisture corn in either rolled or ground form, *J. Dairy Sci.* 80(10), 2487–2496.

Wilkinson, J. M. and Garnsworthy, P. C. (2017). Dietary options to reduce the environmental impact of milk production, *J. Agric. Sci.* 155(2), 334–347.

Williams, Y. J., Pryce, J. E., Grainger, C., Wales, W. J., Linden, N., Porker, M. and Hayes, B. J. (2011). Variation in residual feed intake in Holstein-Friesian dairy heifers in southern Australia, *J. Dairy Sci.* 94(9), 4715–4725.

Yang, W. Z., Beauchemin, K. A. and Rode, L. M. (2001). Effects of grain processing, forage to concentrate ratio, and forage particle size on rumen pH and digestion by dairy cows, *J. Dairy Sci.* 84(10), 2203–2216.

Chapter 8

Improving grassland/forage quality and management to reduce livestock greenhouse gas emissions

Michael O'Donovan, Teagasc, Ireland

1 Introduction

Pasture-based systems of production face multiple challenges, including global food demand, continuing to balance environmental sustainability, and ensuring product quality meets the highest consumer standards of nutrition and health. The challenge ahead for grassland farmers is imposing, but system sustainability and reducing methane emissions by improving management and forage quality is an aspect of pasture farming that needs to be addressed intensely by all elements of the industry. European grasslands sustain a large number of domestic herbivores, 150 million cows and 150 million sheep, which is roughly 15% of the global animal production. Grazers impact the cycling of C and N within pastures via defoliation, excretal returns and mechanical disturbance. They emit CO_2 via their metabolic activity and methane (CH_4) through enteric fermentation. Improving grassland management practise on farms is a key avenue of addressing further reductions of CH_4 emissions at farm level. There are many ways of

http://dx.doi.org/10.19103/AS.2020.0077.11

completing this successfully, and the challenge is to ensure these methods are employed at farm level through:

i optimising the use of home-grown grass;
ii increasing the proportion of grazed grass in the overall diet of the grazing ruminant;
iii extending the grazing season, especially in spring and autumn, thereby reducing the indoor feeding period;
iv adopting a strategy of grazing lower pre-grazing herbage masses during the grazing season (adapting sward morphology);
v grazing swards with higher levels of clover content, targeting levels of >20% in the sward, on average, across the grazing season;
vi improving grass dry matter intake (DMI), milk solids and average daily gain at grass; and
vii measurement of pasture quality at farm level across the growing season and appropriate supplementation patterns based on these figures.

These themes are discussed in the rest of the chapter.

2 Grassland areas and productivity in Europe

Europe maintains one of the highest livestock densities in the world. In 2018, the 27 member states of the European Union (EU-27) produced: 26%, 13%, 22%, 12% and 11% of the world's milk, beef, pork, poultry and eggs, respectively (FAO, 2009). Following considerable growth in the 1960s and 1979s, cattle numbers in Europe have been decreasing since 1980. More recently, reductions in animal production in Europe are related to market developments as well as to changes in agricultural and environmental policies. Between 1970 and 2013, 5.9 million ha of grassland in the EU9 were lost to other purposes (Table 1), equating to a 15.3% reduction in the proportion of permanent grassland. The proportions of permanent grassland reduced significantly in Germany (31.8%), the Netherlands (30.1%), France (28.1%), Denmark (25.9%) and Belgium (21.7%), while the reduction in Italy was much lower (13.3%). There was a small reduction in Ireland (2.2%) and Luxembourg (0.8%), while there was an increase in the United Kingdom (6.4%). Similar reductions in grazing land areas accrued in Austria, Greece, Spain and Hungary.

Grassland productivity will be affected by botanical composition, soil characteristics, climate conditions, altitude, latitude and management (De Vliegher and Carlier, 2007). Lee (1983) reviewed grassland potential productivity data from most European countries, and it was divided into five major geographical/climatic regions:

Table 1 Changes in the proportion of utilised agricultural area in permanent grassland area in EU9 between 1970 and 2013

		1970		2013	
	Total LAA	Permanent grassland (ha)	Proportion (%)	Permanent grassland (ha)	Proportion (%)
Belgium	1 307 900	621 253	47.5	486 539	37.2
Denmark	2 619 340	264 553	10.1	196 451	7.5
France	27 739 430	11 456 385	41.3	8 238 611	29.7
Luxembourg	131 040	67 486	51.5	66 961	51.1
Germany	16 699 580	6 780 029	40.6	4 625 784	27.7
Ireland	4 959 450	4 007 236	80.8	3 917 966	79
Italy	12 098 890	3 823 249	31.6	3 315 096	27.4
The Netherlands	1 847 570	1 104 847	59.8	772 284	41.8
United Kingdom	17 096 170	10 138 029	59.3	10 787 683	63.1
Total	84 499 370	38 263 066	46.9	32 407 374	40.5

Source: Eurostat; authors own calculations.

1 North-west and west Europe;
2 Central Europe;
3 South-east Europe;
4 Mediterranean Europe; and
5 Northern Europe.

Factors that are considered to influence the grassland productivity are altitude, water stress, temperature and aspects such as slope and soil depth.

A coordinated experiment organised under the auspices of the FAO Lowland Grassland Sub-network measured the production and productivity of cutting grassland according to a standardised protocol in 32 European sites. The average annual production ranged from 10 t DM ha^{-1} to 14 t DM ha^{-1} (Peeters and Kopec, 1996). The extremes in grass production were very different ranging from 2 t DM ha^{-1} in Portugal to 20 t DM/ha in Germany (Kiel) (Peeters and Kopec, 1996; Fig. 1). The most productive sites (>15 t DM ha^{-1}) were located on the Atlantic side of Europe between 52°N and 57°N latitude.

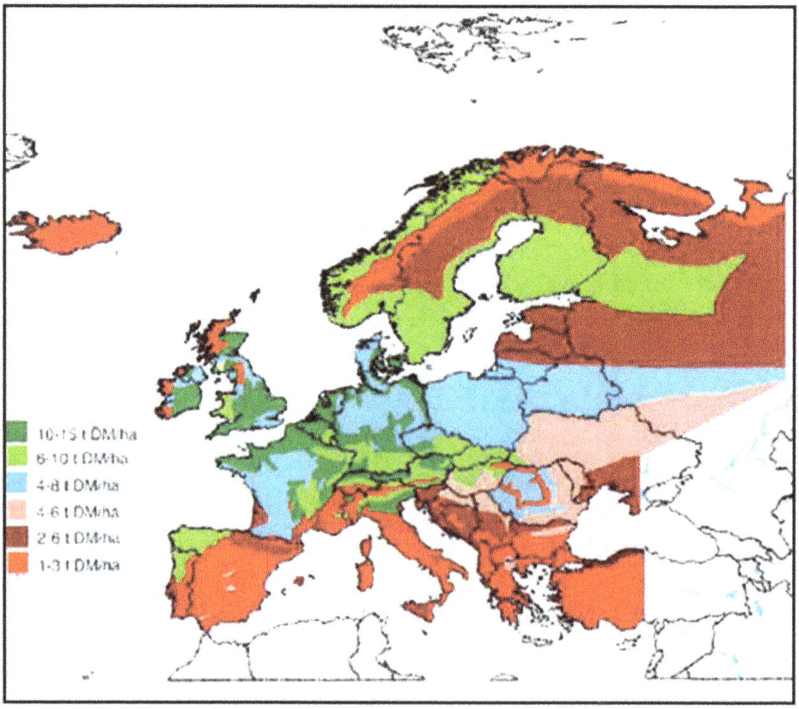

Figure 1 Production potential (annual yields in t DM/ha) of mown and heavily fertilised grasslands (source: Peeters and Kopec, 1996).

These included the Netherlands, Great Britain, Ireland, Belgium, north-western France and northern Germany. The less-productive sites were situated at high or low latitudes in Europe.

The EU (28 countries) currently has a permanent grassland area of about 60 million ha (Eurostat, 2017). Permanent and temporary grasslands represent 40% of the total utilised agricultural area in Europe (Huyghe et al., 2014) and a large acreage of these grasslands is exclusively used as ruminant feed, as either grazed grass or grass silage/hay. This asset of grasslands is extremely important for the human population since ruminants deliver food for humans by converting human-inedible plant biomass into high-quality human edible proteins. By providing feed to ruminants, grasslands contribute to the feeding of member state populations. Grass-based ruminant production delivers a number of other services to society, like carbon (C) sequestration (e.g. Soussana et al., 2010; Conant et al., 2017) and biodiversity (e.g. Isselstein et al., 2005; Van den Pol-van Dasselaar et al., 2019)).

Under climatically and topographically favourable conditions, the European grasslands area has been significantly reduced during the last 30 years (Huyghe et al., 2014). According to the 3rd report of the EU MAES initiative (Mapping of Ecosystems and Ecosystem Services), between 2006 and 2012 the main causes for this process were the conversion of grasslands into arable crops like maize (including for the production of biofuels) and other crops (32% of the lost area), the sprawl of urban areas, economic sites and infrastructures (30%), and the withdrawal from farming (17%) (Erhard et al., 2016). In many countries, the number of dairy cows decreased in the last 30 years but the milk yield of individual cows increased during the same period, with the number of cow reductions mainly driven by the implementation of the milk quota regime. Between 2010 and 2016, however, the bovine population slowly grew again by 1.4% (Eurostat, 2017).

The improvement in individual animal milk production is achieved based on an increasing amount of concentrates and maize in cow rations and declining herbage use from grassland (e.g. Isselstein et al., 2005). More and more farmers have changed to all-year housing and do not provide access to grazing for their cows, for example, and only

- 42% of German dairy cows have access to pasture (Gurrath, 2011); and
- 25% of Danish dairy cows have access to pasture (Van den Pol-van Dasselaar, 2016).

Such a decline in the access to grass has led to increases in milk production costs at farm level, and a worrying dependence on imported feed inside the farm gate; if continued it will lead to more GHG from the ruminant population.

3 The challenge of greenhouse gas emissions from livestock

Greenhouse gas emissions (GHG) from livestock are closely related to ruminant numbers. Factors other than ruminant numbers also have an impact. The size and productivity of animals affects their feed intake and enteric CH_4 emissions. Rarely, if ever, have the levels of grassland management, and the type of grassland offered to the grazing animal, been considered as a mitigation factor for CH_4 production. We know clearly that ruminants produce CH_4 during enteric fermentation of feed and CH_4 and N_2O are released from stored manure. Farm-based studies indicate that there are large differences among farms in terms of animal productivity and environmental impacts. These differences are often related to the management skill of the farmer, technologies applied and environmental conditions. We will discuss the possibilities to reduce CH_4 levels through better grazing management practises and the choice of more appropriate pasture species, given the changes in European grassland area and practise over recent years.

Ruminants lose between 2 and 12% of ingested energy as CH_4. Improvement in forage quality and more specifically forage digestibility has been investigated as a means of enteric CH_4 mitigation (Hristov et al., 2013). Structural carbohydrates have been reported to be more methanogenic than soluble carbohydrates. In ruminants with high feed intakes, reductions in enteric CH_4 emissions per unit intake with increased digestibility of feeds have been reported (Hristov et al., 2013). Greater digestibility is associated with a fermentation profile in the rumen that is unfavourable to CH_4 production. Hristov et al. (2013) stated a more digestible feed is associated with greater intake and production, diluting maintenance energy requirements and resulting in less CH_4 per unit of animal product. The literature is full of studies having evaluated feeds in indoor feeding systems based on concentrates and forage diets. Grazing studies with CH_4 emissions measured are currently scarce but increasing.

4 Grazing management to combat climate change: grazing season

4.1 Grazing season length and impact of climate change

Implementing good grazing management practises to improve the quality of pastures will increase animal productivity and lower CH_4 per unit of product (Boadi et al., 2004). A long grazing season can increase the annual proportion of grazed grass in ruminant diets, which can reduce feed costs and increase profitability (Dillon et al., 2005; Finneran et al., 2012). Grazing is generally positively perceived by consumers when compared to indoor feeding systems

(Van den Pol-van Dasselaar et al., 2019). Phelan et al. (2015) investigated the spatial variation in grazing season lengths from 32 European countries obtained from the results of the EUROSTAT Survey on Agricultural Production Methods (SAPM) and bioclimatic variables for dairy farms. The reference year was 2012 for all countries with the exception of Spain and Portugal which had 2009 as the reference year. Grazing season length was positively correlated with mean temperature during the coldest quarter and negatively correlated with precipitation in the wettest month. Figure 2 illustrates the observed and predicated grazing season lengths for dairy farms in all 32 European countries. The predicted grazing season lengths were longer than observed in Belgium, Estonia, Germany, Hungary and the Netherlands but shorter in Bulgaria, France, Latvia and Lithuania.

4.2 Early season grazing

Grazed grass can be increased in the overall diet of the dairy cow by allowing cow's access to grass early in spring; this is an opportunity for all member states. Many studies have shown an improvement in milk production and composition with this practise (O'Donovan et al., 2004; Kennedy et al., 2006). As well as improving animal performance, early spring grazing can have beneficial effects

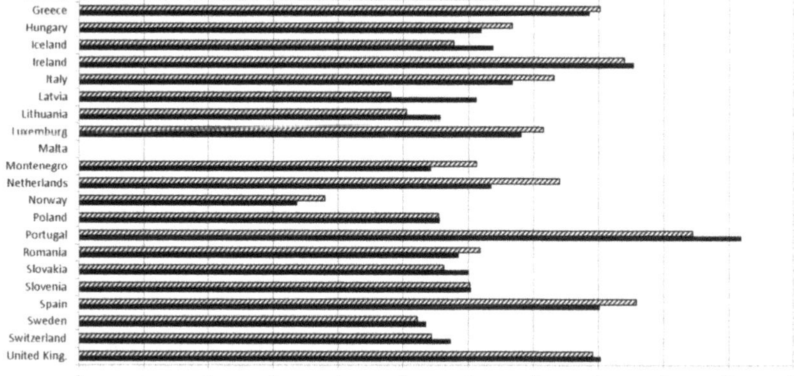

Figure 2 Observed and predicted grazing season length (months) for 32 European countries (source: Phelan et al., 2015).

including increasing grass utilisation, sward quality and simplifying grazing management. Late turnout to grass can lead to under-grazing of pastures for a variety of reasons, for example, excessively high pre-grazing herbage mass, low-grazing stocking rates or poor grass utilisation conditions. O'Donovan et al. (2004) found that early spring grazing can act as a sward conditioner, that is, avoids build-up of excessively high pre-grazing yield. They found high milk production from early grazed swards even with a low grass allowance compared to late-grazed swards, clearly showing the beneficial effects of early grazing on sward structure and quality. The early use of grass reduces the need for supplementary feed, making better use of home-grown feed. Early spring grazing reduces the large requirement for machinery, fuel and fertilisers, and facilitating less GHG production on farms.

O'Neill et al. (2011) and Robertson and Waghorn (2002) found that in early lactation pasture-fed cows produced less CH_4 emissions per day than cows offered a total mixed ration (TMR) diet. The higher DMI exhibited by TMR cows is likely to have caused the increased CH_4 emissions by these cows.

Increasing DMI increases CH_4 production as greater DMI provides a greater intake of fermentable substrate, including both structural and non-structural carbohydrates (Moe and Tyrrell, 1980). In addition, saliva is an important rumen buffer (Owens et al., 1998) and higher saliva production may give rise to increased rumen pH which would maintain favourable rumen fermentation conditions for fibre digestion and methanogenesis (Krause et al., 2002). Pinares-Patiño et al. (2007) found that in their study saliva production was positively correlated with higher daily CH_4 emissions and feed intake.

5 Grazing management to combat climate change: sward structure and quality

5.1 Sward structure characteristics

The presented herbage mass very much dictates the grazing intensity and level of DMI achieved by the grazing animal. Pre-grazing herbage mass also dictates the source from which the animal selects its diet (Wade, 1991). The presented herbage mass influences the sward characteristics presented to the grazing animal. Grazing animals prefer living to dead material, younger to older material, leaf to stem and legume leaves to grass leaves (Leaver, 1985). Michell and Fulkerson (1985) showed that pre-grazing herbage mass increased at the sward base in the latter part of the season, due to stem and dead material accumulation (Mayne et al., 1987). High levels of death and decay accompany poor herbage utilisation. Curran et al. (2010) agreed with those findings, and his work showed that high pre-grazing herbage mass swards had greater accumulations of dead material in the second half of the season due to a lax post-grazing height and low levels of grass utilisation. Korte et al. (1984) suggested that a high grazing intensity reduced the production and development of reproductive tillers. Fulkerson and Donaghy (2001) stated that severe defoliation removes too much water soluble carbohydrate (WSC) storage capacity and reduces regrowth, while under lax defoliation, the loss of DM through leaf senescence and reduced rates of tillering are not compensated for by the increased growth rate that results.

Criteria for determining when to defoliate pastures have been based on rotation length, sward height and pre-grazing herbage mass (HM) (Mayne et al., 2000). Pre-grazing herbage mass takes account of sward height and density, and is an animal sward interface indicator of when the sward is ready to graze. Previous studies by O'Donovan et al. (2004) and Kennedy et al. (2006) showed that high HM swards supported greater stocking rates; however, grazing low HM swards had a positive effect on herbage quality, milk production and grass DMI. Curran et al. (2010) found that low herbage mass swards supported high stocking rates due to a greater number of grazing rotations and improved sward quality due to intense grazing. O'Donovan et al. (2004) found that herbage from early grazed swards (February/March) was of higher quality (increased organic matter digestibility (OMD) and UFL value) relative to late grazed swards (April). This in part reflects seasonal trends in the accumulation of dead herbage. Wade (1991) found sward stem (true and pseudostem) as a barrier to increasing grass DMI, and this is true when cows are forced to graze to low post-grazing residuals. Sheath resistance increases in importance as a barrier to intake as pre-grazing herbage mass and pre-grazing height increase.

Wims et al. (2010) found that dairy cows grazing high pre-grazing herbage mass swards increased their CH_4 production per cow per day (+42 g), per kg of milk yield (+3.5 g/kg), per kg milk solids (+47 g/kg) and per kg grass DMI (+3.1 g/kg). The main difference in these swards was a 10-day difference in rotation length. In both early and late season measurements, Wims et al. (2010) found that cows grazing high HM swards lost a greater proportion of their gross energy intake as CH_4 during both measurement periods (+0.9% and +1% for summer and autumn, respectively). Wims et al. (2010) offered low HM swards to cows which maintained higher grass DMI and milk output; however, while no significant differences were found in CH_4 production, the conclusion was that grazing lower pre-grazing herbage mass swards tended to reduce CH_4 per unit of DMI (−8.2%) and energy corrected milk yield (−10%) compared to grazing high HM swards. Both these studies agreed that offering lower HM swards to grazing cows constituted a viable CH_4 mitigation strategy.

5.2 Maintaining sward quality

Ruminant CH_4 originates from the digestible fraction of the diet rather than the whole diet and fermentation of cell wall carbohydrates (NDF) produces more CH_4 than fermentation of soluble sugars (Moe and Tyrrell, 1980). Blaxter and Clapperton (1965) reported relationships between dietary factors and CH_4 emissions for indoor-fed animals. These authors stated that absolute emissions (g d^{-1}) and CH_4 yield (% of gross energy intake) increase and decrease, respectively, as feed intake increases above maintenance requirements, but both absolute CH_4 emissions and CH_4 yield decreases with increasing digestibility. Plant maturity is the most important factor affecting the morphology and forage quality. As maturity stage increases, the proportion of cell wall (neutral detergent fibre (NDF)) contents increases, whereas the proportion of cell contents decreases. The NDF contents are negatively correlated to rates of digestion and passage and herbage intake (fill effect).

Hammond et al. (2011) found 0.87 of the variation in total enteric CH_4 emissions of grazing sheep was predicted by OM intake, and the relationship between forage chemical composition and total CH_4 emissions and CH_4 yield (g/kg of DM intake) were weak. O'Neill et al. (2011) found that cows grazing a high-quality perennial ryegrass diet had lower CH_4 per unit of feed intake than cows that offered a TMR of lower digestibility. The effect of forage quality on CH_4 emissions most likely depends on the extent of contrast in forage quality between treatments in the completed studies.

Figure 3 shows the impact of grass quality on the actual post-grazing sward height achieved over two grazing seasons by dairy cows grazing sward plots. It is clear that grass varieties with greater grass quality values have subsequently lower post-grazing sward height, meaning improved grass utilisation. Since 2013, Ireland has introduced the Pasture Profit Index (McEvoy et al., 2011), and key traits of this index are seasonal DM yield, pasture quality, silage DM yield and persistency. New traits such as grazing utilisation will be introduced into this index in the next year.

5.3 Grass dry matter intake at pasture

Achieving high grass utilisation consistently takes a considerable amount of grazing management discipline, and its application can be poor on farms. Within the typical range of herbage allowance (HA) in grazing systems, herbage intake increases on average by 0.10–0.15 kg/kg HA at ground level and 0.20–0.25 kg/kg HA above 4–5 cm (Delagarde et al., 2011). This means that the marginal response of pasture utilisation rate when increasing HA is very small (15–25%). There is a consensus that increasing feed intake reduces CH_4 yield as g CH_4/kg DMI although its effect is larger with high- versus low-quality diets

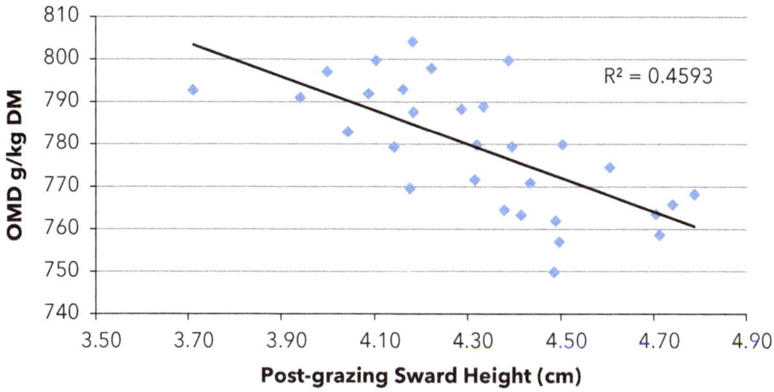

Figure 3 The effect of pasture quality (OMD) on post-grazing sward height over the grazing season (Tubritt, personal comm).

(Blaxter and Clapperton, 1965). Combined with good grassland management, allocating the correct daily HA and pre-grazing HM, CH_4 from grazing dairy cows can be reduced. It is important to maintain a high-quality diet, as possible, when grazing.

5.4 Grass clover swards with dairy cows

Higher animal performance from swards containing white clover can be expected because of its superior nutritional value compared to perennial ryegrass due to higher crude protein content (Butler, 2000), lower structural fibre values (Thomson, 1984) and higher OMD (Wilman and Riley, 1993). White clover is a potential option to reduce CH_4 emissions because of higher grass DMI and higher milk production (McClearn et al., 2019). Grass-white clover has lower aNDF and high voluntary DMI compared to grass-only (Ulyatt, 1970; Enriquez-Hilalgo et al., 2014). Lower aNDF concentration in white clover may lead to rapid ruminal degradation and passage rate, which should lower CH_4 yield as g CH_4/kg DMI.

A balance between the optimum sward white clover content for milk production and pasture production must be achieved to optimize both animal and pasture performance. Within the sward, white clover proportion changes seasonally; to achieve consistent clover content is a real challenge in grazing management. It is difficult to maintain optimum levels of white clover in the sward because of climatic factors (drought, waterlogging, colder soils), poor grazing managements, suboptimal soil fertility and pests. Dineen et al. (2018) completed a meta-analysis from a number of studies which had white clover included in the diet of grazing dairy cows. The mean sward white clover content was 31.6%, mean daily milk yield and milk solids yield per cow were increased by 1.4 kg and 0.12 kg, respectively, milk and milk solids yield were unaffected when cows grazed grass clover compared to grass-only swards.

McClearn et al. (2019) created swards with an average annual sward white clover content of 23.6% and 22.6% in tetraploid perennial ryegrass and white clover swards and diploid perennial ryegrass and white clover, respectively. Milk production did not differ between grass ploidies during a 4-year study, but cows grazing the perennial ryegrass-white clover treatments had significantly greater milk yield (+597 kg/cow per year) and milk solids yields (+48 kg/cow per year) compared with cows grazing the perennial ryegrass-only swards. Increased milk output has been associated with higher herbage nutritive value for perennial ryegrass-white clover swards, especially mid-season, compared to perennial ryegrass-only swards (Soegaard, 1993) and an increase in voluntary herbage DMI (Ribeiro Filho et al., 2005) with numerous studies having shown selective grazing of white clover over perennial ryegrass (Rutter et al., 2004). McClearn et al. (2019) found the difference in milk production from the perennial ryegrass-white clover swards was observed from May onwards in each year. This pattern was consistent with white clover content in the sward increasing as the season progressed and is similar to what Woodward et al. (2001) and Egan et al. (2018) reported. Andrews et al. (2007) suggested that sward-white clover content greater than 20% is required to establish an animal production response. Establishing such white clover proportions on farms will be and is a real challenge. Egan (personal comm) has shown high clover germination rates on farms but low establishment. He found that successful rates of establishment were only 60% across all farms, 12 months after over seeding. Direct reseeding perennial ryegrass-white clover swards is the only reliable way of establishing these swards on farms.

Methane emissions related to gross energy intake of animals-fed legumes are lower than animals-fed grasses (Beauchemin et al., 2008). Some previous research has shown that clover inclusion in pasture can reduce dairy cow CH_4 emissions and others have found no effects (van Dorland et al., 2007). Enriquez-Hilalgo et al. (2014) found that an average annual white clover content of 20% was not sufficient to improve overall sward production, quality or dairy cow productivity. The cows grazing the grass-white clover swards had a tendency to consume more and emitted less CH_4 than cows grazing the grass-only swards.

Structural carbohydrates are fermented at slower rates than non-structural carbohydrates such as starch and sugars to yield more CH_4/unit substrate fermented. Slower rumen fractional outflow rates and higher rumen volumes increase rate of CH_4 emission, most probably by allowing increased digestion of structural carbohydrates and providing a better environment for the growth of methanogens. The higher readily fermentable carbohydrate:structural carbohydrate ratio in white clover compared with perennial ryegrass may decrease the rumen acetate:propionate ratio, which is expected to lower CH_4 yield.

6 Grazing management to combat climate change: legume forages

Legume forages have a substantially higher nutritive value than grasses, biologically fix N and the condensed tannin (CT) containing legumes birdsfoot trefoil and sainfoin (*Onabrychis vicifolia* Scop.) and may have further advantages over non-tannin containing forages such as alfalfa. Sainfoin contains higher concentrations of condensed tannins (50–80 g/kg DM) compared to 5–47 g/kg DM for birdsfoot trefoil. Condensed tannins bind strongly to proteins and it has been proposed that some plants evolved CT production as a chemical defence, first against invasion by pathogenic microorganism, then against being eaten by insects and finally against being eaten by grazing herbivores. Originally it was thought that CT-containing forages were the Lotus species, sulla (*Heddysarium coronarium*) and sainfoin. Newer technologies have shown the presence of CT in grasses, legumes and herbs (Table 2).

Bloat is caused by very high solubility of forage proteins leading to the development of stable foam in the rumen, and it is prevalent in cattle and sheep-fed legumes. Because of their protein-precipitating properties, grazing CT-containing legumes has long been known to eliminate bloat. Recently it has been proposed that the plant CT concentration needed to make forage bloat safe was 5 g CT/kg DM or greater. Most common legumes and grasses used in temperate agriculture have CT concentrations well below this value (Table 2). It will be a challenge to raise CT levels through plant breeding, but it is on the agenda as a breeding goal of the major grass and legume breeding companies worldwide. In temperate systems the challenge will be to integrate such legumes consistently into a sward growth pattern.

Both chicory (*Cichorium intybus L.*) and plantain (Plantago lanceolata L.) are suitable herbs for inclusion into swards managed under intensive grazing because of their high yield potential and ability to maintain sward quality mid-season (Cranston et al., 2015). The use of multispecies swards containing chicory and plantain is of particular interest on sites which are prone to soil moisture deficits, because of their deeper root systems and greater drought tolerance (Lee et al., 2015). Swards containing chicory and plantain can support higher levels of animal performance particularly in summer and autumn because of its ability to maintain sward quality in comparison to grass-white clover swards (Cranston et al., 2015). Grazing management, such as rotation length and post-grazing sward height, affects the growth, persistence and nutritive value of swards containing chicory and plantain (Lee et al., 2015). Multispecies swards containing clover and herbs can persist under grazing for 3–5 years, but are

Table 2 The extractable and bound condensed tannin content of legumes, grasses and herbs fed to ruminants in temperate grazing systems, measured by the butanol-HCl method

Forage	Extractable	Condensed Protein-bound	Tannin (g/kg DM) Fibre-bound	Total
Legumes				
Big trefoil (*Lotus pedunculatus*)	61	14	1	77
Birdsfoot trefoil (*Lotus corniculatus*)	36	9	2	47
Sulla (*Hedysarum coronarium*)	33	9	3	45
Sainfoin (*Onabrychis vicfolia*)	29			
Red clover (*Trifolium pratense*)	0.4	0.6	0.7	1.7
Lucerne (*Medicago sativa*)	0.0	0.5	0.0	0.5
Grasses				
Perennial ryegrass (*Lolium perenne*)	0.8	0.5	0.5	1.8
Herbs				
Chicory (*Chicorium intybus*)	1.4	2.6	0.2	4.2
Sheep's burnet (*Sanguisorba minor*)	1.0	1.4	1.0	3.4

suited to a 3–4-week rotation length with lax grazing to 8 cm (Cranston et al., 2015). These grazing guidelines are very much in contrast with those necessary for modern perennial ryegrass varieties. Their role in more modern grazing systems is currently under evaluation in a number of countries.

7 Grazing management to combat climate change: measurement issues

7.1 CH_4 measurement at grazing

Respiration chambers (RC) have been considered the gold standard for measuring enteric CH_4 emissions from farm animals, but this is only the case if RC are operated properly and recoveries are fixed and preferably close to 100%. Animals in RC must have stable daily feed intake. Approximately 30% of today's CH_4 emissions are as a result of the previous day's DMI. Daily variation in DMI can cause variation in CH_4 emissions.

Compared to housed animals, grazing animals have higher energy requirements due to the added cost of walking and grazing. There is continual debate about measurement methods – the standard method for CH_4 measurement is open-circuit respiration calorimeter chambers – yet this bears no relation to the activity of the grazing animal, when grazing the animal selects their chosen herbage, is exposed to the variances in climate, and has to forage, this is the opposite to the behaviour in an enclosed chamber, where the diet selection is somewhat compromised and supply is regular. Data from chambers cannot be applied to all farm situations; this is why the SF_6 and Green Feed (GF) system need to be more widely used in grazing systems.

Variability with the SF_6 method has been notoriously high but modifications by Deighton et al. (2014) addressed the most important sources of error, and the modified technique produced CH_4 measurements with accuracy similar to measurements using RC. Some of the variation with SF_6 seems intrinsic to the technique because the estimated CH_4 emission rate appears sensitive to factors that affect the proportions of exhaled and eructated air in the air samples and distance of the sampling point from mouth to the mouth/nostrils (Berends et al., 2014). A recently introduced technique for direct measurement of enteric CH_4 emissions is the automated head chamber system, GF, which was developed for spot sampling for exhaled and eructated gases (Zimmerman and Zimmerman, 2012). When used properly with repeated animal measurements, GF can be a reliable technique for measuring enteric CH_4 emissions from ruminants. An important prerequisite for decreasing uncertainty of the measurement when using GF is that all animals visit the unit at times that enable estimation of the diurnal pattern of CH_4 emission over successive 24 h period. For accurate daily emissions estimates, animal visits need to be distributed appropriately over the 24-h feeding cycle and a number of repeated days of measurements are required for each animal. Both GF and SF_6 methods are established techniques, and they can provide accurate estimates of enteric CH_4 emissions when properly used and calibrated. Direct comparisons of both techniques have shown acceptable agreement (Grainger et al., 2007; Huhtanem et al., 2018), and both provide valuable data required for grazing systems.

7.2 Life cycle assessment (LCA) with pasture-based diets

Comparisons of results between life-cycle assessment (LCA) studies is difficult due to differences in computation of life cycle inventories and choice of functional unit (de Boer, 2003; Van der Werf et al., 2009). In many previous LCA studies of grass and confinement systems, CH_4 from enteric fermentation was identified as the main cause of on-farm GHG emission (Casey and Holden,

2005). O'Brien et al. (2012) found that GHG emissions from cultivation of purchased concentrate and forage, emissions associated with the manufacture of inputs, that is, fertiliser used in the production of on-farm forages, were the main contributors to off-farm GHG emissions. Off-farm GHG emissions from confinement systems were more than double the off-farm GHG emission from grass-based system in the O'Brien et al. (2012) study, because of the greater quantity of concentrate feed. Consequently, total GHG emission per unit of milk produced was greater for the confinement system relative to the grass-based system.

Grassland systems can be further improved with strategic use of concentrate feed. Cederberg and Mattison (2000) suggested that the environmental impact of concentrate feed could be reduced by using domestic or regionally produced rapeseed meal rather than imported soya bean meal. Concentrate impacts can be reduced further by lowering the crude protein levels in concentrate by more appropriate diet formulation, in line with the chemical value of the diet, for example, avoiding the use of higher concentrate feeds. This would in turn have positive impacts on reducing N losses and further enhance N-use efficiency. There is much more to be learned about grassland and its seasonal quality fluctuations if there is more emphasis on pasture quality assessment across the growing season.

8 Conclusion

The continual decline in the area of permanent grassland in Europe is not assisting farmers in developing better grassland management practise and many countries are growing their dependence on imported feed (both forage and concentrate). The focus on farm needs to be to optimise the level of home-grown feed in the ruminant's diet.

None of the aspects of grassland management solely will deliver key movements in reducing CH_4 emissions; however, some of the gains that can be made are interacting with one another. High-grazed grass utilisation systems can be effective in this role, not all countries can adapt totally to this system type, but at different stages in the growing season, grass can be capitalised upon. Table 3 shows the effects of new and improved innovation and applications to reduce CH_4 emissions. Some of the improvements are gradual, improvement in animal productivity (based on proper national breeding goals), adapting grassland management, using appropriate grass varieties and clovers. The key aspect of such changes is that they have to be established inside the farm gate, herein is a crucial challenge for the grassland/ruminant sectors. Using home-produced feed, much of the efficiency in CH_4 can be harnessed as set out at the start of this chapter. Diet manipulation through adopting different grazing strategies that improve the forage quality available to the herds is

Table 3 Summary of methane mitigation strategies for grazing animals

Strategy	Potential CH_4 reduction	Technology availability/ feasibility	Production cost benefits
Improving animal productivity	20-30%	Feasible and practical	Increased feed cost Increased milk production Use of fewer animals Less feed per kg of milk
Forage species and maturity	20-25%	Feasible	Increased feed efficiency Increased milk production
Rotational grazing of animals/early grazing	9% or more	Feasible	Increased feed intake Increased milk production
Managed grazing of animals versus confined feeding	25% or more	Feasible needs more investigation	Cheaper feed costs Supplement use Reduced milk fat/ protein content Higher net return
Use of high-quality forages/pastures	25% or more	Feasible	Increased feed intake Increased milk production
Genetic selection (use of high net feed efficiency animals)	21%	Long-term feasibility	Decreased feed intake Increased feed efficiency

Source: Boadi et al. (2004).

readily available to farmers; however, many times it is not clearly advocated by the wider industry. Including grass and legumes in the diet of the grazing ruminant does not require major investment, but an overall change in mind-set. It is clearly a CH_4 mitigation factor and is positively viewed by the wider consumer population.

9 Where to look for further information

While the number of organisations across the world and indeed in Europe completing grazing research combined with measurement of GHG emssions is relatively small, there is a recently assembled research infrastructure consortium in Europe called the Smartcow network (www. Smartcow.eu). Ireland (Teagasc), United Kingdom (SRUC and University of Reading), France (INRAE), Belgium (CRAW), Denmark (Aarhus University), Germany (FBN), the Netherlands (Wageningen University) and Spain (IRTA) are collaborating partners of the consortium, their objective to ensure a commonality in research approach, measurements and protocols in methane research and other research intiatives. This type of infrastructure model will become more important into the future as it will allow a more smooth comparison of research techniques across countries, and this is really important with the techniques involved in GHG measurement at grazing.

10 References

Andrews, M., Scholefield, D., Abberton, M. T., McKenzie, B. A., Hodge, S. and Raven, J. A. (2007). Use of white clover as an laternative to nitrogen fertiliser for dairy pastures in nitrate vulnerable zones in the UK; Productivity, environmental impact and economic considerations. *Annals of Applied Biology* 151(1), 11-23.

Beauchemin, K. A., Kreuzer, M., O'Mara, F. and McAllister, T. A. (2008). Nutritional management for enteric methane abatement; A review. *Australian Journal of Experimental Agriculture* 48(2), 21-27.

Berends, H., Gerrits, W. J. J., France, J., Ellis, J. L., van Zijderveld, S. M. and Dijkstra, J. (2014). Evaluation of the SF6 tracer technique for estimating methane emission rates with references to dairy cows using a mechanistic model. *Journal of Theoretical Biology* 353, 1-8.

Blaxter, K. L. and Clapperton, J. L. (1965). Prediction of the amount of methane production in ruminants; a modelling approach. *Canadian Journal of Animal Science* 81, 563-574.

Boadi, D., Benchaar, C., Chiquette, J. and Masse, D. (2004). Mitigation strategies to reduce enteric methane emissions from dairy cows. Update review. *Canadian Journal of Animal Science* 84(3), 319-335.

Butler, W. R. (2000). Nutritional interactions with reproductive performance in dairy cattle. *Animal Reproduction Science* 60-61, 449-457.

Casey, J. W. and Holden, N. M. (2005). The relationship between greenhouse gas emissions and the intensity of milk production in Ireland. *Journal of Environmental Quality* 34(2), 429-436.

Cederberg, C. and Mattison, B. (2000). Life cycle assessment of milk production - a comparison of conventional and organic farming. *Journal of Cleaner Production* 8(1), 49-60.

Conant, R. T., Cerri, C. E. P., Osborne, B. B. and Paustian, K. (2017). Grassland management impacts on soil carbon stocks: a new synthesis. *Ecological Applications: A Publication of the Ecological Society of America* 27(2), 662-668.

Cranston, L. M., Kenyon, P. R., Morris, S. T. and Kemp, P. D. (2015). A review of the use of chicory, plantain, red clover and white clover in a sward mix for increased sheep and beef production. *Journal of New Zealand Grasslands* 77, 89-94.

Curran, J., Delaby, L., Kennedy, E., Murphy, J. P., Boland, T. M. and O'Donovan, M. (2010). Sward characteristics, grass dry matter intake and milk production performance are affected by pre grazing herbage mass and pasture allowance. *Livestock Science* 127(2-3), 144-154.

de Boer,I. J. M. (2003). Environmental impact assessment of conventional and organic milk production. *Livestock Production Science* 80(1-2), 69-77.

De Vliegher, A. and Carlier, L. (2007). The effect of the age of grassland on yield, botanical composition and nitrate content in the soil under grazing conditions. *Grassland Science in Europe* 13, 51-54.

Deighton, M. H., Williams, S. R. O., Hannah, M. C., Eckard, R. J., Boland, T. M., Wales, W. J. and Moate, P. J. (2014). A modified sulphur hexafluoride tracer technique enables accurate determination of enteric methane emissions from ruminants. *Animal Feed Science and Technology* 197, 47-63.

Delagarde, R., Faverdin, P., Baratte, C. and Peyraud, J. L. (2011). GrazeIn; a model of herbage intake and milk production for grazing dairy cows. 2. Prediction of intake under rotational and continuously stocked grazing management. *Grass and Forage Science* 66, 61-77.

Dillon, P., Roche, J. R., Shalloo, L. and Horan, B. (2005). Optimising financial return from grazing in temperate pastures. Utilisation of grazed grass in temperate animals systems. In: Murphy, J. J. (Ed.), *Proceedings of the a Satellite Workshop of the XXth International Grassland Congress, Cork, Ireland*. Wageningen Academic Publishers, The Netherlands, 131-148.

Dineen, M., Delaby, L., Gilliland and McCarthy, B. (2018). Metaanalysis of the effect of white clover inclusion in perennial ryegrass swards on milk production. *Journal of Dairy Science* 101, 1804-1816.

Egan, M., Galvin, N. and Hennessy, D. (2018). Incorporating white clover (Trifolium repens L.) into perennial ryegrass (Lolium perenne L.) swards receiving varying levels of nitrogen fertiliser: effects on milk and herbage production. *Journal of Dairy Science* 101(4), 3412-3427.

Enriquez-Hilalgo, D.,Gilliland, T. J., Deighton, M. H., O'Donovan, M. and Hennessy, D. (2014). Milk production and enteric methane emissions by dairy cows grazing fertilised perennial ryegrass pasture with or without inclusion of white clover. *Journal of Dairy Science* 97(3), 1400-1412.

Erhard, M., Teller, A., Maes, J., Meiner, A., Berry, P., Smith, A., Eales, R., Papadopoulou, L., Bastrup-Birk, A., Ivits, A., Royo Gelabert, E., Dige, G., Petersen, J.-E., Reker, J., Cugny-Seguin, M., Kristensen, P., Uhel, R., Estreguil, C., Fritz, M., Murphy, P., Banfield, N., Ostermann, O., Abdul Malak, D., Marín, A., Schröder, C., Conde, S., Garcia-Feced, C., Evans, D., Delbaere, D., Naumann, S., Davis, M., Gerdes, H., Graf, A., Boon, A., Stoker, B., Mizgajski, A., Santos Martin, F., Jol, A., Lükewille, A., Werner, B., Romao, C., Desaulty, D., Larsen, F. W., Louwagie, G., Zal, N., Gawronska, S. and Christiansen, T. (2016) *Mapping and Assessment of Ecosystems and Their Services*. 3rd Report - Final, March 2016. Technical Report - 2016 - 095. European Union, European Environment Agency, 183 pp.

Eurostat (2017). *Agriculture, Forestry and Fishery Statistics* (2017 edn.), 165 pp.

FAO (Food and Agricultural Organisation of the United Nations) (2009). Global agriculture towards 2050. High Level Expert Forum Issues Paper. Food and Agriculture Organization, Rome.

Finneran, E., Crosson, P., O'Kiely, P., Shalloo, L., Forristal, D. and Wallace, M. (2012). Stochastic simulation of the cost of home produced feeeds for ruminant livestock systems. *The Journal of Agricultural Science* 150, 123-139.

Fulkerson, W. J. and Donaghy, D. J. (2001). Plant-soluble carbohydrate reserves and senescense-key criteria for developing and effective grazing management system for ryegrass-based pastures;a review. *Australian Journal of Experimental Agriculture* 41(2), 261–275.

Grainger, C. T., Clarke, S. M., McGinn, M. J., Auldist, K. A., Beauchemin, M. C., Hannah, G. C., Waghorn, G. C., Clark, H. and Eckard, R. J. (2007). Methane emissions from dairy cows measured using the sulphur hexafluoride (SF_6) tracer and chamber techniques. *Journal of Dairy Science* 90, 2755–2766.

Gurrath, P. (2011). *Landwirtschaft auf einen Blick*. Statistisches Bundesamt, Wiesbaden, Germany.

Hammond, K. J., Hoskin, S. O., Burke, J. L., Waghorn, G. C., Koolaard, J. P. and Muetzel, S. (2011). Effects of feeding fresh white clover (Trifolium repens) or perennial ryegrass (Lolium perenne) on enteric methane emissions from sheep. *Animal Feed Science and Technology* 166–167, 398–404.

Hristov, A. N., Oh, J., Firkins, J. L., Dijkstra, J., Kebreab, E., Waghorn, G., Makkar, H. P. S., Adesogan, A. T., Yang, W., Lee, C., Gerber, P. J., Henderson, B. and Tricarico, J. M. (2013). Mitigation of methane and nitrous oxide emissions from animal operations; I. A review of enteric Methans emissions mitigation options. *Journal of Animal Science* 91(11), 5045–5069.

Huhtanem, P., Ramin, M. and Hristov, A. N. (2018). Comparison of methane production measured by the Greenfeed system and predicted by empirical equations. *Journal of Dairy Science*, https://doi.org/10.3168/jds.2017-14218.

Huyghe, C., De Vliegher, A., van Gils, B. and Peeters, A. (2014). Grasslands and herbivore production in Europe and effects of common policies. Quae Éditions, Versailles, 287 pp.

Isselstein, J., Jeangros, B. and Pavlů, V. (2005). Agronomic aspects of biodiversity targeted management of temperate grasslands in Europe - a review. *Agronomy Research* 3, 139–151.

Kennedy, E., O'Donovan, M., Murphy, J. P., O'Mara, F. P. and Delaby, L. (2006). The effect of spring of initial spring grazing date and subsequent stocking rate on the grazing management, grass dry matter intake and milk production of dairy cows in summer. *Grass and Forage Science* 61(4), 375–384.

Korte, C. J., Watkin, B. R. and Harris, W. (1984). Effects of timing and intensity of spring grazings on reproductive development, tillering and herbage productionof perennial ryegrass dominant pasture. *New Zealand Journal of Agricultural Research* 27(2), 135–149.

Krause, K. M., Combs, D. K. and Beauchemin, K. A. (2002). Effects of forage particle size and grain fermentability in midlactation cows. II. ruminal pH and chewing activity. *Journal of Dairy Science* 85(8), 1947–1957.

Leaver, J. D. (1985). Milk production from grazed temperate grassland. *Journal of Dairy Research* 52(2), 313–344.

Lee, J. (1983). The spatial distribution of grassland production in Europe. In: *Proceedings of the 9th General Meeting of the European Grassland Federation* (Rothamsted), 11–20.

Lee, J. M., Hemmingson, N. R., Minnee, E. M. K. and Clark, C. E. F. (2015). Management strategies for chicory (Cichorium intybus) and plantain (Plantago lanceolata): impact on dry matter yield, nutritive characteristics and plant density. *Crop and Pasture Science* 66(2), 168–183.

Mayne, C. S., Newberry, R. D., Woodcock, S. C. F. and Wilkins, R. J. (1987). Effect of grazing severity on grass utilisation and milk production of rotationally grazed dairy cows. *Grass and Forage Science* 42(1), 59–72.

Mayne, C. S., Wright, I. A. and Fisher, G. E. J. (2000). Grassland management under grazing and animal response. In: Hopkins, A. (Ed.), *Grass: Its Production and Utilisation*. Centre for Agriculture and Biosciences, New York, 247–291.

McClearn, B., Gilliland, T. J., Delaby, L., Guy, C., Dineen, M., Coughlan, F. and McCarthy, B. (2019). Milk production per cow and per hectare of spring calving dairy cows grazing swards differing in *Lolium perenne* L. ploidy and *Trifolium repens* L. composition. *Journal of Dairy Science* 102(9), 8571–8585.

McEvoy, M., O'Donovan, M. and Shalloo, L. (2011). Development and application of an economic ranking index for perennial ryegrass cultivars. *Journal of Dairy Science* 94(3), 1627–1639.

Michell, P. and Fulkerson, W. J. (1985). Effect of level of utilization in spring on pasture composition in summer and on milk production in sping and summer. In: Philips, T. J. (Ed.). *The Proceedings Conference*. Albury-Wodong, Australia, 66–67.

Moe, P. W. and Tyrrell, H. F. (1980). Methane production in dairy cattle. In: *Energy metabolism: Proceedings of the 8th Symposium Energy Metabolic*. Butterworths, London, 59.

O'Brien, D., Shalloo, L., Patton, J., Buckley, F., Grainger, C. and Wallace, M. (2012). A life cycle assessment of seasonal grass based and confinement dairy farms. *Agricultural Systems* 107, 33–46.

O'Donovan, M., Delaby, L. and Peyraud, J. L. (2004). Effect of time initial grazing date and subsequent stocking rate on pasture production and dairy cows performance. *Animal Research* 53(6), 489–502.

O'Neill, B. F., Deighton, M. H., O'Loughlin, B. M., Mulligan, F. J., Boland, T. M., O'Donovan, M. and Lewis, E. (2011). Effects of a perennial ryegrass diet or total mixed ration diet offered to spring calving Holstein-Friesian dairy cows on methane emissions. *Journal of Dairy Science* 94, 1941–1951.

Owens, F. N., Secrist, D. S., Hill, W. J. and Gill, D. R. (1998). Acidosis in cattle: a review. *Journal of Animal Science* 76(1), 275–286.

Peeters, A. and Kopec, S. (1996). Production and productivity of cutting grasslands in temperature climates of Europe. *Grassland Science in Europe* 1, 59–73.

Phelan, P., Morgan, E. R., Rose, H., Grant, J. and O'Kiely, P. (2015). Predications of future grazing season length for European dairy, beef and sheep farms based on regression with bioclimatic variables. *Journal of Agricultural Science* 154(5), 765–781 10.1017/S0021859615000830.

Pinares-Patiño, C. S., Waghorn, G. C., Machmüller, A., Vlaming, B., Molano, G., Cavanagh, A. and Clark, H. (2007). Methane emissions and digestive physiology of non-lactating dairy cows fed pasture forage. *Canadian Journal of Animal Science* 87(4), 601–613.

Ribeiro Filho, H. M. N., Delarge, R. and Peyraud, J. L. (2005). Herbage intake and milk yield of dairy cows grazing perennial ryegrass swards of white clover/perennial ryegrass at low and medium herbage allowances. *Animal Feed Science and Technology* 119, 13–27.

Robertson, L. J. and Waghorn, G. (2002). Dairy industry perspectives on methane emissions and production from cattle fed pasture or total mixed rations in New Zealand. *Proceedings of the New Zealand Grassland Association* 62, 213–218.

Rutter, S. M., Orr, R. J., Yarrow, N. H. and Champion, R. A. (2004). Dietary preference of dairy cows grazing ryegrass and while clover. *Journal of Dairy Science* 87(5), 1317–1324.
Soegaard, K. (1993). Nutritive value of white clover. In: *White Clover in Europe: State of the Art*. RUER Technical Series 29. Complied by J. Framen. Food and Agriculture Organization, Rome, Italy, 17–23.
Soussana, J. F., Tallec, T. and Blanfort, V. (2010). Mitigating the greenhouse gas balance of ruminant production systems through carbon sequestration in grasslands. *Animal: An International Journal of Animal Bioscience* 4(3), 334–350.
Thomson, D. (1984). The nutritive value of white clover. *Forage Legumes*, 78–92.
Ulyatt, M. J. (1970). Evaluation of pasture quality under New Zealand conditions. *Proceedings of the New Zealand Grassland Association* 32, 61–68.
Van den Pol-van Dasselaar, A. (2016). *Kijken met een weide blik*. Publicatienummer, 16-002PP. Aeres University of Applied Sciences, Dronten, 47 pp.
Van den Pol-van Dasselaar, A., Bastiaansen-Aantjes, B., O'Donovan, F. and Huyghe, M. (2019). *Grassland Use in Europe – A Syllabus for Young Farmers*. Editions Quae. Available at: https://www.quae-open.com/produit/123/9782759231461/grassland-use-in-europe.
Van der Werf, H., Kanyarushoki, C. and Michael Scott, C. (2009). An operational method for the evaluation of resource use and environmental impacts of dairy farms by life cycle assessment. *Journal of Environment Management* 90, 2643–3652.
van Dorland, H. A., Wettstein, H.-R., Leuenberger, H. and Kreuzer, M. (2007). Effect of supplementationof fresh and ensiled clovers to ryegrass on nitrogen loss and methane emission of dairy cows. *Livestock Science* 111(1–2), 57–69.
Wade, M. H. (1991). Factors affecting the availability of vegetative *Lolium perenne* to grazing cows with special reference to sward characteristics, stocking rate and grazing method. PhD Thesis. University of Rennes.
Wilman, D. and Riley, J. A. (1993). Potential nutritive value of a wide range of grassland species. *The Journal of Agricultural Science* 120(1), 43–50.
Wims, C. M., Deighton, M. H., Lewis, E., O'Loughlin, B., Delaby, L., Boland, T. M. and O'Donovan, M. (2010). The effect of pre grazing herbage mass on methane production, dry matter intake and milk production of grazing dairy cows during the mid-season period. *Journal of Dairy Science* 93(10), 4976–4985.
Woodward, S. L., MacDonald, K. A., Carter, W. A., Eerens, J. P. J. and Crush, J. R. (2001). Milksolids production from different combinations of perennial ryegrass and while clover cultivars. II. Milksolids production and farm productability. *Proceedings of the New Zealand Grassland Association* 63, 97–102.
Zimmerman, P. R. and Zimmerman, R. S. (2012). Method and system for monitoring and reducing ruminant methane production. United States Pat. No US20090288606 A1. P. R. Zimmerman, assignee.

Chapter 9

The use of plant bioactive compounds to reduce greenhouse gas emissions from farmed ruminants

Cécile Martin, Vincent Niderkorn, Gaëlle Maxin, INRAE, France; Jessie Guyader, INRAE-ADM NEOVIA, France; and Maguy Eugène and Diego P. Morgavi, INRAE, France

1 Introduction

Livestock farming activities account for 14.5% of global greenhouse gas (GHG) emissions of anthropogenic origin (Gerber et al., 2013). Concerning the ruminant livestock sector, the largest contribution is from cattle and sheep mainly in the form of methane (CH_4) and nitrous oxide (N_2O) emissions, which represent, in carbon equivalent, 44% and 30% of GHG emissions, respectively. Methane has a digestive (enteric) origin in ruminants and is mostly eliminated into the atmosphere by eructation. During the microbial fermentation process of feeds in the rumen (bacteria, protozoa and fungi), hydrogen (H_2) is produced and is immediately used by archaea methanogens to reduce carbon dioxide into CH_4. Nitrous oxide is produced in the soil during microbial processes (nitrification and denitrification) of urinary nitrogen (N) (urea and ammonia) excreted by ruminants (De Klein and Eckard, 2008; Selbie et al., 2015). Both enteric CH_4 emissions and urinary N waste represent loss of dietary energy (2-15%) and N (75-95%) ingested, which could be otherwise available for animal production (Hristov et al., 2013b). Therefore, decreasing enteric CH_4 emissions and N excretion from ruminants is important for reducing the environmental impact of ruminant production and for improving feed efficiency and the sustainability of this sector.

Compounds produced by the secondary metabolic processes of plants have been used for medicinal purposes by humans since antiquity (Wink, 2015).

http://dx.doi.org/10.19103/AS.2020.0077.13

Research on the use of compounds in animal production increased when the use of antibiotics as growth promoters was banned in Europe and other parts of the world in the mid-2000s. The potential use of plant bioactive compounds in animal nutrition to reduce CH_4 emissions and N waste is the subject of renewed interest as they are seen as a natural alternative to chemical additives and are well perceived by consumers. Some plant compounds have marked biological activity and, depending on their concentration in ruminant diets, can have positive or negative effects on animal responses. Plant bioactive compounds are promoted as improving health (are antiparasitic, reduce bloating and are antioxidant) and performance (N use efficiency). Conversely, they can decrease intake and diet digestibility (Mueller-Harvey, 2006) and can be toxic to animals (Reed, 1995).

We reviewed the current information on the use of plant bioactive compounds in ruminant nutrition to promote livestock farming that is not only more environment-friendly and efficient in the use of feed but also compliant with consumer demands for quality and safety in animal products. We focussed on the potential of plant bioactive compounds to mitigate enteric CH_4 production in ruminants and, when information is available, N waste. The main families of compounds considered as plant lipids are secondary compounds that are tannins, saponins, halogenated compounds and essential oils. Priority was given to information from *in vivo* studies by exploring the ability of plant compounds to positively modulate not only ruminant responses but also their mechanisms of action and utilization on farms. We selected two case studies showing the value of growing and using sainfoin forage in ruminant diets to decrease GHG emissions at the farm scale and combining dietary strategies with different modes of action to increase enteric CH_4 abatement. Future research on the use of plant bioactive compounds to reduce GHG emissions from farmed ruminants is also considered.

2 Families of plant bioactive compounds

2.1 Lipids

Lipids have a high nutritional value. The primary reason to use lipids in the diet of ruminants was to increase the potential production of animals and to improve the quality of meat and dairy products in terms of fat content and composition. However, an excessive dietary fat supplementation (> 7% dry matter [DM]) can affect microbial fermentation and fibre digestion in the rumen (Palmquist and Jenkins, 2017). The variable effects of lipids on ruminal fermentation are usually attributed to differences in their lipid structure (Bayat et al., 2018; Vargas et al., 2020). One factor is their degree of unsaturation because unsaturated fatty acids (from oleaginous oils or seeds and by-products such as residues from food processing plants) inhibit fermentation more than saturated fatty acids.

Commercial inert lipids (e.g. calcium salts of saturated fatty acids) are rumen bypass fats, which do not affect fibre digestion in the rumen at normal levels of supplementation in the diet.

Lipids are a proven dietary strategy for reducing CH_4 emissions from ruminants whose effectiveness depends on many factors, such as the dose, the source and the mode of distribution (Martin et al., 2010; Beauchemin et al., 2020).

Meta-analyses agree that the CH_4-mitigating effect of lipids are dose-dependent (Giger-Reverdin et al., 2003; Eugène et al., 2008; Grainger and Beauchemin, 2011; Doreau et al., 2011); the decrease in CH_4 emissions (g/kg DM intake [DMI]) varies between 1% and 7% per 10 g/kg of fat added DM in the diet. For low doses of added dietary lipids (< 2% DM), the mitigating effect was not systematic in cattle (Chung et al., 2011; Veneman et al., 2015). For high doses of added dietary lipids, the decrease is linear in cattle with rapeseed oil (0%, 5.4% and 9.5% added lipids; Jentsch et al., 1972) and coconut oil (0%, 1.3%, 2.7% and 3.3% added lipids; Hollmann et al., 2012) or quadratic in sheep with coconut oil (0%, 3.5% and 7% added lipids; Machmüller and Kreuzer, 1999). In dairy cows, many trials have shown the decreasing effect of lipids on methanogenesis (Martin et al., 2016; Bayat et al., 2018; van Gastelen et al., 2017) with different forage-based diets. The dose-response effect of extruded linseed (0%, 1.8%, 3.6% and 5.4% added lipids) on CH_4 emissions was more substantial with a corn silage-based diet compared to a hay-based diet (Martin et al., 2016). This more marked effect on methanogenesis was related to the adverse effects of lipids on animal performance (intake, digestibility and milk yield) with high doses of lipid supplementation.

The form of presentation of lipids also greatly influenced CH_4 output from dairy cows: inhibition of methanogenesis increased with the theoretical availability of linseed lipids in the rumen (oil > extruded seed > whole seed) (Martin et al., 2008). In practical conditions, extruded linseed is the most used form, because it is more readily available, easy to use, and less costly than oil and more efficient than crude linseed.

Concerning the source of lipids, conclusions of meta-analyses are not consensual: Grainger and Beauchemin (2011) did not find an effect of the source of lipids on CH_4 emissions, whereas medium chain and polyunsaturated fatty acids were reported to be more potent than others, according to the study of Doreau et al. (2011). In addition, the CH_4-mitigating effect of extruded linseed (2–3% added lipids) persisted for up to 1 year in dairy cows fed diets based on grazed pasture (80%) or grass silage (60%) (Martin et al., 2011). The persistency of this effect is very important for practical use.

The modes of action of lipids in the mitigation of ruminal methanogenesis are multiple (Martin et al., 2010). A common effect for all lipids is that when carbohydrates are substituted by lipids, as lipids are not fermented in the rumen, they do not contribute to H_2 production, unlike the carbohydrates they replace. Moreover, lipids have a toxic effect on some microbial populations (Popova et al., 2011; Vargas et al., 2020), more particularly on H_2-producing microbes (cellulolytic bacteria and protozoa), and reduce the metabolic activity of archaea methanogens by limiting H_2 availability and consequently CH_4 production. Ruminal biohydrogenation of unsaturated lipids may also help decrease methanogenesis, but the H_2 sink function of this biohydrogenation process was estimated to be negligible (i.e. 1–2%) based on stoichiometric (Czerkawski, 1986) and modelling (Mills et al., 2001; Giger-Reverdin et al., 2003) approaches.

Adding fat supplements in a proper dose to ruminant diets is a real opportunity to persistently reduce enteric CH_4 emissions without altering animal performance. This dietary strategy can be immediately implemented on commercial farms, especially if the quality of the meat and milk produced is improved, as is the case with unsaturated lipid sources. Most of the time, lipids are added in mixed diets as oil, oilseeds or food by-products, which limits their utilization for grazing ruminants. Notwithstanding, the use of lipids as feed ingredients in ruminant diets is relatively costly; if they are not locally produced, they may have a higher carbon footprint, which should be considered before adopting this approach. The combination of lipids with other dietary strategies has the potential to further reduce enteric CH_4 emissions (see Section 3).

2.2 Secondary compounds

2.2.1 Tannins

Among the polyphenolic compounds, special emphasis has been placed on the effects of tannins as they may act at several levels to reduce GHG emissions from ruminants.

First, the well-known ability of tannins to bind dietary proteins and reduce rumen proteolysis results in an increase of N duodenal flow and a shift from urinary to faecal N excretion (Aufrère et al., 2008; Theodoridou et al., 2010). As urinary N deposition results in N_2O emissions that are much higher than those arising from faecal N deposition (Luo and Kelliher, 2010), incorporating tannins in ruminant diets has great potential to decrease these emissions. In addition, a direct application of tannin extract (rather than feeding it) to barns reduces urease activity, thereby decreasing ammonia loss from dairy barn floors (Powell et al., 2011). Finally, if tannins are from forage legume species, such as sainfoin (*Onobrychis viciifolia*), birdsfoot trefoil (*Lotus corniculatus*) or sulla (*Hedysarum coronarium*), their ability to fix and transfer atmospheric N into the soil reduces the use of N fertilizers, which are a source of N_2O emissions through microbial nitrification and denitrification processes (Bouwman, 1996).

Second, abundant literature reports show that tannins in ruminant diets decrease enteric CH_4 emissions (review of Piluzza et al., 2014). The effect of these molecules on methanogenesis is highly variable between studies according to their nature (condensed or hydrolyzable), their chemical structure (molecular weight), especially the dose ingested by the animal, and the form of presentation. Given the extremely diverse structure of tannins in the plant kingdom, efforts have recently been made to understand their structure/activity relationships to facilitate their applicability (Mueller-Harvey et al., 2019). This aim was achieved, thanks to remarkable progress in the chemical analysis of both condensed (Zeller, 2019) and hydrolyzable tannins (Engström et al., 2019).

Using purified hydrolyzable (from chestnut and sumach) and condensed tannins (from mimosa and quebracho), Jayanegara et al. (2015) showed *in vitro* that hydrolyzable tannins have a greater effect in reducing CH_4 emissions with a less detrimental effect on digestibility than condensed tannins. Rira et al. (2019) reported the same conclusion with tropical forages tested *in vitro*: hydrolysable tannin-rich sources (*Acacia nilotica*) were more effective in supressing methanogenesis than condensed tannins-rich sources (*Calliandra calothryrsus* and *Leucaena leucocephala*). In addition, a combination of these plants did not highlight synergies between these two types of tannins. The type of molecular interaction that drives the protein-binding capacity of tannins is highly related to the structure of both tannins and proteins. Protein precipitation increases consistently with the mean degree of polymerization and tends to be higher with prodelphinidin-rich condensed tannins due to a greater number of potential hydrogen-bond participants available to interact with proteins (Zeller, 2019). The ability of hydrolyzable tannins to form insoluble complexes with the protein seems to be related not only to their molecular weight (oligomers are superior to monomers) but also to the type and number of functional groups (e.g. galloyl groups) in monomers (Engström et al., 2019).

Using purified condensed tannins of different structures from eight plants, Huyen et al. (2016b) showed *in vitro* that the proportion of prodelphinidins in condensed tannins had the largest effect on CH_4 production and fermentation characteristics, followed by the average polymer size. Other *in vitro* studies have shown that condensed tannins with a high degree of polymerization are more potent in lowering CH_4 production and the diversity and abundance of rumen methanogens (Hatew et al., 2016; Saminathan et al., 2016). Similarly, Baert et al. (2016) investigated *in vitro* how the degree of oligomerization of purified ellagitannins, an important family of hydrolyzable tannins, can influence their ability to alter ruminal fermentation including CH_4 production. They showed that large oligomers have more detrimental effects on gas production and volatile fatty acids (VFA) than small oligomers, while being similarly effective in their ability to decrease CH_4 production.

The meta-analysis conducted by Jayanegara et al. (2012), including data from a total of 30 experiments (both *in vitro* and *in vivo*), helped to partly clarify the underlying mode of action of tannins on methanogenesis. These authors reported that related CH_4 reduction is associated with reduced OM digestibility, especially fibre, because of a decreased number/activity or impaired substrate adhesion of fibrolytic microbes. This inhibitor effect of tannins on fibrolysis was more marked for condensed tannins than for hydrolysable tannins in dairy ewes (Buccioni et al., 2015). Recently, Costa et al. (2018) reported in sheep that gram-positive specialized fibrolytic bacteria (*R. albus, R. flavefaciens* and *B. proteoclasticus*) were more affected by condensed tannins than gram-negative bacteria (*F. succinogenes, S. ruminantium* and *P. bryantii*), with a decrease in rumen volatile fatty acids concentration, mostly acetate.

Methane emissions also declined when expressed per kg of digested organic matter (DOM), suggesting that other mechanisms account for the anti-methanogenic activity of tannins (Jayanegara et al., 2012). Tannins have been shown to directly inhibit H_2 using methanogens in the rumen of sheep (Liu et al., 2011) and beef cattle (Yang et al., 2017). This direct effect of tannins on methanogens microbiota, without affecting fibre digestion, would be more specific to hydrolysable tannins as reviewed by Vasta et al. (2019).

The potential of tannins to reduce methanogenesis has been widely reviewed in both *in vitro* and *in vivo* studies, thus highlighting the large variability of data (reviews of Piluzza et al., 2014 and Vasta et al., 2019). Archimède et al. (2016) observed a linear relationship between the condensed tannins content of three tropical-rich plants (*Glyricidia sepium, Leucaena leucocephala and Manihot esculenta*) and CH_4 reduction *in vivo*. The potential of mitigation ranged between 13% and 36% in sheep fed a forage diet containing between 1.5% and 4.0% DM of condensed tannins. The authors also reported better palatability (and intake) of tannin-rich tropical plants and a strong decrease in CH_4 emissions in animals fed the plants as pellets. Concerning hydrolysable tannins, their potential of mitigation ranged between 10% and 25% with sheep (Liu et al., 2011) and 11–30% with beef cattle (Yang et al., 2017) fed diets containing 1–3% DM of tannins from chestnut and pure hydrolysable tannins, respectively.

Few general equations of CH_4 prediction concerning tannins, mostly derived from *in vitro* trials, are available because of the diversity of the chemical analysis methods and the types of tannins (Jayanegara et al., 2012). We conducted a quantitative review of the literature by meta-analysis to assess the specific effect of tannins (condensed or hydrolyzable) on *in vivo* CH_4 emissions in ruminants (Eugène et al., 2019). Using an existing database (Methafour, INRA, 2018) on the effect of forages fed to ruminants on CH_4 emissions, we were able to significantly improve the accuracy of Eq. [1] based on the animal-feeding level and forage diet composition to predict CH_4 emissions, by taking into account the tannin content of the forage diets as mentioned in Eq. [2]:

- Eq. [1] for forage diets

$$CH_4 / DOM = 34.95 - 4.05FL + 0.027NDF - 0.010DOM$$
$$(n = 412, nexp = 153, RMSE = 3.1)$$

- Eq. [2] for forage diets containing tannins

$$CH_4 / DOM = 34.26 - 3.96FL + 0.027NDF - 0.008DOM - 1.72Log_{10}(1 + TAN)$$
$$(n = 398, nexp = 147, RMSE = 3.1)$$

where CH_4/DOM is the CH_4 production (g/kg DOM), FL is the feeding level (kg DM intake, % liveweight), NDF is the NDF content (g/kg DM), DOM is the DOM content (g/kg DM) and TAN is the tannin content (g/kg DM) of the diet, which is transformed on a logarithmic basis to account for its largely abnormal distribution (Sauvant et al., 2018).

The coefficients of regression of all variables remain stable between Eq. [1] and Eq. [2], highlighting the specific effect of tannins on methanogenesis. Based on current scientific knowledge, we propose to use the coefficient of TAN in Eq. [2] to evaluate the average quantitative effect of tannins *in vivo* on CH_4 emissions in all types of diets (Fig. 1). Our results confirm that CH_4 mitigation increases with the dose of tannins in the diets (Jayanegara et al.,

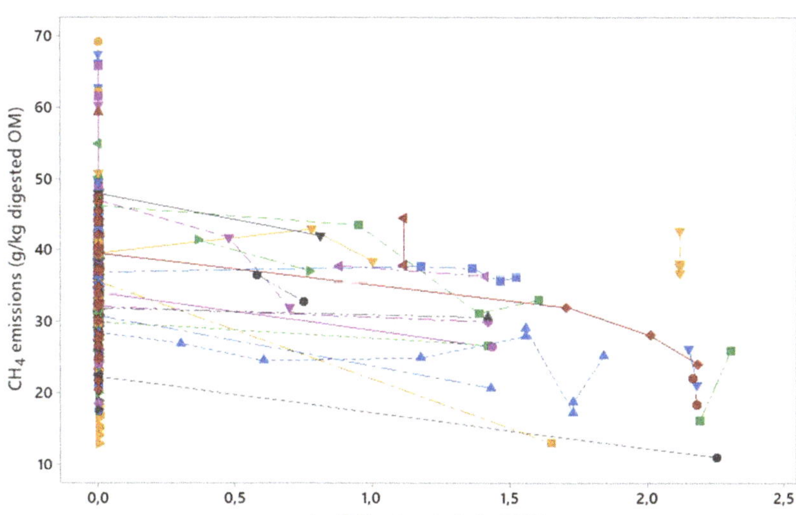

Figure 1 Relationship between methane emissions (g/kg DOM) and tannin content (g/kg DM) in the diet.

2012). Unfortunately it is not possible to give a minimum threshold of tannins content to observe an effect on methanogenesis because it is modulated by the FL, NDF and DOM contents of the diet.

Tannins are consumed by animals as plants or added as an extract to rations. The use of fodder-containing tannins is particularly relevant for grazing ruminants since many forage legumes are rich in tannins. New insights into the chemical structure of tannins help to explain the inconsistencies of the effects on protein-binding ability and on CH_4 emissions reported in the literature. Despite this progress, there are still questions to address concerning the use of tannin-rich resources on farms. Considering the variability in tannin activity, one solution could be the production of batches of feeds or plant extracts analysed before their marketing. Also, in the context of agro-ecological ruminant production systems, the use of integrated solutions such as tannin-containing legumes offers opportunities to act at multiple levels of GHG production (see Section 3).

2.2.2 Saponins

Saponins are secondary metabolites present in seeds, leaves and roots of a broad variety of plants. They are usually classified into two major classes, triterpenoids (soy, pea, garlic, sunflower and panama bark) and steroid glycosides (e.g. oat, eggplant, tomato, yucca and fenugreek), but Vincken et al. (2007) refined their classification in 11 main categories based on their carbon skeleton. Despite some negative effects upon feeding in animal nutrition (increased membrane permeability of erythrocytes and enterocytes, or impaired animal production and reproduction; reviews of Addisu and Assefa, 2016; Francis et al., 2007), saponins can have beneficial effects on rumen fermentation and animal health when used in a proper dosage. Among others, saponins can decrease *in vivo* degradability of feed protein, avoid N accumulation and increase efficiency of microbial protein synthesis in the rumen (Francis et al., 2007; Patra and Saxena, 2009). In addition, saponins from *Quillaja saponaria* (known as quillaja saponins), *Yucca schidigera* (known as yucca saponins) and *Camellia sinensis* or *assamica* (known as tea saponins) have been extensively studied for their mitigating effect on rumen methanogenesis. Other saponin sources have shown interesting CH_4-mitigating impacts: mangosteen peel powder (Wanapat et al., 2014) and alfalfa saponins (Klita et al., 1996), but these effects need to be confirmed.

The underlying mechanism mainly involves an inhibitory effect towards rumen microbes and more particularly protozoa, which produce large amounts of H_2 and are known to live in symbiosis with methanogenic archaea (Guyader et al., 2014; Morgavi et al., 2010). The membrane-disrupting activity of saponins would explain their toxic effect on protozoa through the formation of complexes with sterols present in the protozoal cell wall, thereby inducing cell lysis (Morgavi et al., 2010).

The CH_4 mitigation potential of saponins depends on the dose and source of saponins (Patra and Saxena, 2010). Most studies on this subject have been conducted *in vitro*. In a meta-analysis combining 23 studies, Jayanegara et al. (2014) reported a linear inhibiting dose-response effect of saponins towards methanogenesis (tested dosage between 0% and 0.6% DM). Compared with quillaja and tea saponins, yucca saponins induced the greatest reduction in CH_4 expressed as mL per unit of incubated substrate. However, when expressed as a percentage of total gas produced, all tested saponin sources were statistically similar and produced less CH_4 than the control. However, *in vivo* results are not as clear. Using up to 1.4% DM of quillaja saponins in the diet, Pen et al. (2007) and Holtshausen et al. (2009) did not observe a difference in CH_4 emissions of sheep and dairy cows, respectively. Patra and Saxena (2009) summarized published papers studying the *in vivo* effect of saponins, including yucca saponins, on fermentation parameters. Only two articles out of five showed a significant reduction (−14% in Santoso et al., 2004; −7% in Wang et al., 2009) in methanogenesis with sheep fed a diet containing 1.3% DM of yucca saponins. Similarly, the effect of tea saponins seems to be highly variable. With similar dosages comprising between 0.4% and 0.8% DM, three articles reported a decrease in CH_4 yield (g/kg DMI) on adult sheep (−17% in Yuan et al., 2007; −26% in Zhou et al., 2011) and on lambs (−69% in Mao et al., 2010). More recent papers showed an absence of effect on steers (Li and Powers, 2012) and non-lactating cows (Guyader et al., 2015), or even an increase in CH_4 emissions on lactating dairy cows (+18% in Guyader et al., 2017) with the same dosages. Concerning the effect of tea saponins on other digestion parameters, data are scarce but mostly indicate an absence of effect on nutrient digestibility or N balance. However, milk yield in dairy cows (Guyader et al., 2017) and average daily weight gain in growing steers (Li and Powers, 2012) decreased (−18% and −80%, respectively) as a consequence of lower intake in both studies (−12% and −27%, respectively). However, 0.4% tea saponin in a Chinese wild rye-based diet did not affect feed intake or growth of lambs (Mao et al., 2010).

Four main reasons might explain the variable effect of saponins on methanogenesis and limit their utilization in animal nutrition. The quality of saponins is an important criterion for their CH_4-mitigating efficiency (Jayanegara et al., 2014). Plant maturity, geographical area of production and extraction methods are three parameters affecting the final concentration and quality of saponins (Li and Powers, 2012). Denaturation of saponins might also be possible during pelleting processes. Indeed, a modification of the miscellaneous structure of quillaja saponins was observed after heating from 20°C to 60°C (Mitra and Dungan, 1997). Guyader et al. (2015) assumed that denaturation was one possible explanation for the lack of effect on CH_4 production of tea saponins fed to non-lactating dairy cows.

The transient effect of saponins on rumen microbes is another limitation for their utilization. After adaptation, rumen bacteria are able to separate the active compound of saponins (sapogenin) from the sugar moiety, leading to their inactivation (Ramos-Morales et al., 2017). Newbold et al. (1997) supplemented sheep with saponins from foliage of an African multipurpose tree, *Sesbania sesban*. Protozoa concentrations dropped by 60% after 4 days, but the population recovered after 10 days. However, the time for adaptation of rumen microbes seems to be dependent upon the source of saponins: the anti-protozoal effect of saponins from *Sapindus rarak* was persistent over 105 days in sheep (Wina et al., 2006). The chemical modification of their structure to avoid microbiota adaptation may maximize the CH_4-mitigating potential of saponins (Ramos-Morales et al., 2017).

The CH_4-mitigating response seems to be dependent on the composition of the basal diet. For instance, in a study with young Holstein males, Wang et al. (2019) concluded that changes in the ruminal microbial community with tea saponin supplementation were different between alfalfa-, hay- or soybean hull-based diets. Given that the protozoal community is strongly affected by the basal diet, Patra and Saxena (2009) assumed that the diet-dependent effect of saponins is related to their selectivity for specific protozoal species.

Finally, the last challenge in saponin utilization is their impact on feed intake. Reduced intake has been reported following dietary supplementation with tea saponins in lactating dairy cows (Guyader et al., 2017) and steers (Li and Powers, 2012). Nevertheless, this drawback is not systematic: among the 43 papers compiled by Patra and Saxena (2009), who did not include recent articles on tea saponins, only two reported a decrease in feed intake with yucca saponin supplementation.

The effects of saponin supplementation in the diets of ruminants are highly contrasted. The conditions in which yucca, quillaja and tea saponins reduce CH_4 emissions from ruminants must be refined (optimal dose, long-term persistency). In addition, given the wide variety of saponin structures, screening of other plants might highlight the beneficial effect of new sources available for localized markets. Before adoption by farmers, the potential effects of saponins on digestion efficiency and zootechnical performance should also be investigated in depth.

2.2.3 Halogenated compounds

Halogenated products (e.g. bromoform, dibromomethane, dichloromethane, bromochloroacetic acid, etc.) exist naturally in seaweed at different concentrations, and much more in red and brown algae than in green ones.

These compounds may be produced as defense against disease and marine herbivores, anti-oxidants or by-products of metabolic processes (Keng et al., 2020).

Different macroalgae have been shown to decrease *in vitro* CH_4 production effectively (Dubois et al., 2013; Machado et al., 2014; Kinley and Fredeen, 2015). Among 20 tropical species screened, the red macroalgae, *Asparagopsis taxiformis*, was identified as the most efficient (Machado et al., 2014). Low doses (2% OM incubated) of *A. taxiformis* almost eliminated *in vitro* CH_4 production (Machado et al., 2016a), without any effect on forage digestibility (Kinley et al., 2016) and without compromising other fermentation parameters at a 5% OM supplementation rate (Roque et al., 2019a).

The CH_4-mitigating effect of red seaweed *Asparagopsis* spp. *(A. taxiformis and A. armata)* was recently confirmed in three *in vivo* trials. Li et al. (2016) reported a consistent (over a 72-day period) and dose-dependent reduction in CH_4 emissions (−50% to −80%) when adding *A. taxiformis* at 1–3% of diet OM, respectively. In dairy cows, adding *A. armata* at 0.5% and 1% of diet OM reduced CH_4 emissions (−26% and −67%, respectively) over 21 days while compromising animal performances (milk yield and intake) only at the high dose (Roque et al., 2019a). A recent experiment in feedlot beef cattle, *A. taxiformis* was tested in a high grain diet at three inclusion levels (0.05%, 0.10% and 0.20% of diet OM) over a 90-day period (Kinley et al., 2020). Steers receiving 0.10% and 0.20% *A. taxiformis* demonstrated decreased CH_4 emissions up to −40% and −98% and demonstrated weight gain improvements of +53% and +42%, respectively. There was no negative effect on daily feed intake, feed conversion efficiencies or rumen function, and no residues or changes in meat-eating quality were detected.

Bromoform is the most abundant natural product in *Asparagopsis taxiformis* and thus has been identified as the compound involved in CH_4 reduction, even if a combination of the different compounds may play a role in this reduction (Machado et al., 2016b). Halogenated compounds in *Asparagopsis taxiformis* appear to act as structural analogues of coenzyme M and thus inhibit the final step of the methanogenesis pathway (Liu et al., 2011). It has been shown that the decrease in abundance of methanogens in the rumen was positively correlated with the decrease of methanogenesis and increase in H_2 emissions (Machado et al., 2018; Roque et al., 2019a). Emissions of bromoform into the atmosphere may occur during the growth of seaweed or during desiccation processes (Keng et al., 2020), which would prevent - or at least greatly hamper - the farming of red seaweed on a commercial basis.

Macroalgae have a tremendous potential to inhibit methanogenesis in ruminants at low doses of supplementation. *Asparagopsis* spp. are the most effective species. Further investigations are required to confirm a long-term persistency effect on methanogenesis and long-term safety in animal responses before adoption by farmers. In addition, widespread use of red seaweed for animal nutrition raises concerns about their contribution to biogenic halocarbon emissions and their impact on the environment (i.e. ozone depletion related to bromoform). The carbon footprint of each step of algae production (harvesting, drying process, delivery, etc.) also needs to be considered for upstream emissions.

2.2.4 Essential oils

In addition to the compounds considered in previous sections, there are other plant bioactive compounds, collectively known as 'essential oils', that have the potential to mitigate CH_4 and ammonia production in ruminants (Cobellis et al., 2016). The name essential oil is not specific; it mainly comprises a diverse group of terpene and phenylpropene compounds as well as organo-sulphur compounds (Benchaar and Greathead, 2011). There are thousands of compounds that are qualified as essential oils, and many of them have been tested *in vitro* (reviewed by Calsamiglia et al., 2007; Hart et al., 2008; Benchaar and Greathead, 2011; Cobellis et al., 2016). However, for multiple reasons only a handful of these compounds have been pursued in animal studies. Many of the compounds tested decreased methanogenesis through a general reduction in feed degradation and fermentation in the rumen and, therefore, are not further considered in this chapter. The effect of some compounds was observed at high doses not compatible with their incorporation (as an extract or as the plant containing the active component) in the diet. In addition, for some compounds or plants, there are issues of toxicity, palatability, cost and availability that preclude their utilization even for experimental purposes. Further, the majority of *in vivo* studies have tested the effects of essential oils on general production parameters and only a handful of them included measurements of enteric CH_4 emissions (Cobellis et al., 2016). In this section, we will focus on those plant secondary compounds that were tested *in vivo* for their anti-methanogenic activity.

Most metabolites tested to reduce CH_4 production in ruminants are naturally produced by plants to fend off microbial invasion. Compounds that are effective *in vitro*, such as eugenol, which is abundant in the essential oil of clove, and cinnamaldehyde, which is abundant in cinnamon (Macheboeuf et al., 2008; Patra and Yu, 2012), did not reduce CH_4 emissions when tested on dairy cows (Benchaar, 2015; Benchaar et al., 2015). Carvacrol is a monoterpene with a phenol ring structure that is abundant in oregano and thyme. Oregano (*Origanum vulgare*) leaves fed to lactating dairy cows at doses of 250, 500 or 750 g/d decreased CH_4 production by up to −40% for the medium dose (Tekippe et al., 2011; Hristov et al., 2013a). But the CH_4 measurements were done up to 8 h after feeding, and the authors noted that 24-h continuous measurement is needed to validate the results. In another study on dairy cows, the use of oregano extract mixed into the diet at 0.056% DM tended to reduce CH_4 yield (g/kg DMI) by −22% (Kolling et al., 2018).

Flavonoids are a class of plant secondary compounds that have antimicrobial, anti-inflammatory and anti-oxidative functions. These compounds have been extensively studied in animal nutrition (Olagaray and Bradford, 2019). In studies with ruminants, supplementation of diets with flavonoids from mulberry decreased CH_4 yield (g/kg DMI) by −11% in sheep (Ma et al., 2017). The main flavonoids of mulberry are quercetin glycosides and rutin, a glucorhamnoside of quercetin (Ju et al., 2018). In contrast, the use of pure rutin or rutin contained in buckwheat seeds did not have any effect on CH_4 emissions in dairy cows (Stoldt et al., 2016). Catechins, flavonoids contained in green tea leaves, decreased CH_4 emissions (g/kg digestible DMI) in dairy cows by −18% (Kolling et al., 2018). Green tea also contains saponins that may have a synergistic effect in reducing CH_4. Notwithstanding, a commercial purified catechin extract linearly decreased CH_4 emissions in sheep by 7–13% (Aemiro et al., 2016). Catechins have known antimicrobial activities including a toxic effect on protozoa (Aemiro et al., 2016), but they are also known H_2 sinks that can compete with CH_4 production in the rumen environment (Becker et al., 2014). For flavonoids, in general, it is noted that those that have anti-inflammatory functions in the host animal are not effective in reducing CH_4 emissions or modulating microbial fermentation in the rumen (Olagaray and Bradford, 2019).

Sinigrin is a glucosinolate found in some plants of the Brassicaceae family, such as black mustard and horseradish, which is naturally converted to allyl isothiocyanate when the plants are processed (Mohammed et al., 2004). The latter compound is responsible for the strong flavour of horseradish and low palatability if used as a feed additive (Mohammed et al., 2004). A coated additive would avoid the problem of palatability and provide a gradual release of the sinigrin. A cyclodextrin-coated horseradish oil added to the diet of steers decreased CH_4 emissions by −19%. Although the mechanism of action is not well understood, the number of methanogens also decreased significantly. A parallel *in vitro* study showed a large increase in H_2 associated with CH_4 reduction (Mohammed et al., 2004), similar to that observed with specific inhibitors of methanogens or methanogenesis, such as garlic or halogenated compounds.

A particular mention is made for organo-sulphur compounds from garlic (*Allium sativum*). Sulphur compounds in garlic have both general antimicrobial properties and are specific inhibitors of the enzyme hydroxymethylglutaryl-CoA (HMG-S-CoA) reductase, which is essential for the production of the cell wall of archaea methanogens. These compounds remarkably reduce CH_4 production *in vitro* (reviewed by Hart et al., 2008; Benchaar and Greathead, 2011). However, there are few reports describing the *in vivo* use of garlic oil or diallyl disulphide, the main component of garlic oil. A decrease of about

−8% (g CH_4/kg digestible OM intake) was reported in sheep supplemented with garlic extract (Ma et al., 2016), whereas garlic leaves, which are normally discarded after harvesting the bulbs, reduced emissions by −10% (g CH_4/kg DMI) in sheep (Panthee et al., 2017).

Garlic oil combined with linseed oil reduced emissions in lambs, but the effect cannot be ascribed solely to garlic oil (Saro et al., 2018). Similarly, dried garlic combined with mangosteen peel rich in tannins and saponins reduced CH_4 emissions in cattle (Manasri et al., 2012), but the effect is confounded. In contrast, no effect was observed in a study with diallyl disulphide, garlic oil or raw garlic (Klevenhusen et al., 2011; Patra et al., 2011). More recently, a commercial mixture of garlic and citrus extracts (Mootral) was tested in dairy cows with positive results (Roque et al., 2019b; Vrancken et al., 2019). These results are encouraging but should be confirmed with further studies. For instance, in the work of Roque et al (2019b) the reduction in CH_4 was observed in the last week of the 12-week study, but not before. In addition to the product mentioned previously, there are several commercial products based on mixtures of essential oils that have been tested for their CH_4-reducing activity. The most tested are CRINA Ruminants (DSM; mixture of essential oil components) and Agolin Ruminant (Agolin; mixture of coriander oil, eugenol, geranyl acetate and geraniol, among others) and XTRACT Ruminant (Pancosma; mixture of cinnamon, cloves and capsicum oleoresin from chili peppers). The first product showed no effect on CH_4 emissions in beef cattle (Beauchemin and McGinn, 2006; Tomkins et al., 2015). The effect of Agolin Ruminant on dairy cows was recently evaluated in a meta-analysis (Belanche et al., 2020). A total of 23 *in vivo* experiments and on-farm studies were identified in which the additive was supplemented at 1 g/d/cow. Of these, nine had records of enteric CH_4 that showed an average decrease of −8.8% in CH_4 production (g/d), of −12.9% in CH_4 yield (g/kg DMI) and −9.9% in CH_4 intensity (g/kg milk) without a negative effect on feed digestibility or milk yield. The effects were observed only after an initial period of adaptation of at least 4 weeks.

Anacardic acid is an akylphenolic compound that is found in the shell of the cashew nut. It has antimic activity, particularly against gram-positive bacteria. In the rumen, it decreased the numbers of H_2- and formate-producing bacteria such as *Ruminococcus flavefaciens*, *Butyrivibrio fibrisolvens* and *Treponema bryantii*, whereas succinate-producing bacteria such as *Prevotella* spp., *Selenomonas ruminantium*, *Anaerovibrio lipolytica* and *Succinivibrio dextrinosolvens* increased (Shinkai et al., 2012; Konda et al., 2019). Methanogen numbers also decrease with changes in the community composition (Shinkai et al., 2012; Kang et al., 2018). The use of cashew nut shell liquid (CNSL) as a feed additive reduced CH_4 emissions in Holstein cows by −19% and −38% in a dose-dependent manner (Shinkai et al., 2012). The anti-methanogenic effect of CNSL was also observed in Thai native cattle and buffaloes (Konda et al., 2019). These changes in CH_4 were observed along with increases in propionate and decreases in acetate in the rumen. The technical grade CNSL (t-CNSL) is the main by-product of the cashew industry that does not contain anacardic acid, as during the production process it is converted into cardanol. The utilization of t-CNSL was not as effective at reducing enteric CH_4 emissions in dairy cows (Branco et al., 2015), suggesting that anacardic acid is the main active component in CNSL.

Essential oils used to mitigate CH_4 emissions and N waste have been extensively studied mainly *in vitro*, but validation of the results *in vivo* is still scarce. The use of cocktails of molecules in most *in vivo* studies makes it difficult to identify active molecules and to understand the mechanisms of action. Work still needs to be done to demonstrate the effectiveness of certain essential oils and to consider their synergistic or antagonistic interaction with other compounds. For most compounds, there is also a need to assess their efficacy in the long term, not only in reducing CH_4 emissions but also in the production and welfare of animals, to facilitate adoption by farmers.

3 Case studies

3.1 Sainfoin, a traditional forage legume containing condensed tannins

Several methods can be used to deliver tannins to the animals, using crude extracts from plants, by-products, wood or whole plants. Agro-ecological and local solutions that integrate several dimensions of ruminant nutrition can decrease GHG emissions at multiple levels, while improving protein self-sufficiency and reducing inputs such as fertilizers and drug treatments (Soussana et al., 2015). In this context, one option is to include legume species containing tannins in animal diets. Recently, a European multidisciplinary research consortium of agronomists, plant breeders, ruminant nutritionists, veterinarians and experts in tannin chemistry focussed on sainfoin, a traditional forage legume (Mueller-Harvey et al., 2019). Here, we present the specific results of this project regarding the potential of sainfoin to decrease GHG emissions (Fig. 2).

A first interesting result was obtained at the field level, where symbiotic N fixation by sainfoin was shown to be comparable with major N-fixing species such as white and red clover without treatment with a commercial rhizobia product (Malisch et al., 2017). The authors concluded that sainfoin has great potential for cropping grass-legume mixtures with increased forage yields, especially when cutting frequency and N fertilizer input are low. These results indicate that sainfoin is relevant to local production of high-protein forage without applying excessive N fertilization leading ultimately to N_2O emissions.

When ensiled with grass, sainfoin preserves silage quality via increased fermentation intensity and reduction in protein degradation in the silos (Copani

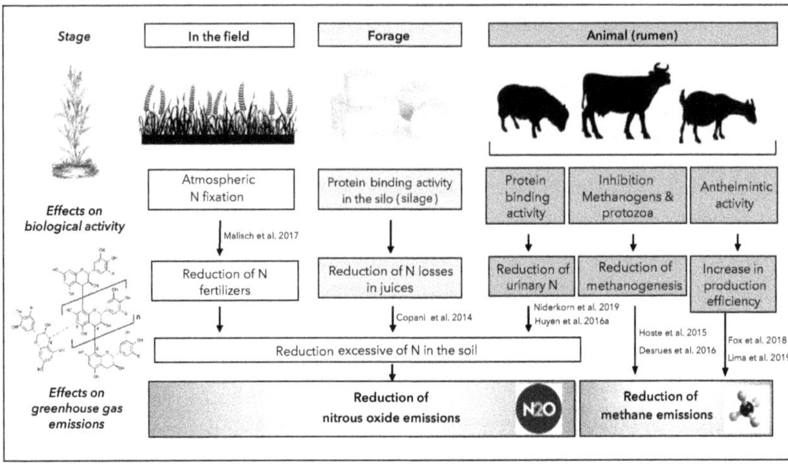

Figure 2 Multiple effects of growing and using sainfoin in ruminant diets to decrease greenhouse gas emissions.

et al., 2014). This was shown by a lower proportion of soluble N and ammonia (relative to total N) in silage compared to grass silage, which reduces N losses in fermentation juices and decreases the adverse impacts on the environment.

At the animal level, the decrease in CH_4 yield (g/kg DMI) and changes in N partition when sainfoin is incorporated in the diet were consistently observed in sheep (Niderkorn et al., 2019) and dairy cows (Huyen et al., 2016a). A particularly interesting result was obtained in dairy cows when a sainfoin-containing diet reduced CH_4 yield and diet digestibility of fibre but improved milk yield compared to the same diet in which sainfoin was replaced by grass. The authors hypothesized that sainfoin may redirect metabolism towards body protein accretion at the expense of body fat (Huyen et al., 2016a). Sainfoin-condensed tannins were shown to have anthelmintic activities in both small (Hoste et al., 2015) and large ruminants (Desrues et al., 2016), showing that a large spectrum of these compounds counteracts infection by gastrointestinal nematodes. This effect may help to decrease GHG emissions by animals, as shown by recent results indicating that parasitism increases CH_4 emissions in sheep (Fox et al., 2018; Lima et al., 2019).

3.2 Additive effect of different anti-methanogenic dietary strategies, a proof of concept

Lipids have emerged as a persistent option for mitigating enteric CH_4 emissions from ruminants (Doreau et al., 2014). However, their potential mitigation is moderate (~20%) if used at a suitable dose avoiding negative effects on animal performance (see Section 3.1).

As proof of concept, we tested whether it was possible to increase the CH_4-mitigation potential of lipids (linseed oil) by combining them with another dietary strategy (nitrate) with a different mode of action on the metabolism of H_2 in the rumen. In a meta-analysis, we reported that lipids may be relevant in reducing H_2 production (via reduction of protozoa), whereas nitrate may stimulate H_2 consumption (H_2 sink) by a competitive pathway to methanogenesis (Guyader et al., 2014). We assumed that simultaneous manipulation of H_2 production and H_2 utilization allows a greater reduction in CH_4 emissions than when acting on a single pathway. To test this hypothesis, we tested the effect of linseed oil and nitrate fed alone or in combination on CH_4 emissions and digestive processes in non-lactating cows.

The daily kinetics of CH_4 emission measurements clearly showed an additive effect of the dietary strategies (−32% reduction for linseed + nitrate vs. −17% and −22% reduction for linseed and nitrate fed alone, respectively). Linseed oil supplementation reduced CH_4 emissions throughout the day compared to the control diet, while nitrate had a transient but marked action for 3 h post-feeding. Combination of the strategies cumulated the two modes of actions (Fig. 3).

In addition, we showed that linseed oil plus nitrate fed to lactating cows for 2 months induced a constant reduction of CH_4 emissions (−29% g/kg DMI), without any effect on digestibility of nutrients, N balance and milk performance. This persistent effect showed the absence of adaptation of rumen microbiota. However, the energetic benefits from the decreased CH_4 emissions did not appear beneficial for dairy cows (Guyader et al., 2016).

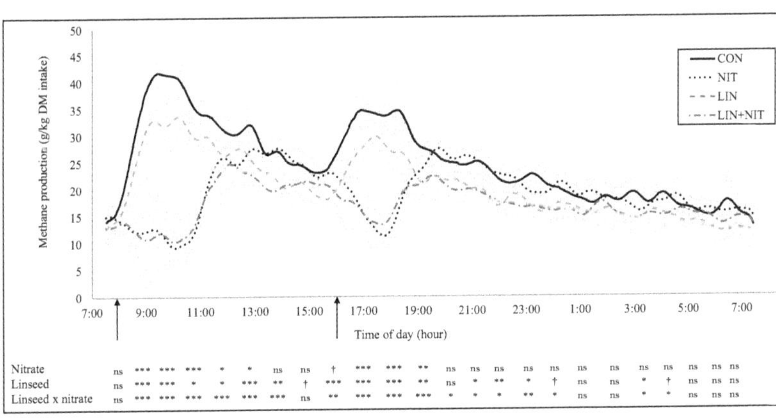

Figure 3 Daily methane production pattern of non-lactating cows fed four different diets containing linseed oil and calcium nitrate alone or in association (n = 4). Treatments consisted of control diet (CON), CON plus 3% calcium nitrate (NIT), CON plus 4% linseed oil (LIN) and CON plus 4% linseed oil and 3% calcium nitrate (LIN+NIT). The arrows indicate time of feeding. Error bars indicate SD. Adapted from Guyader et al. (2015).

Our work confirmed the initial working hypothesis that combining dietary strategies with different mechanisms of action to reduce H_2 availability in the rumen reduces methanogenesis more markedly than when lipids are fed individually. This opens up a range of possibilities for designing new strategies to increase CH_4 abatement (Beauchemin et al., 2020).

4 Outstanding questions and future trends in research

Considering the current health crisis, integrating animal production into a 'one health' approach is more relevant than ever: it s important to consider health care for humans, animals and the Earth in a systemic and integrated way at local, national and global levels. In this context, the use of plant bioactive substances from local resources in animal nutrition is a strong 'card to play' for promoting efficient and safe livestock farming to feed populations, while minimizing its environmental impact.

Many feed resources contain lipids and secondary compounds that are likely, if used properly, to improve animal performances and health, decrease enteric CH_4 emissions and N waste, and improve the quality of animal products. However, many questions have to be addressed before widespread application at the farm level. One demand of stakeholders concerns the possibility of standardized resources with a guaranteed content in active principle. This implies further research to characterize active molecules and their mode of action in order to design and evaluate new feeding strategies that are more efficient in minimizing GHG emissions from ruminants.

Evaluation of new resources based only on the traditional feed value is not sufficient. A multicriteria approach of these resources, as well as practices (crop growing, conservation, processing, feed delivery), is required to consider all animal responses, without neglecting the evaluation of the cost-benefit ratio for farmers. Another challenge is to develop resources with valuable properties for pasture-based systems in order to better integrate the context of agro-ecological ruminant production.

5 Where to look for further information

5.1 *Key articles or books*

- Vasta, V., Daghio, M., Cappucci, A., Buccioni, A., Serra, A., et al. 2019. Invited review: Plant polyphenols and rumen microbiota responsible for fatty acid biohydrogenation, fiber digestion, and methane emission: Experimental evidence and methodological approaches. *J. Dairy Sci.* 102, 3781–3804. https://doi.org/10.3168/jds.2018-14985.
- Mueller-Harvey, I., Bee, G., Dohme-Meier, F., Hoste, H., Karonen, M., Kölliker, R., Lüscher, A., Niderkorn, V., Pellikaan, W. F., Salminen, J. P., Skøt,

L., Smith, L. M. J., Thamsborg, S. M., Totterdell, P., Wilkinson, I., Williams, A. R., Azuhnwi, B. N., Baert, N., Grosse Brinkhaus, A., Copani, G., Desrues, O., Drake, C., Engström, M., Fryganas, C., Girard, M., Huyen, N. T., Kempf, K., Malisch, C., Mora-Ortiz, M., Quijada, J., Ramsay, A., Ropiak, H. M., Waghorn, G. C. 2018. Benefits of condensed tannins in forage legumes fed to ruminants: importance of structure, concentration and diet composition. Invited review. *Crop Science*, 59, 861–885. https://doi.org/10.2135/cropsci2017.06.0369.

- Cobellis, G., Trabalza-Marinuccia, M., Yu, Z., 2016. Critical evaluation of essential oils as rumen modifiers in ruminant nutrition: A review. *Sci. Total Environ.* 545–546, 556–568. http://dx.doi.org/10.1016/j.scitotenv.2015.12.103.
- Hristov, A. N., Oh, J., Firkins, J. L., Dijkstra, Kebreab, J. E., Waghorn, G., et al. 2013b. Special topics–Mitigation of methane and nitrous oxide emissions from animal operations: I. A review of enteric methane mitigation options. *J. Anim. Sci.* 91, 5045–5069. https://doi.org/10.2527/jas.2013-6583.
- Beauchemin, K. A., Ungerfeld, E. M., Eckard, R. J., Wang, M. 2020. Review: Fifty years of research on rumen methanogenesis: Lessons learned and future challenges for mitigation. *Animal*, 14:S1, s2–s16. https://doi.org/10.1017/S1751731119003100.

5.2 Key conferences

- International Symposium on the Nutrition of Herbivores, Clermont-Ferrand, FRA (2018-09-02–2018-09-06). Proceedings of the 10th International Symposium on the Nutrition of Herbivores in *Advances in Animal Biosciences*, 9(3), 337–786. doi:10.1017/S2040470018000146.
- International Symposium on Ruminant Physiology, Leipzig, DEU (2019-09-03–2019-09-06). Proceedings of the XIIIth International Symposium on Ruminant Physiology in *Advances in Animal Biosciences*, 10(3), 369–649. doi:10.1017/S2040470019000037.
- Greenhouse Gases and Animal Agriculture conference, Iguassu, BRA (2019-08-04–2019-08-10). Proceedings of the VIIth Greenhouse Gas and Animal Agriculture Conference. http://www.ggaa2019.org/sites/default/files/proceedings-ggaa2019.pdf.

5.3 Major international research projects and networks

- LegumePlus (2012-2015): European project aiming to optimize plant polyphenols in legumes for ruminant nutrition and health plus environmental sustainability (Project Number PITN-GA-2011-289377).

- Pro YoungStock (2018–2021): European CORE Organic Co-fund Project aiming to collect, develop and assess natural feeding strategies increasing dairy livestock welfare (Preject FiBL 50090).
- SmartCow (2018–2022): European project on infrastructures for increased research capability and innovation in the European cattle sector. Joint research activities are focussed on improving the quality and ethics of research services through advances in the capabilities to investigate feed efficiency and emissions in cattle at a large scale and to valorize data from sensors monitoring nutrition, health and behaviour.

6 References

Addisu, S. and Assefa, A. 2016. Role of plant containing saponin on livestock production: a review. *Adv. Biol. Res.* 10:309–314. doi:10.5829/idosi.abr.2016.309.314.

Aemiro, A., Hanada, M., Umetsu, K. and Nishida, T. 2016. The effect of sunphenon 30S-O on methane emission, nutrient intake, digestibility and rumen fermentation. *Anim. Feed Sci. Tech.* 214:34–43. doi:10.1016/j.anifeedsci.2016.02.007.

Archimède, H., Rira, M., Barde, D. J., Labirin, F., Marie-Magdeleine, C., Calif, B., Periacarpin, F., Fleury, J., Rochette, Y., Morgavi, D. P. and Doreau, M. 2016. Potential of tannin-rich plants, *Leucaena leucocephala*, *Glyricidia sepium* and *Manihot esculenta*, to reduce enteric methane emissions in sheep. *J. Anim. Physiol. Anim. Nutr.* 100(6):1149–1158.

Aufrère, J., Dudilieu, M. and Poncet, C. 2008. In vivo and in situ measurements of the digestive characteristics of sainfoin in comparison with Lucerne fed to sheep as fresh forages at two growth stages and as hay. *Animal* 2(9):1331–1339.

Baert, N., Pellikaan, W. F., Karonen, M. and Salminen, J. P. 2016. A study of the structure-activity relationship of oligomeric ellagitannins on ruminal fermentation in vitro. *J. Dairy Sci.* 99(10):8041–8052.

Bayat, A. R., Tapio, I., Vilkki, J., Shingfield, K. J. and Leskinen, H. 2018. Plant oil supplements reduce methane emissions and improve milk fatty acid composition in dairy cows fed grass silage-based diets without affecting milk yield. *J. Dairy Sci.* 101(2):1136–1151. doi:10.3168/jds.2017-13545.

Beauchemin, K. A. and McGinn, S. M. 2006. Methane emissions from beef cattle: effects of fumaric acid, essential oil, and canola oil. *J. Anim. Sci.* 84(6):1489–1496. doi:10.2527/2006.8461489x.

Beauchemin, K. A., Ungerfeld, E. M., Eckard, R. J. and Wang, M. 2020. Review: fifty years of research on rumen methanogenesis: lessons learned and future challenges for mitigation. *Animal* 14(S1):s2–s16. doi:10.1017/S1751731119003100.

Becker, P. M., van Wikselaar, P. G., Franssen, M. C. R., de Vos, R. C. H., Hall, R. D. and Beekwilder, J. 2014. Evidence for a hydrogen-sink mechanism of (+)catechin-mediated emission reduction of the ruminant greenhouse gas methane. *Metabolomics* 10(2):179–189. doi:10.1007/s11306-013-0554-5.

Belanche, A., Newbold, C. J., Morgavi, D. P., Bach, A., Zweifel, B. and Yáñez-Ruiz, D. R. 2020. A meta-analysis describing the effects of the essential oils blend agolin ruminant on performance, rumen fermentation and methane emissions in dairy cows. *Animals* 10(4):620. doi:10.3390/ani10040620.

Benchaar, C. 2015. Diet supplementation with cinnamon oil, cinnamaldehyde, or monensin does not reduce enteric methane production of dairy cows. *Animal* 10(3):418–425. doi:10.1017/S175173111500230X.

Benchaar, C. and Greathead, H. 2011. Essential oils and opportunities to mitigate enteric methane emissions from ruminants. *Anim. Feed Sci. Tech.* 166–167:338–355. doi:10.1016/j.anifeedsci.2011.04.024.

Benchaar, C., Hassanat, F. and Petit, H. V. 2015. Dose-response to eugenol supplementation to dairy cow diets: methane production, N excretion, ruminal fermentation, nutrient digestibility, milk production, and milk fatty acid profile. *Anim. Feed Sci. Tech.* 209:51–59. doi:10.1016/j.anifeedsci.2015.07.027.

Bouwman, A. F. 1996. Direct emission of nitrous oxide from agricultural soils. *Nutr. Cycl. Agroecosyst.* 46(1):53–70. doi:10.1007/BF00210224.

Branco, A. F., Giallongo, F., Frederick, T., Weeks, H., Oh, J. and Hristov, A. N. 2015. Effect of technical cashew nut shell liquid on rumen methane emission and lactation performance of dairy cows. *J. Dairy Sci.* 98(6):4030–4040. doi:10.3168/jds.2014-9015.

Buccioni, A., Pauselli, M., Viti, C., Minieri, S., Pallara, G., Roscini, V., Rapaccini, S., Marinucci, M. T., Lupi, P., Conte, G. and Mele, M. 2015. Milk fatty acid composition, rumen microbial population, and animal performances in response to diets rich in linoleic acid supplemented with chestnut or quebracho tannins in dairy ewes. *J. Dairy Sci.* 98(2):1145–1156. doi:10.3168/jds.2014-8651.

Calsamiglia, S., Busquet, M., Cardozo, P. W., Castillejos, L. and Ferret, A. 2007. Invited review: essential oils as modifiers of rumen microbial fermentation. *J. Dairy Sci.* 90(6):2580–2595. doi:10.3168/jds.2006-644.

Chung, Y. -H., He, M. L., McGinn, S. M., McAllister, T. A. and Beauchemin, K. A. 2011. Linseed suppresses enteric methane emissions from cattle fed barley silage, but not from those fed grass hay. *Anim. Feed Sci. Technol.* 166–167:321–329.

Cobellis, G., Trabalza-Marinucci, M. and Yu, Z. 2016. Critical evaluation of essential oils as rumen modifiers in ruminant nutrition: a review. *Sci. Total Environ.* 545–546:556–568. doi:10.1016/j.scitotenv.2015.12.103.

Copani, G., Ginane, C., Le Morvan, A. and Niderkorn, V. 2014. Bioactive forage legumes as a strategy to improve silage quality and minimise nitrogenous losses. *Anim. Prod. Sci.* 54(10):1826–1829.

Costa, M., Alves, S. P., Cappucci, A., Cook, S. R., Duarte, A., Caldeira, R. M., McAllister, T. A. and Bessa, R. J. B. 2018. Effects of condensed and hydrolyzable tannins on rumen metabolism with emphasis on the biohydrogenation of unsaturated fatty acids. *J. Agric. Food Chem.* 66(13):3367–3377. doi:10.1021/acs.jafc.7b04770.

Czerkawski, J. W. 1986. *An Introduction to Rumen Studies*, Pergamon Press, Oxfordshire and New York.

De Klein, C. A. M. and Eckard, R. J. 2008. Targeted technologies for nitrous oxide abatement from animal agriculture. *Aust. J. Exp. Agric.* 48(2):14–20.

Desrues, O., Peña-Espinoza, M., Hansen, T. V. A., Enemark, H. L. and Thamsborg, S. M. 2016. Anti-parasitic activity of pelleted sainfoin (Onobrychis viciifolia) against Ostertagia ostertagi and Cooperia Oncophora in calves. *Parasites Vectors* 9(1):329.

Doreau, M., Bamière, L., Pellerin, S., Lherm, M. and Benoit, M. 2014. Mitigation of enteric methane for French cattle: potential extent and cost of selected actions. *Anim. Prod. Sci.* 54(9):1417–1422. doi:10.1071/AN14207.

Doreau, M., Martin, C., Eugène, M., Popova, M. and Morgavi, D. P. 2011. Leviers d'action pour réduire la production de méthane entérique par les ruminants. *INRA Prod. Anim.* 24(5):461–474. doi:10.20870/productions-animales.2011.24.5.3278.

Dubois, B., Tomkins, N., Kinley, R. D., Bai, M., Seymour, S., Paul, N. and de Nys, R. 2013. Effect of tropical algae as additives on rumen in vitro gas production and fermentation characteristics. *Am. J. Plant Sci.* 4:34–43. doi:10.4236/ajps.2013.412A2005.

Engström, M. T., Arvola, J., Nenonen, S., Virtanen, V. T. J., Leppä, M. M., Tähtinen, P. and Salminen, J.-P. 2019. Structural features of hydrolysable tannins determine their ability to form insoluble complexes with bovine serum albumin. *J. Agric. Food Chem.* 67(24):6798–6808.

Eugène, M., Benchaar, C., Chiquette, J. and Masse, D. 2008. Meta-analysis on the effects of lipid supplementation on methane emissions and milk performance of lactating dairy cows. *Can. J. Anim. Sci.* 84:331–334.

Eugène, M., Doreau, M., Archimède, H., Giger-Reverdin, S. and Sauvant, D. 2019. Modelling by meta-analysis enteric methane emissions from ruminants fed forages supplemented or not with tannins. International Workshop on Modelling Nutrient Digestion and Utilization in Farm Animals, Itamambuca Eco Resort, Brazil.

Fox, N. J., Smith, L. A., Houdijk, J. G. M., Athanasiadou, S. and Hutchings, M. R. 2018. Ubiquitous parasites drive a 33% increase in methane yield from livestock. *Int. J. Parasitol.* 48(13):1017–1021. doi:10.1016/j.ijpara.2018.06.001.

Francis, G., Kerem, Z., Makkar, H. P. S. and Becker, K. 2007. The biological action of saponins in animal systems: a review. *Br. J. Nutr.* 88(6):587–605. doi:10.1079/BJN2002725.

Gerber, P. J., Hristov, A. N., Henderson, B., Makkar, H., Oh, J., Lee, C., Meinen, R., Montes, F., Ott, T., Firkins, J., Rotz, A., Dell, C., Adesogan, A. T., Yang, W. Z., Tricarico, J. M., Kebreab, E., Waghorn, G., Dijkstra, J. and Oosting, S. 2013. Technical options for the miti-gation of direct methane and nitrous oxide emissions from livestock–a review. *Animal* 7(S2):220–234. doi:10.1017/S1751731113000876.

Giger-Reverdin, S., Morand-Fehr, P. and Tran, G. 2003. Literature survey of the influence of dietary fat composition on methane production in dairy cattle. *Livest. Prod. Sci.* 82(1):73–79. doi:10.1016/S0301-6226(03)00002-2.

Grainger, C. and Beauchemin, K. A. 2011. Can enteric methane emissions from ruminants be lowered without lowering their production? *Anim. Feed Sci. Technol.* 166-167:308–320. doi:10.1016/j.anifeedsci.2011.04.021.

Guyader, J., Doreau, M., Morgavi, D. P., Gérard, C., Loncke, C. and Martin, C. 2016. Long-term effect of linseed plus nitrate fed to dairy cows on enteric methane emission and nitrate and nitrite residuals in milk. *Animal* 10(7):1173–1181. doi:10.1017/S1751731115002852.

Guyader, J., Eugène, M., Doreau, M., Morgavi, D. P., Gérard, C. and Martin, C. 2017. Tea saponin reduced methanogenesis in vitro but increased methane yield in lactating dairy cows. *J. Dairy Sci.* 100(3):1845–1855. doi:10.3168/jds.2016-11644.

Guyader, J., Eugène, M., Meunier, B., Doreau, M., Morgavi, D. P., Silberberg, M., Rochette, Y., Gerard, C., Loncke, C. and Martin, C. 2015. Additive methane-mitigating effect between linseed oil and nitrate fed to cattle. *J. Anim. Sci.* 93(7):3564–3577. doi:10.2527/jas2014-8196.

Guyader, J., Eugène, M., Nozière, P., Morgavi, D. P., Doreau, M. and Martin, C. 2014. Influence of rumen protozoa on methane emissions in ruminants: a meta-analysis approach. *Animal* 8(11):1816–1825. doi:10.1017/S1751731114001852.

Hart, K. J., Yáñez-Ruiz, D. R., Duval, S. M., McEwan, N. R. and Newbold, C. J. 2008. Plant extracts to manipulate rumen fermentation. *Anim. Feed Sci. Tech.* 147(1-3):8-35.

Hatew, B., Stringano, E., Mueller-Harvey, I., Hendriks, W. H., Carbonero, C. H., Smith, L. M. J. and Pellikaan, W. F. 2016. Impact of variation in structure of condensed tannins from sainfoin (O nobrychis viciifolia) on in vitro ruminal methane production and fermentation characteristics. *J. Anim. Physiol. Anim. Nutr.* 100(2):348-360.

Hollmann, M., Powers, W. J., Fogiel, A. C., Liesman, J. S., Bello, N. M. and Beede, D. K. 2012. Enteric methane emissions and lactational performance of Holstein cows fed different concentrations of coconut oil. *J. Dairy Sci.* 95(5):2602-2615.

Holtshausen, L., Chaves, A. V., Beauchemin, K. A., McGinn, S. M., McAllister, T. A., Odongo, N. E., Cheeke, P. R. and Benchaar, C. 2009. Feeding saponin-containing Yucca schidigera and Quillaja Saponaria to decrease enteric methane production in dairy cows. *J. Dairy Sci.* 92(6):2809-2821. doi:10.3168/jds.2008-1843.

Hoste, H., Torres-Acosta, J. F., Sandoval-Castro, C. A., Mueller-Harvey, I., Sotiraki, S., Louvandini, H., Thamsborg, S. M. and Terrill, T. H. 2015. Tannin containing legumes as a model for nutraceuticals against digestive parasites in livestock. *Vet. Parasitol.* 212(1-2):5-17. doi:10.1016/j.vetpar.2015.06.026.

Hristov, A. N., Lee, C., Cassidy, T., Heyler, K., Tekippe, J. A., Varga, G. A., Corl, B. and Brandt, R. C. 2013a. Effect of Origanum vulgare L. leaves on rumen fermentation, production, and milk fatty acid composition in lactating dairy cows. *J. Dairy Sci.* 96(2):1189-1202. doi:10.3168/jds.2012-5975.

Hristov, A. N., Oh, J., Firkins, J. L., Dijkstra, J., Kebreab, E., Waghorn, G., Makkar, H. P., Adesogan, A. T., Yang, W., Lee, C., Gerber, P. J., Henderson, B. and Tricarico, J. M. 2013b. Special topics–mitigation of methane and nitrous oxide emissions from animal operations: I. A review of enteric methane mitigation options. *J. Anim. Sci.* 91(11):5045-5069. doi:10.2527/jas.2013-6583.

Huyen, N. T., Desrues, O., Alferink, S. J. J., Zandstra, T., Verstegen, M. W. A., Hendriks, W. H. and Pellikaan, W. F. 2016a. Inclusion of sainfoin (Onobrychis viciifolia) silage in dairy cow rations affects nutrient digestibility, nitrogen utilization, energy balance, and methane emissions. *J. Dairy Sci.* 99(5):3566-3577.

Huyen, N. T., Fryganas, C., Uittenbogaard, G., Mueller-Harvey, I., Verstegen, M. W. A., Hendriks, W. H. and Pellikaan, W. F. 2016b. Structural features of condensed tannins affect in vitro ruminal methane production and fermentation characteristics. *J. Agric. Sci.* 154(8):1474-1487.

INRA. 2018. *INRA Feeding System for Ruminants*, Wageningen Academic Publishers, Wageningen, the Netherlands.

Jayanegara, A., Goel, G., Makkar, H. P. S. and Becker, K. 2015. Divergence between purified hydrolysable and condensed tannin effects on methane emission, rumen fermentation and microbial population in vitro. *Anim. Feed Sci. Technol.* 209:60-68.

Jayanegara, A., Leiber, F. and Kreuzer, M. 2012. Meta-analysis of the relationship between dietary tannin level and methane formation in ruminants from in vivo and in vitro experiments. *J. Anim. Physiol. Anim. Nutr.* 96(3):365-375.

Jayanegara, A., Wina, E. and Takahashi, J. 2014. Meta-analysis on methane mitigating properties of saponin-rich sources in the rumen: influence of addition levels and plant sources. *Asian Australas. J. Anim. Sci.* 27(10):1426-1435.

Jentsch, W., Wittenburg, H. and Schiemann, R. 1972. Die Verwertung der Futterenergie fur die Milchproduction. 4. *Unters. Verwertung Futterenergie Rapsoleinsatz Arch. Tierernahr.* 22:697-720.

Ju, W. T., Kwon, O. C., Kim, H. B., Sung, G. B., Kim, H. W. and Kim, Y. S. 2018. Qualitative and quantitative analysis of flavonoids from 12 species of Korean mulberry leaves. *J. Food. Sci. Technol.* 55(5):1789–1796. doi:10.1007/s13197-018-3093-2.

Kang, S., Suzuki, R., Suzuki, Y., Koike, S., Nagashima, K. and Kobayashi, Y. 2018. Rumen responses to dietary supplementation with cashew nut shell liquid and its cessation in sheep. *Anim. Sci. J.* 89(11):1549–1555. doi:10.1111/asj.13100.

Keng, F. S., Phang, S., Abd Rahman, N., Leedham Elvidge, E. C., Malin, G. and Sturges, W. T. 2020. The emission of volatile halocarbons by seaweeds and their response towards environmental changes. *J. Appl. Phycol.* 32(2):1377–1394. doi:10.1007/s10811-019-02026-x.

Kinley, R. D., de Nys, R., Vucko, M. J., Machado, L. and Tomkins, N. W. 2016. The red macroalgae Asparagopsis taxiformis is a potent natural antimethanogenic that reduces methane production during in vitro fermentation with rumen fluid. *Anim. Prod. Sci.* 56(3):282–289. doi:10.1071/AN15576.

Kinley, R. D. and Fredeen, A. H. 2015. In vitro evaluation of feeding North Atlantic stormtoss seaweeds on ruminal digestion. *J. Appl. Phycol.* 27(6):2387–2393. doi:10.1007/s10811-014-0487-z.

Kinley, R. D., Martinez-Fernandez, G., Matthews, M. K., de Nys, R., Magnusson, M. and Tomkins, N. W. 2020. Mitigating the carbon footprint and improving productivity of ruminant livestock agriculture using a red seaweed. *J. Clean. Prod.* 259. doi:10.1016/j.jclepro.2020.120836.

Klevenhusen, F., Zeitz, J. O., Duval, S., Kreuzer, M. and Soliva, C. R. 2011. Garlic oil and its principal component diallyl disulfide fail to mitigate methane, but improve digestibility in sheep. *Anim. Feed Sci. Tech.* 166–167:356–363.

Klita, P. T., Mathison, G. W., Fenton, T. W. and Hardin, R. T. 1996. Effects of alfalfa root saponins on digestive function in sheep. *J. Anim. Sci.* 74(5):1144–1156.

Kolling, G. J., Stivanin, S. C. B., Gabbi, A. M., Machado, F. S., Ferreira, A. L., Campos, M. M., Tomich, T. R., Cunha, C. S., Dill, S. W., Pereira, L. G. R. and Fischer, V. 2018. Performance and methane emissions in dairy cows fed oregano and green tea extracts as feed additives. *J. Dairy Sci.* 101(5):4221–4234. doi:10.3168/jds.2017-13841.

Konda, S., Onodera, R., Kanchanasatit, E., Boonsaen, P., Sawanon, S., Nagashima, K., Suzuki, Y., Koike, S. and Kobayashi, Y. 2019. Effect of cashew nut shell liquid feeding on fermentation and microbiota in the rumen of Thai native cattle and swamp buffaloes. *Livest. Sci.* 226:99–106. doi:10.1016/j.livsci.2019.06.011.

Li, W. and Powers, W. 2012. Effects of saponin extracts on air emissions from steers. *J. Anim. Sci.* 90(11):4001-4013. doi:10.2527/jas.2011-4888.

Li, X., Norman, H. C., Kinley, R. D., Laurence, M., Wilmot, M., Bender, H., de Nys, R. and Tomkins, N. 2016. Asparagopsis taxiformis decreases enteric methane production from sheep. *Anim. Prod. Sci.* 58(4). doi:10.1071/AN15883.

Lima, PdM. T., Crouzoulon, P., Sanches, T. P., Zabré, G., Kabore, A., Niderkorn, V., Hoste, H., Amarante, A. F. Td, Costa-Júnior, L. M., Abdalla, A. L. and Louvandini, H. 2019. Effects of Acacia mearnsii supplementation on nutrition, parasitological, blood parameters and methane emissions in Santa Inês sheep infected with Trichostrongylus colubriformis and Haemonchus contortus. *Exp. Parasitol.* 207:107777.

Liu, H., Wang, J., Wang, A. and Chen, J. 2011. Chemical inhibitors of methanogenesis and putative applications. *Appl. Microbiol. Biotechnol.* 89(5):1333–1340. doi:10.1007/s00253-010-3066-5.

Luo, J. and Kelliher, F. M. 2010. Partitioning of animal excreta N into urine and dung and developing the N_2O inventory. Report for the MAF Policy (MAF POL 0910-11528 09-03), Wellington.

Ma, T., Chen, D. D., Tu, Y., Zhang, N. F., Si, B. W., Deng, K. D. and Diao, Q. Y. 2016. Effect of supplementation of allicin on methanogenesis and ruminal microbial flora in Dorper crossbred ewes. *J. Anim. Sci. Biotechnol.* 7(1). doi:10.1186/s40104-015-0057-5.

Ma, T., Chen, D. D., Tu, Y., Zhang, N. F., Si, B. W. and Diao, Q. Y. 2017. Dietary supplementation with mulberry leaf flavonoids inhibits methanogenesis in sheep. *Anim. Sci.* 88(1):72–78. doi:10.1111/asj.12556.

Machado, L., Magnusson, M., Paul, N. A., de Nys, R. and Tomkins, N. 2014. Effects of marine and freshwater macroalgae on in vitro total gas and methane production. *PLoS ONE* 9(1):e85289. doi:10.1371/journal.pone.0085289.

Machado, L., Magnusson, M., Paul, N. A., Kinley, R., de Nys, R. and Tomkis, N. 2016a. Dose-response effects of Asparagopsis taxiformis and Oedogonium sp. on in vitro fermentation and methane production. *J. Appl. Phycol.* 28(2):1443–1452. doi:10.1007/s10811-015-0639-9.

Machado, L., Magnusson, M., Paul, N. A., Kinley, R., de Nys, R. and Tomkis, N. 2016b. Identification of bioactives from the red seaweed Asparagopsis taxiformis that promote antimethanogenic activity in vitro. *J. Appl. Phycol.* 28(5):3117–3126. doi:10.1007/s10811-016-0830-7.

Machado, L., Tomkis, N., Magnusson, M., Midgley, D. J., de Nys, R. and Rosewarne, C. P. 2018. In vitro Response of Rumen microbiota to the Antimethanogenic Red Macroalga Asparagopsis taxiformis. *Microb. Ecol.* 75(3):811–818.

Macheboeuf, D., Morgavi, D. P., Papon, Y., Mousset, J.-L. and Arturo-Schaan, M. 2008. Dose-response effects of essential oils on in vitro fermentation activity of the rumen microbial population. *Anim. Feed Sci. Tech.* 145(1–4):335–350.

Machmüller, A. and Kreuzer, M. 1999. Methane suppression by coconut oil and associated effects on nutrient and energy balance in sheep. *Can. J. Anim. Sci.* 79(1):65–72.

Malisch, C. S., Suter, D., Studer, B. and Lüscher, A. 2017. Multifunctional benefits of sainfoin mixtures: effects of partner species, sowing density and cutting regime. *Grass Forage Sci.* 72(4):794–805.

Manasri, N., Wanapat, M. and Navanukraw, C. 2012. Improving rumen fermentation and feed digestibility in cattle by mangosteen peel and garlic pellet supplementation. *Livest. Sci.* 148(3):291–295. 10.1016/j.livsci.2012.06.009

Mao, H. L., Wang, J. K., Zhou, Y. Y. and Liu, J. X. 2010. Effects of addition of tea saponins and soybean oil on methane production, fermentation and microbial population in the rumen of growing lambs. *Livest. Sci.* 129(1–3):56–62. doi:10.1016/j.livsci.2009.12.011.

Martin, C., Ferlay, A., Mosoni, P., Rochette, Y., Chilliard, Y. and Doreau, M. 2016. Increasing linseed supply in dairy cow diets based on hay or corn silage: effect on enteric methane emission, rumen microbial fermentation, and digestion. *J. Dairy Sci.* 99(5):3445–3456. doi:10.3168/jds.2015-10110.

Martin, C., Morgavi, D. P. and Doreau, M. 2010. Methane mitigation in ruminants: from microbe to the farm scale. *Animal* 4(3):351–365. doi:10.1017/S1751731109990620.

Martin, C., Pomiès, D., Ferlay, A., Rochette, Y., Martin, B., Chilliard, Y., Morgavi, D. P. and Doreau, M. 2011. Methane output and rumen microbiota in dairy cows in response to longterm supplementation with linseed or rapeseed of grass silage or pasture based diets. *Proc. N. Z. Soc. Anim. Prod.* 71:243–247.

Martin, C., Rouel, J., Jouany, J. P., Doreau, M. and Chilliard, Y. 2008. Methane output and diet digestibility in response to feeding dairy cows with crude linseed, extruded linseed or linseed oil. *J. Anim. Sci.* 86(10):2642–2650.

Mills, J. A., Dijkstra, J., Bannink, A., Cammell, S. B., Kebreab, E. and France, J. 2001. A mechanistic model of whole-tract digestion and methanogenesis in the lactating dairy cow: model development, evaluation, and application. *J. Anim. Sci.* 79(6):1584–1597.

Mitra, S. and Dungan, S. R. 1997. Micellar properties of Quillaja saponin. 1. Effects of temperature, salt, and pH on solution properties. *J. Agric. Food Chem.* 45(5):1587–1595.

Mohammed, N., Ajisaka, N., Lila, Z. A., Hara, K., Mikuni, K., Hara, K., Kanda, S. and Itabashi, H. 2004. Effect of Japanese horseradish oil on methane production and ruminal fermentation in vitro and in steers. *J. Anim. Sci.* 82(6):1839–1846.

Morgavi, D. P., Forano, E., Martin, C. and Newbold, C. J. 2010. Microbial ecosystem and methanogenesis in ruminants. *Animal* 4(7):1024–1036. doi:10.1017/S1751731110000546.

Mueller-Harvey, I. 2006. Unravelling the conundrum of tannins in animal nutrition and health. *J. Sci. Food Agric.* 86(13):2010–2037.

Mueller-Harvey, I., Bee, G., Dohme-Meier, F., Hoste, H., Karonen, M., Kölliker, R., Lüscher, A., Niderkorn, V., Pellikaan, W. F., Salminen, J., Skøt, L., Smith, L. M. J., Thamsborg, S. M., Totterdell, P., Wilkinson, I., Williams, A. R., Azuhnwi, B. N., Baert, N., Brinkhaus, A. G., Copani, G., Desrues, O., Drake, C., Engström, M., Fryganas, C., Girard, M., Huyen, N. T., Kempf, K., Malisch, C., Mora-Ortiz, M., Quijada, J., Ramsay, A., Ropiak, H. M. and Waghorn, G. C. 2019. Benefits of condensed tannins in forage legumes fed to ruminants: importance of structure, concentration, and diet composition. *Crop Sci.* 59(3):861–885.

Newbold, C. J., El Hassan, S. M., Wang, J., Ortega, M. E. and Wallace, R. J. 1997. Influence of foliage from African multipurpose trees on activity of rumen protozoa and bacteria. *Br. J. Nutr.* 78(2):237–249.

Niderkorn, V., Copani, G., Martin, C., Maxin, G., Torrent, A., Anglard, F., Rochette, Y. and Ginane, C. 2019. Effects of including bioactive legumes in grass silage on digestion parameters, nitrogen balance and methane emissions in sheep. *Grass Forage Sci.* 74(4):626–635. doi:10.1111/gfs.12454.

Olagaray, K. E. and Bradford, B. J. 2019. Plant flavonoids to improve productivity of ruminants–a review. *Anim. Feed Sci. Tech.* 251:21–36. 10.1016/j.anifeedsci.2019.02.004

Palmquist, D. L. and Jenkins, T. C. 2017. A 100-year review: fat feeding of dairy cows. *J. Dairy Sci.* 100(12):10061–10077. doi:10.3168/jds.2017-12924.

Panthee, A., Matsuno, A., Al-Mamun, M. and Sano, H. 2017. Effect of feeding garlic leaves on rumen fermentation, methane emission, plasma glucose kinetics, and nitrogen utilization in sheep. *J. Anim. Sci. Technol.* 59(1):14. 10.1186/s40781-017-0139-3

Patra, A. K., Kamra, D. N., Bhar, R., Kumar, R. and Agarwal, N. 2011. Effect of Terminalia chebula and Allium sativum on in vivo methane emission by sheep. *J. Anim. Physiol. Anim. Nutr. (Berl)* 95(2):187–191. doi:10.1111/j.1439-0396.2010.01039.x.

Patra, A. K. and Saxena, J. 2009. The effect and mode of action of saponins on the microbial populations and fermentation in the rumen and ruminant production. *Nutr. Res. Rev.* 22(2):204–219. doi:10.1017/S0954422409990163.

Patra, A. K. and Saxena, J. 2010. A new perspective on the use of plant secondary metabolites to inhibit methanogenesis in the rumen. *Phytochemistry* 71(11–12):1198–1222.

Patra, A. K. and Yu, Z. 2012. Effects of essential oils on methane production and fermentation by, and abundance and diversity of, rumen microbial populations. *Appl. Environ. Microbiol.* 78(12):4271–4280. doi:10.1128/aem.00309-12.

Pen, B., Takaura, K., Yamaguchi, S., Asa, R. and Takahashi, J. 2007. Effects of Yucca schidigera and Quillaja Saponaria with or without β 1-4 galacto-oligosaccharides on ruminal fermentation, methane production and nitrogen utilization in sheep. *Anim. Feed Sci. Technol.* 138(1):75–88. doi:10.1016/j.anifeedsci.2006.11.018.

Piluzza, G., Sulas, L. and Bullitta, S. 2014. Tannins in forage plants and their role in animal husbandry and environmental sustainability: a review. *Grass Forage Sci.* 69(1):32–48. doi:10.1111/gfs.12053.

Popova, M., Morgavi, D. P., Doreau, M. and Martin, C. 2011. Production de méthane et interactions microbiennes dans le rumen. *INRA Prod. Anim.* 24(5):447–460. doi:10.20870/productions-animales.2011.24.5.3277.

Powell, J. M., Aguerre, M. J. and Wattiaux, M. A. 2011. Tannin extracts abate ammonia emissions from simulated dairy barn floors. *J. Environ. Qual.* 40(3):907–914. doi:10.2134/jeq2010.0492.

Ramos-Morales, E., de la Fuente, G., Nash, R. J., Braganca, R., Duval, S., Bouillon, M. E., Lahmann, M. and Newbold, C. J. 2017. Improving the antiprotozoal effect of saponins in the rumen by combination with glycosidase inhibiting iminosugars or by modification of their chemical structure. *PLoS ONE* 12(9):e0184517. doi:10.1371/journal.pone.0184517.

Reed, J. D. 1995. Nutritional toxicology of tannins and related polyphenols in forage legumes. *J. Anim. Sci.* 73(5):1516–1528. doi:10.2527/1995.7351516x.

Rira, M., Morgavi, D. P., Genestoux, L., Djibiri, S., Sekhri, I. and Doreau, M. 2019. Methanogenic potential of tropical feeds rich in hydrolyzable tannins1,2. *J. Anim. Sci.* 97(7):2700–2710.

Roque, B. M., Salwen, J. K., Kinley, R. and Kebreab, E. 2019a. Inclusion of Asparagopsis armata in lactating dairy cows' diet reduces enteric methane emission by over 50 percent. *J. Clean. Prod.* 234:132–138. doi:10.1016/j.jclepro.2019.06.193.

Roque, B. M., Van Lingen, H. J., Vrancken, H. and Kebreab, E. 2019b. Effect of Mootral–a garlic- and citrus-extract-based feed additive–on enteric methane emissions in feedlot cattle. *Transl. Anim. Sci.* 3(4):1383–1388. 10.1093/tas/txz133

Saminathan, M., Sieo, C. C., Gan, H. M., Abdullah, N., Wong, C. M. V. L. and Ho, Y. W. 2016. Effects of condensed tannin fractions of different molecular weights on population and diversity of bovine rumen methanogenic archaea in vitro, as determined by high-throughput sequencing. *Anim. Feed Sci. Technol.* 216:146–160.

Santoso, B., Mwenya, B., Sar, C., Gamo, Y., Kobayashi, T., Morikawa, R. and Takahashi, J. 2004. Effect of Yucca schidigera with or without nisin on ruminal fermentation and microbial protein synthesis in sheep fed silage- and hay-based diets. *Anim. Sci.* 75(6):525–531. doi:10.1111/j.1740-0929.2004.00223.x.

Saro, C., Hohenester, U. M., Bernard, M., Lagrée, M., Martin, C., Doreau, M., Boudra, H., Popova, M. and Morgavi, D. P. 2018. Effectiveness of interventions to modulate the rumen microbiota composition and function in pre-ruminant and ruminant lambs. *Front. Microbiol.* 9:1273. 10.3389/fmicb.2018.01273

Sauvant, D., Giger-Reverdin, S. and Eugène, M. 2018. Émissions de méthane entérique. In: Nozière, P., Sauvant, D., Delaby, L. and INRA, (Eds.) *Alimentation des ruminants. Apports nutritionnels - Besoins et réponses des animaux - Rationnement - Tables des*

valeurs des aliments. Federal Railroad Administration, Versailles, Editions Quae, pp. 243–247.
Selbie, D. R., Buckthought, L. E. and Shepherd, M. A. 2015. The challenge of the urine patch for managing nitrogen in grazed pasture systems. *Adv. Agron.* 129:229–292.
Shinkai, T., Enishi, O., Mitsumori, M., Higuchi, K., Kobayashi, Y., Takenaka, A., Nagashima, K., Mochizuki, M. and Kobayashi, Y. 2012. Mitigation of methane production from cattle by feeding cashew nut shell liquid. *J. Dairy Sci.* 95(9):5308–5316. doi:10.3168/jds.2012-5554.
Soussana, J. F., Tichit, M., Lecomte, P. and Dumont, B. 2015. Agroecology: integration with livestock. In: Agroecology for Food Security and Nutrition Proceedings of the FAO International Symposium, 18–19 September 2014, Rome, Italy. Rome: FAO, 225–249. ISBN 978-92-5-108807-4. (International Symposium on Agroecology for Food Security and Nutrition, Rome, Italie, 18 Septembre 2014/19 Septembre 2014.) Available at: http://www.fao.org/3/a-i4729e.pdf.
Stoldt, A. K., Derno, M., Das, G., Weitzel, J. M., Wolffram, S. and Metges, C. C. 2016. Effects of rutin and buckwheat seeds on energy metabolism and methane production in dairy cows. *J. Dairy Sci.* 99(3):2161–2168. doi:10.3168/jds.2015-10143.
Tekippe, J. A., Hristov, A. N., Heyler, K. S., Cassidy, T. W., Zheljazkov, V. D., Ferreira, J. F. S., Karnati, S. K. and Varga, G. A. 2011. Rumen fermentation and production effects of Origanum vulgare L. leaves in lactating dairy cows. *J. Dairy Sci.* 94(10):5065–5079. doi:10.3168/jds.2010-4095.
Theodoridou, K., Aufrère, J., Andueza, D., Pourrat, J., Le Morvan, A., Stringano, E., Mueller-Harvey, I. and Baumont, R. 2010. Effects of condensed tannins in fresh sainfoin (Onobrychis viciifolia) on in vivo and in situ digestion in sheep. *Anim. Feed Sci. Technol.* 160(1–2):23–38.
Tomkins, N. W., Denman, S. E., Pilajun, R., Wanapat, M., McSweeney, C. S. and Elliott, R. 2015. Manipulating rumen fermentation and methanogenesis using an essential oil and monensin in beef cattle fed a tropical grass hay. *Anim. Feed Sci. Tech.* 200:25–34. doi:10.1016/j.anifeedsci.2014.11.013.
van Gastelen, S., Visker, M. H. P. W., Edwards, J. E., Antunes-Fernandes, E. C., Hettinga, K. A., Alferink, S. J. J., Hendriks, W. H., Bovenhuis, H., Smidt, H. and Dijkstra, J. 2017. Linseed oil and DGAT1 K232A polymorphism: effects on methane emission, energy and nitrogen metabolism, lactation performance, ruminal fermentation, and rumen microbial composition of Holstein-Friesian cows. *J. Dairy Sci.* 100(11):8939–8957. doi:10.3168/jds.2016-12367.
Vargas, J. E., Andrés, S., López-Ferreras, L., Snelling, T. J., Yáñez-Ruíz, D. R., García-Estrada, C. and López, S. 2020. Dietary supplemental plant oils reduce methanogenesis from anaerobic microbial fermentation in the rumen. *Sci. Rep.* 10(1):1613. doi:10.1038/s41598-020-58401-z.
Vasta, V., Daghio, M., Cappucci, A., Buccioni, A., Serra, A., Viti, C. and Mele, M. 2019. Invited review: plant polyphenols and rumen microbiota responsible for fatty acid biohydrogenation, fiber digestion, and methane emission: experimental evidence and methodological approaches. *J. Dairy Sci.* 102(5):3781–3804. doi:10.3168/jds.2018-14985.
Veneman, J. B., Muetzel, S., Hart, K. J., Faulkner, C. L., Moorby, J. M., Perdok, H. B. and Newbold, C. J. 2015. Does dietary mitigation of enteric methane production affect rumen function and animal productivity in dairy cows? *PLoS ONE* 10: e0140282.

Vincken, J. P., Heng, L., de Groot, A. and Gruppen, H. 2007. Saponins, classification and occurrence in the plant kingdom. *Phytochemistry* 68(3):275-297. doi:10.1016/j.phytochem.2006.10.008.

Vrancken, H., Suenkel, M., Hargreaves, P. R., Chew, L. and Towers, E. 2019. Reduction of enteric methane emission in a commercial dairy farm by a novel feed supplement. *Open J. Anim. Sci.* 09(3):286-296. doi:10.4236/ojas.2019.93024.

Wanapat, M., Chanthakhoun, V., Phesatcha, K. and Kang, S. 2014. Influence of mangosteen peel powder as a source of plant secondary compounds on rumen microorganisms, volatile fatty acids, methane and microbial protein synthesis in swamp buffaloes. *Livest. Sci.* 162:126-133. doi:10.1016/j.livsci.2014.01.025.

Wang, B., Ma, M. P., Diao, Q. Y. and Tu, Y. 2019. Saponin-induced shifts in the rumen microbiome and metabolome of young cattle. *Front. Microbiol.* 10. doi:10.3389/fmicb.2019.00356.

Wang, C. J., Wang, S. P. and Zhou, H. 2009. Influences of flavomycin, ropadiar, and saponin on nutrient digestibility, rumen fermentation, and methane emission from sheep. *Anim. Feed Sci. Technol.* 148(2-4):157-166. doi:10.1016/j.anifeedsci.2008.03.008.

Wina, E., Muetzel, S. and Becker, K. 2006. The dynamics of major fibrolytic microbes and enzyme activity in the rumen in response to short- and long-term feeding of Sapindus rarak saponins. *J. Appl. Microbiol.* 100(1):114-122. doi:10.1111/j.1365-2672.2005.02746.x.

Wink, M. 2015. Modes of action of herbal medicines and plant secondary metabolites. *Medicines* 2(3):251-286. doi:10.3390/medicines2030251.

Yang, K., Wei, C., Zhao, G. Y., Xu, Z. W. and Lin, S. X. 2017. Effects of dietary supplementing tannic acid in the ration of beef cattle on rumen fermentation, methane emission, microbial flora and nutrient digestibility. *J. Anim. Physiol. Anim. Nutr.* 101(2):302-310. doi: 10.1111/jpn.12531. Epub 2016 Jun 8. PMID: 27272696.

Yuan, Z. P., Zhang, C. M., Zhou, L., Zou, C. X., Guo, Y. Q., Li, W., Liu, J. and Wu, Y. 2007. Inhibition of methanogenesis by tea saponin and tea saponin plus disodium fumarate in sheep. *J. Anim. Feed Sci.* 16(Suppl. 2):560-565.

Zeller, W. E. 2019. Activity, purification, and analysis of condensed tannins: current state of affairs and future endeavors. *Crop Sci.* 59(3):886-904.

Zhou, Y. Y., Mao, H. L., Jiang, F., Wang, J. K., Liu, J. X. and McSweeney, C. S. 2011. Inhibition of rumen methanogenesis by tea saponins with reference to fermentation pattern and microbial communities in Hu sheep. *Anim. Feed Sci. Technol.* 166-167:93-100. doi:10.1016/j.anifeedsci.2011.04.007.

Chapter 10

The use of feed supplements to reduce livestock greenhouse gas emissions: direct-fed microbials

Natasha Doyle, Teagasc Moorepark Food Research Centre, Ireland; Philiswa Mbandlwa, University College Cork, Ireland; Sinead Leahy and Graeme Attwood, AgResearch Limited, New Zealand; Bill Kelly, Ashhurst, New Zealand; Collin Hill and R. Paul Ross, Teagasc Moorepark Food Research Centre and University College Cork, Ireland; and Catherine Stanton, Teagasc Moorepark Food Research Centre, University College Cork and VISTAMILK SFI Centre – Teagasc, Ireland

1 Introduction

The agricultural sector contributes approximately 24% of all global greenhouse gas (GHG) emissions (IPCC, 2014). The main routes of production for GHG emissions are enteric fermentation and manure management (Haque, 2018). The main gases produced are methane and nitrous oxide, and to a lesser extent, carbon dioxide (McMichael et al., 2007). In 2010, total anthropogenic methane and nitrous oxide emissions accounted for approximately 20% and 5% of all emissions to date, respectively, based on the fifth assessment report (IPCC, 2014).

At a global level, livestock annually produce around 80 million tonnes (Tg) of enteric methane (Patra, 2012). Of these 80 million tonnes of enteric methane,

http://dx.doi.org/10.19103/AS.2020.0077.14

an estimated 18.9 Tg are attributed to dairy cattle, 55.9 Tg to beef cattle, and 9.5 Tg to sheep and goats (Hook et al., 2010). Of all livestock, ruminants are the main producers of enteric methane (Lassey et al., 2007). These animals contain a four-chambered stomach, which generates methane primarily via eructation and belching as a result of the complex microbiological fermentation that occurs in the rumen. This enteric fermentation involves degradation of cellulose and other macromolecules (Boadi et al., 2004) to allow for adsorption into the bloodstream. The large and diverse microbial population ferments the polymers to volatile fatty acids (VFAs), carbon dioxide (CO_2) and methane (Kataria, 2015).

Due to methanogenesis, the ruminant suffers a loss of ingested feed-derived energy of approximately 6–14% depending upon the diet (Johnson and Johnson, 1995). Rather than losing resources to an ineffective microbial process, this energy could instead be used by the animal to produce better-quality milk for its own development (Tapio et al., 2017). It has been predicted that reducing methane generation in the rumen would mean that more energy would be retained by the animal, thereby enhancing its nutritional efficiency (Yang et al., 2016).

Methane production by ruminants is influenced by various factors such as the physical and chemical characteristics of the feed, the feeding schedule and the feed additives. Methane is derived from ingested feed and therefore diet composition, while intake can be used to manipulate fermentation by altering the microbial interactions through feed additives, that is, direct fed microbials (DFMs). This would also have the potential to positively impact animal production. Manipulating host diet may reduce methane emissions by decreasing fermentation of organic matter, therefore shifting the site of fermentation of organic matter from the rumen to the intestine, consequently diverting hydrogen away from methane production (McGinn et al., 2004). For the cattle industry, reducing methane losses can represent an improvement in feed efficiency. Therefore, mitigating methane losses from cattle has both long-term environmental and short-term economic benefits (McGinn et al., 2004). There is a challenge, however, to maintain a balance between productivity, household food security and environmental preservation (Wright et al., 2011).

2 Methane and agriculture

Methane is a prominent GHG which is found in natural wetlands, rice fields, livestock and biomass burning. It is emitted through human activities such as the production and transport of coal, natural gases and oil, as well as naturally, through animal fermentations and gas deposits, such as peatlands. As the second most abundant GHG, it has been projected that one tonne of methane will absorb 34 times more thermal energy than one tonne of carbon dioxide, over a 100-year period (IPCC, 2013). Due to its thermal conductivity, methane levels have the potential to influence climate change on short-time scales (Rice et al., 2016).

Since the beginning of the industrial revolution, the levels of methane observed in the atmosphere have increased more than two-fold, with a continued 1–2% rise per annum since the 1980s, as measured by the National Oceanic and Atmospheric Administration (Singh et al., 2018). Realising the catastrophic threat posed by climate change, the 2015 Paris Agreement, under the United Nations Framework Convention on Climate Change (UNFCCC) and backed by 195 countries, aims to limit the increase in global average temperatures to below 2°C and, where/if possible, limit it to 1.5°C. It is expected that methane emissions from domesticated ruminants will decline in developed countries, due to an ever-growing trend towards an animal-free diet. However, factors such as population growth, rising incomes and commercialisation of previously small-scale farms will result in increased methane production in developing countries (EPA, 2014). In this way, global methane emissions from enteric fermentations are estimated to increase 32% by 2020 (EPA, 2013).

The microbial composition of the rumen, the fore-stomach of the ruminant animal, has a major influence on the feed digestion and the release of end products, such as methane, into the environment. The rumen is home to a vast array of protozoa, anaerobic fungi, anaerobic bacteria and archaea. This diverse array of microorganisms are responsible for the degradation of lignocellulose, which is used as an energy source for the animal. Short-chain fatty acids (SCFAs) are produced from these soluble sugars, and absorbed into the rumen epithelium, resulting in by-products of hydrogen (H_2), carbon dioxide (CO_2), formate and methyl-containing compounds. These by-products are important substrates for methane-forming archaea.

Due to the high microbial diversity within the rumen, methane is formed by many types of methanogens, each using distinct metabolic pathways and precursors. Although methane production can also occur in the lower gastrointestinal tract, a surprising 89% of methane emitted from ruminants is produced in the rumen itself and exhaled through the mouth and nose (Hook et al., 2010). In general, methanogenic archaea use H_2 + CO_2, formate, methylated C1 compounds, or acetate as energy and carbon sources for growth (Deppenmeier, 2002). The majority of rumen methanogens have been shown to belong to the *Methanobrevibacter* genus, accounting for 74% of all archaea (Henderson et al., 2015). When combined with *Methanosphaera* spp. and two *Methanomassiliicoccaceae*-affiliated groups, there are five dominant methanogen groups that comprise 89.2% of the community (Henderson et al., 2015).

Methanogenesis is a complex process dependent upon a range of microbes, which contribute either indirectly by creating the appropriate environment required for the growth of methanogens or directly by producing the substrates used by methanogens. The production of methane in cattle is also influenced by diet composition (ingredient and chemical), feed intake, and digestibility (Hristov et al., 2018). It has long been established that an increase in concentrate levels in the diet results in a decrease in methane emission as a proportion of energy intake or expressed by unit of animal product such as milk or meat (Wanapat et al., 2015).

High-starch diets have been shown to decrease methane emissions in ruminants better than fibrous diets, such as those containing beet pulps. Non-structural carbohydrates such as starch and sugars are associated with higher ruminal fermentation rates and accelerated feed turnover which cause a change in the rumen physico-chemicals and a shift in the microbial population. A shift in VFA (volatile fatty acid) production from acetate towards propionate occurs with the development of starch-fermenting microbes. This results in lower methane production because the relative proportion of ruminal hydrogen sources declines whereas that of hydrogen sinks increases. As propionate production and methanogenesis are competing pathways, starch-fermenting bacteria can compete with methanogens for hydrogen, therefore less methane would be produced in the rumen (Moss et al., 2000). For this reason, maize silage or whole-crop silage can reduce methane production in the rumen (Haque, 2018).

Manipulating feed in a manner which will improve feed utilisation and ameliorate product yields while reducing methane emissions will be beneficial for farm production and preferable for the environment.

3 Nitrous oxide and carbon dioxide in agriculture

Nitrous oxide (N_2O) is a potent GHG with a 100-year global warming potential 298 times greater than carbon dioxide (EPA, 2017). Currently, the main sources of anthropogenic N_2 emissions are agriculture, industry, biomass burning and indirect emissions from reactive nitrogen, leaching and atmospheric degradation (Reay et al., 2012). When considering direct agricultural emissions, 38% is attributed to N_2O, 32% to methane from ruminants, 12% from biomass burning, 11% from rice production and 7% from manure management (Bellarby et al., 2008). Livestock-related nitrous oxide emissions are estimated to total between 1 and 2 million tonnes of nitrous oxide-N each year, mainly due to animal waste. Nitrous oxide from synthetic fertilisers, manure applications and crop residues left on farms account for over 40% of total agricultural emissions (WRI, 2014). Nitrous oxide is an intermediate gas for both nitrification (transformation from ammonium to nitrate) and denitrification (the biological reduction of nitrate to N_2 gas), and these processes are both facilitated by microbial action (Mosier et al., 1998). The amount of N_2O released depends on the system and duration of waste management.

Nitrogenous fertilisers and manure pits combine to drive the growth of these emissions. Fertilisers are in general applied in excess and not fully absorbed by the plants themselves, which leads to only 50% recovery of fertilizer N in global crop production (Eickhout et al., 2006). Consequently, a great proportion accumulates in soil and is either lost directly as nitrous oxide, or leaches into water courses, enhancing downstream, indirect N_2O emissions. The amount lost will greatly depend on many other factors such as climate, soil and management practices (Brentrup et al., 2004; Eickhout et al., 2006).

Fertilisers containing N compounds consume up to 10 times more energy and consequently result in more GHG emissions, than fresh manure which is a low C-emitting alternative. Fertilisers are commonly used in agriculture, with the production of fertilisers emitting ~1.2% of the world's total GHGs (Wood and Cowie, 2004). Efficacy in the manufacturing of fertilisers can contribute to a significant reduction in nitrous oxide levels. Improvements would be related to greater energy efficiency in ammonia production plants, introduction of new nitrous oxide reduction technologies and other general energy-saving measures in manufacturing. With an increasing population and a demand for greater food production, N_2O emissions are likely to continue to rise in the coming decades (Reay et al., 2012).

Carbon dioxide is a colourless, odourless gas, released through natural processes such as respiration and volcanic eruptions, as well as human activities of deforestation, land-use changes and the burning of fossils fuels. Increasing concentrations of atmospheric CO_2 and other radioactive greenhouse gases, will ultimately lead to profound effects in the ecosystem. CO_2 does not break down easily in the atmosphere and can persist for several centuries.

Carbon sequestration refers to the process by which atmospheric CO_2 is transferred to soil or vegetation (Teagasc, 2017). The earth's soils contain approximately 1500 Pg (Picogram) of carbon, making it the largest surface of terrestrial carbon (C) (Post et al., 1990). Agricultural soil can act as both a source and sink of atmospheric CO_2 because it not only produces C but can also store C in soil and vegetation (Baah-Acheamfour et al., 2016; Paustian et al., 2000). Soil organic carbon (SOB) is influenced by the physical and chemical environments of the soil (e.g. moisture, temperature, aeration, pH and nutrient availability), the characteristics of the organic matter (i.e. susceptibility to microbial decay) and the physical accessibility of the organic matter to microbes (Paustian et al., 2000). Reconstructions of global land-use change suggest that terrestrial ecosystems have contributed as much as half of the increases in CO_2 emissions from human activity in the past two centuries (Post et al., 1990; Houghton and Skole, 1990).

Current knowledge suggests that agricultural soils have the capabilities to act as CO_2 sinks, however, this is dependant on changes in management practices. Management adaptation strategies which may be incorporated to reduce CO_2 emissions from agricultural soils include: (1) reduced tillage (2) cropping intensification and increased production efficiency (Paustian et al., 2000).

Arguments in favour of using agricultural soil C sequestration as a mitigation option are that additional benefits such as improving soil and water quality, reducing erosion, enhancing better soil fertility and crop production will rise from increasing soil organic matter (Paustian et al., 2000). However, carbon sequestration may lead to important water and nutrient depletion and increased soil salinity and acidity (Jackson, 2005). Although sequestration reduces the levels of CO_2 in our atmosphere, these negative effects on crop yield are not favoured, with an ever-increasing world population and demand for food.

4 Direct-fed microbials (DFMs)

Direct-fed microbials (DFMs) refer to microorganisms which are used to supplement feed to exert a beneficial effect on the animal. They contain live, viable cells, rather than additives which may only contain bacterial constituents. The term probiotic and DFM can be used interchangeably. The US Food and Drug Administration (FDA) authority define DFMs as 'products that are purported to contain live (viable) microorganisms (bacteria and/or yeast)'. DFMs are regulated as feed ingredients by the American Association of Feed Control Officials (AAFCO) and the FDA. DFMs are provided to the ruminant in the form of a bolus or mixed in with feed (Khan and Oh, 2015). According to USDA's National Animal Health Monitoring System's (NAHMS) Dairy 2007 study, 20% of dairy and heifer operations used DFMs for preventative purposes, an increase from 14.4% in 2002. In the European Union, approximately 20 microbial feed additives are authorized for use (Meieregger et al., 2010).

DFMs have been shown to limit gastrointestinal infections and provide optimally regulated environments in the digestive tract (Seo et al., 2010). DFMs detoxify toxic compounds, modulate the innate immune system, and maintain optimal gut movement and mucosal integrity of the intestine (Kumar et al., 2015). These effects have been mainly shown in pre-ruminants, where their benefits include a reduction in the incidence of diarrhoea, a decrease in faecal shedding of coliforms, promotion of ruminal development, improved feed efficiency, increased body weight gain and reduction in morbidity (Krehbiel et al., 2003).

In adult ruminants, there is little research available in relation to the efficacy of DFMs containing lactic acid bacteria (LAB). The use of yeast as a DFM has shown varied results. *Saccharomyces cerevisiae* was shown to reduce methane by 6–10% with varying concentrations (Lila et al., 2004). Adversely, the use of three bacterial DFM treatments of *Propionibacterium freudenreichii* 53-W, *Lactococcus pentosus* D31 and *Lactococcus bulgaricus* D1 did not alter ruminal fermentation and failed to reduce methane emissions in lactating primiparous cows on a high-starch or high-fibre diet (Jeyanathan et al., 2019).

Due to the restriction of using antibiotics as animal supplements, DFM use has become more popular due to their potential to influence the rumen environment and enhance feed efficiency. There have been several recent studies (Table 1) which investigated the use of DFMs as methane mitigators, applied alone or in combination with other treatment methods.

4.1 Types of DFMs

Many types of DFMs exist, ranging from bacterial, to yeast and fungal sources. Natural methods for reducing GHGs also exist, such as seaweed and other organic feed supplements. The type of DFM used varies on its effectiveness and intended use. This review aims to outline both the most common DFMs used, as well as mentioning new feed supplement strategies that are also aiming to reduce our GHG levels.

4.1.1 Lactic acid bacteria

Lactic acid bacteria (LAB) are an order of gram-positive bacteria which have a G+C content below 55mol% (König et al., 2009) are acid-tolerant, generally non-sporulating, non-respiring, either rod or coccus shaped bacteria that share common metabolic and physiological characteristics. The LAB group is usually reserved for the genera such as *Lactobacillus, Leuconostoc, Pediococcus* and *Streptococcus*, as well as others.

LAB produce a variety of inhibitory compounds such as organic acids, hydrogen peroxide and ethanol, which may offer them a competitive advantage in the ruminal ecosystem by inhibiting pathogenic microbial species (McAllister and Newbold, 2008). In addition, LAB produce antimicrobial peptides such as bacteriocins, ribosomially synthesised proteins produced by a bacterium of one strain, which are active against those of a closely related strain (Yang et al., 2014). Bacteriocins are deemed safe, since they are non-hazardous to eukaryotic cells.

LAB are good candidates to use as DFM because they are environmentally robust and have a number of mechanisms whereby, they may alter or influence neighbouring microbial communities with a beneficial effect on the animal (McAllister et al., 2011). LAB used as DFMs may produce lactic acid, which results in a lower pH in the rumen environment.

4.1.2 Lactic acid-utilizing bacteria

Lactic acid-utilizing bacteria such as *Megasphaera elsedenii, Propionibacterium shermanii* and *P. jensenii* have also been proposed as DFMs and have been used

Table 1 Detailing the use of DFM as feed additives and their effect on enteric methane production

Study	Type of Diet	DFM used	Effect on methane	Effect on animal welfare/ productivity
Oh et al. (2019) Effects of *Saccharomyces cerevisiae*-based DFM and exogenous enzyme products on enteric methane emission and productivity in lactating dairy cows.	The basal diet consisted of (dry matter basis) 60% forage and 40% concentrates and contained 16.5% crude protein and 32.0% neutral detergent fibre.	*S. cerevisae* (SDM) and exogenous enzyme product (ENZ).	No effect.	SDM increased milk yield by 2 kg/d without affecting DMI or feed efficiency. Supplementation of the diet with ENZ did not affect DMI, milk yield or feed efficiency.
Meller et al. (2019) Potential roles of nitrate and live yeast culture in suppressing methane emission and influencing ruminal fermentation, digestibility and milk production in lactating Jersey cows.	Ground corn.	Live yeast culture and nitrate.	Nitrate decreased methane by 17% but decreased dry matter intake by 10% (from 19.8 to 17.8 kg/d) such that methane:dry matter intake ratio numerically decreased by 8%.	Milk and milk fat production were not affected, but NO_3− decreased milk protein from 758 to 689 g/d.
Deng et al. (2018) Ruminal fermentation, nutrient metabolism and methane emissions of sheep in response to dietary supplementation with *Bacillus licheniformis*.	Total mixed ration (TMR).	Spore-forming B*acillus licheniformis*.	Daily methane production in the treatment groups was lower than in the control.	Dietary *B. licheniformis* supplementation effectively increased energy and protein utilisation in the sheep.

Jeyanathan et al. (2019) Bacterial DFMs fail to reduce methane emissions in primiparous lactating dairy cows.	Cows were randomly divided into two groups that were fed a corn silage-based, high-starch diet (HSD) or a grass silage-based, high-fibre diet (HFD).	DFM treatments: *Propionibacterium freudenreichii* 53-*Lactobacillus pentosus* D31 *Lactobacillus bulgaricus D1*.	No mitigating effect of DFM was observed on methane emissions in dairy cows.	The effect of DFM on milk fatty acid composition was negligible. *Propionibacterium* and *L. pentosus* DFMs tended to increase body weight gain of cows.
Ellis et al. (2016) DFM inoculated silage with LAB.	Treatment with one of four grouping, with inoculant long term and short term. Diets consisted of grass silage and concentrate (75:25 on a dry matter basis).	*Lactobacillus plantarum*, *Lactococcus lactis* and *Lactobacillus buchneri*.	Methane levels were not affected.	Dry matter intake, energy, milk and fat composition were not affected.
Thota et al. (2017) Effect of probiotic supplementation on nutrient digestibilities, growth performance and enteric methane emissions in Deccani ram lambs.	12 Deccani ram lambs of uniform body weight (16.5±0.64 kg with 130.11±3.00 days of age) were randomly allotted to two treatments in a completely randomized design. Animals were fed basal diet (chopped sorghum stover), concentrate and chopped green fodder.	*S. cerevisiae47*, *S. boulardii*, *L. acidophilus* and *P. freudenreichii*	Mean enteric methane treatment was 21.9% less than control group.	Feed efficiency of the animals was improved.

successfully to decrease concentrations of lactate and maintain ruminal pH. Since propionate is the major precursor for gluconeogenesis in early lactation dairy cows (Reynolds et al., 2003), increments of propionate production in the rumen result in increases of hepatic glucose production (Stein et al., 2006), providing more substrates for lactose synthesis, improving energetic efficiency and reducing ketosis (Weiss et al., 2008). In addition, increased propionate may reduce hydrogen available for methane production in the rumen.

4.1.3 Yeast

The most commonly used DFM is the anaerobic yeast *Saccharomyces cerevisae* and the filamentous fungus *Aspergillus oryzae* (Phillipeau et al., 2017). Traditionally, yeast products were used as feed additives to improve animal health and welfare, thereby improving animal performance and its effect as a probiotic is well-established (Darabighane et al., 2019). As such, supplementation with yeast may indirectly reduce methane production per protein (milk and meat) produced through enhancing ruminal fibre degradation and overall feed conversion efficiency (Bayat et al., 2015).

Yeast can modify rumen fermentation in a manner that can potentially reduce methane formation by decreasing rumen pH or favouring the production of certain VFAs such as acetate, propionate and butyrate (Chung et al., 2011; Iqbal et al., 2008). This is dependent on the diet offered to the animal (Islam and Lee, 2019). It has been suggested that live yeast promote the use of hydrogen by ruminal acetogens and drive the fermentation process towards acetate production instead of methane formation (Kataria, 2015). Additional proposed mechanisms by which yeast reduce methane is by reducing protozoan numbers. High populations of protozoa generally are associated with higher ruminal ammonia concentrations and increased methane production.

5 Direct-fed microbials (DFMs) and greenhouse gas (GHG) reduction

Of all agriculture GHGs, methane is the most impactful. DFMs have been explored in terms of methane reduction; however, research is still lacking for the use of DFMs against CO_2 and N_2O.

5.1 Methane

There is an increasing interest in exploring the use of naturally occurring feed additives. DFMs are already used to improve productivity and health of ruminant livestock, therefore they can be used as a possible option to reduce methane (Jeyanathan et al., 2014). Although the exact mechanism for methane reduction by DFMs has not been elucidated, it is thought that DFMs are responsible for the redirection of H_2 away from the methanogenesis pathway, as well as decreased production of H_2 during feed fermentation (Jeyanathan et al., 2011). Rumen methanogens produce methane by reducing CO_2 using H_2; therefore H_2 is a limiting substrate for methanogenesis. The amount of hydrogen produced in the rumen is highly dependent on the diet and type of rumen microbes as the microbial fermentation of feeds produces different end products that are not equivalent in terms of hydrogen output (Mirzaei-Aghsaghali and Maheri-Sis, 2011).

Proposed mechanisms of action include: (1) increased butyrate or propionate production, which may result in reduced methane production due to the utilisation of metabolic H_2 by acetogenic bacteria to produce acetate (Lila et al., 2004); (2) decrease in the number of ciliate protozoa in the rumen (Broucek, 2018); high populations of protozoa generally are associated with higher ruminal ammonia concentrations and increased methane production; this suggests that protozoa themselves or associated bacteria actively degrade dietary proteins and are methanogenic; (3) increase in lactic acid-utilizing bacteria, resulting in a reduction of lactic acid, leading to a more stable ruminal environment (Boadi et al., 2004).

Although the mechanism of action is not fully understood, studies have been performed on a broad range of DFMs for methane mitigation, with varied results. It is proposed that because some strains of yeast increase rumen bacterial growth (Chaucheyras-Durand et al., 2008), less methane may be produced due to a shift in partitioning of hydrogen between microbial cells and fermentation products (Newbold and Rode, 2006).

The scientific literature points to the idea that although the concept of using DFMs as mitigators of enteric methane emission is not novel, scientists have not yet elucidated the exact mechanism by which this occurs. Doyle et al. (2019) summarises studies which have used LAB as DFM to reduce methane, and although methane inhibition *in vitro* is evident, these seem to yield mixed results where efficacy tends to be strain-specific and dependent on delivery mode. Furthermore, a limited number of available animal trials make it difficult to draw an outright conclusion, therefore more research is needed to identify whether the use of DFM supplements can the use of DFM supplements can be used effectively to mitigate methane generation in ruminant livestock.

The use of DFMs as feed additives and their effect on enteric methane production has been tested with a range of diets, organisms and ruminants, of which is summarised in Table 1. Many trials incorporate the use of a combination of DFMs, such as yeast and LAB (Thota et al., 2017).

Yeast such as *A. oryzae* has been seen to reduce methane by 50% which was correlated directly to a 45% decrease in the population of protozoa by

Frumhloz et al. (1989). More recently, Mwenya (2004) showed promising results when yeast was added as part of feed for sheep, where a reduction ($P<0.05$) in methane production (L per day) was observed in the supplemented diets compared to control. Conversely, Martin and Nisbet (1990) found that adding *S.cerevisae* and *A.oryzae* resulted in increased methane production.

Methane reduction by yeast may be strain-dependent and variable in the impact on the rumen, as indicated in a study by Chung et al. (2011), where 15 non-lactating Holstein cows were fed a diet containing two strains on *S.cerevisae* at $1x10^{10}$ CFU per cow daily. Results indicated that yeast strain 1 did not affect methane emission intensity compared with the control treatment, while strain 2 lowered methane emissions but increased the risk of acidosis. More research is needed to screen yeast strains which can confer a probiotic benefit when used in feed and reduce methane simultaneously.

In general, studies investigating the effect of *S.cerevisae* on enteric methane production have shown differing conclusions. A meta-analysis compiled by Darabighane et al. (2019) comparing 46 scientific publications on dairy and beef cattle, concluded that supplementation of yeast does not significantly reduce methane production or methane/DMI. A comparison of 11 studies of yeast- recipient versus control group indicated that the addition of yeast, when it reduced methane emissions was not statistically significant (standard mean deviation (SMD) = −0.051; P = 0.792). The authors of this meta-analysis suggested that the results should be interpreted with caution because they were based on a small number of studies. A compilation of studies investigating the effect using *S. cerevisiae* as a DFM for reducing rumen methanogenesis showed inconsistent results between *in vitro* and *in vivo* studies (Jeyanathan et al., 2014). This could be attributed to the discrepancies in experimental conditions.

In terms of LAB/LUB, a meta-analysis completed by Krehbiel et al. (2003), indicated a generally positive trend for the improved health of young cattle treated with DFM inoculants. The use of bacteriocin like particles from LAB and LUB is of particular interest. Bacteriocins have antimicrobial proteinaceous properties which are ubiquitous in nature and are produced by a range of gram-positive and gram-negative organisms. Bovicin, a bacteriocin produced by *Streptococcus bovis*, has been found to reduce methane in vitro by up to 50% (Lee et al., 2002). Bacteriocins can be useful in directly inhibiting methanogens and/or reducing H_2 to other reductive microorganisms (Garsa et al., 2019).

Nisin is composed of 34 amino acids with two structural domains and is classed as a lantibiotic. This bacteriocin produced by *Lactococcus lactis* species, with GRAS (generally regarded as safe) status, has also shown methane reduction potential. However, Nisin supports the propionate production and depicts selective activity against gram-positive bacteria of rumen origin; therefore, complete exploration must be carried out before its use as an animal additive (Garsa et al., 2019). *Megasphaera elsdenii* (one of the most important lactate-utilizing species in the rumen) has been reported to establish a successful DFM product, but ruminal pH and fermentation have been inconsistent (Klieve et al., 2003; Henning et al., 2010)

Bacterial DFMs do not always result in positive results. Jeyanathan et al. (2019) reported that DFM treatment results in no mitigation effect, as well as no effect on milk composition (Table 1). Similarly, it was seen that in a study using eight Holstein Friesian dairy cows, treated with both long- and short-term LAB inoculations of *Lactobacillus plantarum, Lactococcus lactis* and *Lactobacillus buchneri*, no methane reduction was observed (Ellis et al., 2016). Variances in strain and dose, as well as difference in basal silage conditions may be responsible for the lack of effects observed. There have been other attempts to inoculate the rumen with LAB and fungi (*Candida kefyr*) along with nitrate supplementation to both control methanogenesis and possibly prevent nitrite formation, but no consistent animal data, have been reported (Takahashi, 2011). Although fermentation of lactate to VFA would help prevent a decreased ruminal pH, introduction of lactate-producing DFM would require careful scrutiny in situations in which sub-acute rumen acidosis might occur (Hristov et al., 2013).

5.2 Nitrous oxide and carbon dioxide

Diet can play a fundamental role in manure emissions, as it influences the volume and composition of manure. In particular, diet affects the amount, form and partition of N excretion between urine and faeces, and the amount of fermentable organic matter excreted (Hristov et al., 2013). Reducing ruminally degradable protein concentration can reduce NH_3 (Ammonia) emissions from manure, through a marked reduction of urinary urea excretion, NH_3 concentration and potentially N_2O emissions from dairy manure (Forabosco et al., 2017). Feed additives and dietary manipulation options targeting nitrous oxide emissions are mostly studied in isolation, but can have unexpected synergistic or antagonistic effects. Further research in this field needs to be carried out to test this.

6 Strengths and challenges of direct-fed microbials (DFMs)

6.1 Consumer acceptance

Realising the GHG mitigation potential of agriculture is ultimately dependent on farm-level decisions based on how adoption will benefit the individual farmer (Chandra et al., 2016). Mitigation options that both reduce GHG emissions and increase farm productivity, that is, cost-effective practices, are more likely to be adopted (Smith et al., 2007, 2008) than practices which would negatively affect farmers' income.

6.2 Expenditure

Studies have shown that DFMs as probiotics offer economic advantages, with a cost savings of 1-5%. Initiatives such as Pasture Base Ireland aims to help farmers make better decisions around grassland management, thus ensuring that the grass offered to the animals is of the highest quality resulting in reduced methane emissions (Wims et al., 2010). This will reduce methane emissions by minimising the amount of silage and supplemental feed in the diet and improving feed quality (Analysis Ireland, 2020–2030). Table 2 summarises the possible benefits of using commercially available DFMs, which in the long run will provide financial benefits.

Table 2 Composition and benefits of commercially available DFM products

DFM name	Bacterial #'s	CFUs	Benefits
Generator D	Nine species of live bacteria	24 billion CFU per 2 g	Replaces digestive organisms, higher milk and meat production
Generator Elite	Fourteen species of live bacteria	264 billion CFU	Replaces beneficial organisms, supports appetite, helps maintain pH, aids DM intake
Generator ProSacc	Nine species of live bacteria	2.27 trillion CFUs	Lactic acid-producing bacteria, enzymes to break down feed, B vitamins
Generator ULTRA	Eight species of live bacteria	68.25 billion per feeding	Enzyme-producing bacteria, purified digestive enzymes
Generator PYK	Eight species of live bacteria	480 billion CFUs	Digestive supplement, ideal as a calf starter
ProP169	Propionibacterium freudenreicheii strain P169	60 billion CFUs per feeding	Increase rumen propionate yield
Bovine DFM Powder	Unknown	300 billion CFUs per pound	Improves palatability of feed and rumen function, increased fibre digestion, feed intake and milk/meat production
Blended Fermentation Product Series - COMBO	Five species of live bacteria	100 million - 4 billion CFUs per gram	Source of amino acids, cellular proteins, vitamins (B), minerals

6.3 Inconsistent results

There is inadequate evidence of the direct enteric methane mitigating effect of yeast and other DFMs. DFMs function as probiotics which stabilize pH and stimulate rumen function. These have been reported to result in significant improvement in animal productivity and enhanced feed efficiency. This ultimately reduces methane emission intensity. However the issue of maintaining antimethanogenesis in the rumen has not been fully studied and it seems that DFMs would have to be continuously administered to maintain their efficacy (Jeyanathan et al., 2011).

6.4 Alternative hydrogen sinks

It has been suggested that rumen function will be disrupted if methane production is significantly decreased by directly inhibiting methanogenic archaea without the provision of alternative hydrogen sinks (McAllister and Newbold, 2008). Of all the hydrogen sinks present in the rumen, methane production is the most important in that the activity of rumen methanogens lowers H_2 concentration to about 1 μM, which permits a more rapid fermentation of the feed, giving the animal access to more VFA in a given time (Wolin, 1979). Conversely, high concentrations of H_2 in the rumen slows the microbial activity of feed fermenters, potentially slowing down the conversion to VFAs (Buddle et al., 2011). It is therefore important to introduce a system that will ensure that the accumulation of hydrogen in the rumen is minimised.

It has been proposed that inhibiting methanogenesis could favour microbial biomass production as an alternative hydrogen sink (Henderson et al., 2015). Inhibiting methanogenesis could have consequences on microbial ATP generation. In anaerobic systems like the rumen, part of the negative Gibbs energy change associated with fermentation is used to generate ATP through substrate level- and electron transport-linked phosphorylation. Changes in the H_2 sink levels could also affect such energy generation pathways. Inhibiting methanogens without providing alternative electron acceptors commonly results in an overall reduction in feed digestibility and often, reduced animal productivity (Beauchemin, 2009).

7 Other methane mitigation methods

7.1 Rumen manipulation

Biochemical pathways and microorganisms that play a role in methanogenesis directly or indirectly can be targeted and modulated. The amount, type and rate of fermentation of dietary carbohydrates affect both the total amount and proportions of individual VFAs formed and, ultimately, the amount of methane produced. Methylotrophs are microorganisms that can use single carbon organic compounds such as methane and methanol (Jeyanathan et al., 2015). These are key players in the carbon cycle and can be used to compete against methanogens for substrates (Iguchi et al., 2015). Acetogens are microorganisms that generate acetate as an end product of anaerobic respiration, directly producing the H_2 required for methanogenic growth. Targeting acetogens to reduce their population levels in the rumen has been shown to divert hydrogen away from methane production during ruminal fermentation (McGinn et al., 2004). Nitrate can act as an alternative H_2 sink, which has the potential to reduce rumen methanogenesis. Nitrate-reducing bacteria such as *W.succinogenes* and *S.ruminantium* are present in concentrations of 10^6 cells/mL of rumen fluid (Jeyanathan et al., 2011); however, to compete with methanogenesis, they must be present in higher concentrations. Sulphate-reducing bacteria (SRB) can act both in a competitive or co-operative association with methanogens, depending on levels of sulphate available. When sulphate levels are unlimited, SRB compete with methanogens for H_2, formate and acetate. In depleted sulphate environments, SRB co-operate with methanogens by acting as net producers of H_2.

7.2 Methane inhibitors

Bromochloromethane (BCM) has been found to decrease methane production in ruminants and is considered one of the most effective mitigative strategies for methane inhibition (Martinez-Fernandez et al., 2016). The BCM dramatically increases H_2 expelled without affecting DM intake and feed digestibility. However, the use of BCM has been restricted due to its ozone-depleting capacity. Recently, another chemical, 3-nitrooxypropanol (3NOP), which has been developed synthetically has been shown to reduce methane emission levels by up to 30% without negatively affecting animal performance (Hristov et al., 2015). This non-toxic compound has shown promise as a feed additive as some studies have reported increase in animal weight gain in dairy cows (Haisan et al., 2014 and Hristov et al., 2015) while consistently decreasing methane production. 3NOP is a structural analogue which inhibits methanogens by binding to the active site of the enzyme methyl-coenzyme M reductase which is responsible for the final step in methanogenesis pathway (Martinez-Fernandez et al., 2018). This chemical methane inhibitor is still under development and requires registration as a zoo-technical feed additive under European Union legislation (Rooke et al., 2016).

In terms of its practicality as methane inhibitor on-farm, administering 3 NOP at doses of 111 mg/kg diet DM or less resulted in 21% reduction in methane. However, there was a large degree of uncertainty (95% confidence interval, 13–29%) due to the variation in dose and method of administering 3NOP. Current view of the manufacturers is that the method for administration would appear to be critical and currently recommendations include dosing of a total mixed ration to ensure coupling of feed intake to intake of methane inhibitor (Rooke et al., 2016).

The product could be supplied as a premix to be incorporated into diets on farm. The daily recommended dose of 106 mg/kg diet DM is small enough that it may be practicable to administer the compound as a bolus into the rumen to release 3NOP over an extended time period and therefore compatible with the grazing situation.

7.3 Archael phage

Bacteriophages, or phage, are viruses which attack bacteria. Although not yet isolated from the rumen, *Siphroviridae* phages have been reported to infect *Methanobrevibacter*, *Methanococcus* and *Methanobacterium* species (McAllister and Newbold, 2008). Phages are highly specific and their use as methane biocontrol agents would likely not disrupt the normal microbial composition of the rumen ecosystem.

7.4 Natural feed supplements

Recently there have been several studies investigating the effects of seaweed (or macroalgae) as a livestock feed supplement and its potential to reduce methane production in ruminants. Antimethanogenic compounds are known to exist naturally in seaweed. The tropical red algae species of *Asparagopsis taxiformis* produces the halogenated antimethanogenic compounds, bromoform and dibromochloromethane which have been shown to reduce methane production by up to 80% when fed to sheep (Li et al., 2016). Another red algae species, *Asparagopsis armata*, has been demonstrated to reduce methane emissions by around 20% in dairy cattle, which was increased to around 60% reduction with an increased inclusion level of 1% (Roque et al., 2019).

8 Conclusion

DFMs are a viable method for the reduction of GHG emissions in livestock. DFMs may be used in conjunction with existing technologies, that is, vaccines, and small molecule inhibitors. For this to work in on-farm operations, there needs to be a balance between the reduction of GHGs, cost-effectiveness, animal welfare and animal productivity. Essentially, improving feed quality and the overall efficiency of dietary nutrient can be an effective way of decreasing the intensity of GHG emissions from livestock.

9 Acknowledgements

METHLAB, refining DFMs and silage inoculants for reduction of methane emissions from ruminants has been funded by FACCE ERA-GAS, an EU ERA-NET Cofund programme, whereby national money is pooled to fund transnational projects, and the European Commission also provides co-funding for the action. FACCE ERA-GAS is the ERA-NET Cofund for Monitoring and Mitigation of Greenhouse Gases from Agri- and Silvi-culture, and comprises funding agencies and project partners from 19 organisations across 13 European countries. Teagasc, the Agriculture and Food Development Authority in Ireland, is the overall coordinator of the ERA-NET. METHLAB was one of 10 successful projects to be funded from 79 proposals from across Europe, the United States and New Zealand. The New Zealand contribution to the METHLAB project is funded by the New Zealand Government in support of the objectives of the Livestock Research Group of the Global Research Alliance on Agricultural Greenhouse Gases.

10 References

Baah-Acheamfour, M., Carlyle, C. N., Lim, S. S., Bork, E. W. and Chang, S. X. (2016). Forest and grassland cover types reduce net greenhouse gas emissions from agricultural soils. *Science of the Total Environment* 571, 1115-1127.

Bayat, A. R., Kairenius, P., Stefański, T., Leskinen, H., Comtet-Marre, S., Forano, E., Chaucheyras Durand, F. and Shingfield, K. J. (2015). Effect of camelina oil or live yeasts (*Saccharomyces cerevisiae*) on ruminal methane production, rumen fermentation, and milk fatty acid composition in lactating cows fed grass silage diets. *Journal of Dairy Science* 98(5), 3166-3181.

Beauchemin, K. (2009). Dietary mitigation of enteric methane from cattle. *CAB Reviews: Perspectives in Agriculture, Veterinary Science, Nutrition and Natural Resources* 4, 1-18

Bellarby, J., Foereid, B., Hastings, A. and Smith, P. (2008). *Cool Farming: Climate Impacts of Agriculture and Mitigation Potential*. Greenpeace International, Amsterdam, The Netherlands.

Boadi, D., Benchaar, C., Chiquette, J. and Masse, D. (2004). Mitigation strategies to reduce enteric methane emissions from dairy cows: update review. *Canadian Journal of Animal Science* 84(3), 319-335.

Brentrup, F., Küsters, J., Kuhlmann, H. and Lammel, J. (2004). Environmental impact assessment of agricultural production systems using the life cycle assessment methodology. I. theoretical concept of a lca method tailored to crop production. *European Journal of Agronomy* 20(3), 247-264.

Broucek, J. (2018). Options to methane production abatement in ruminants: a review. *Journal of Animal and Plant Sciences* 28, 348-364.

Buddle, B. M., Denis, M., Attwood, G. T., Altermann, E., Janssen, P. H., Ronimus, R. S., Pinares-Patiño, C. S., Muetzel, S. and Wedlock, D. N. (2011). Strategies to reduce methane emissions from farmed ruminants grazing on pasture. *The Veterinary Journal* 188(1), 11-17.

Chandra, N., Lal, S., Venkataramani, S., Patra, P. K. and Sheel, V. (2016). Temporal variations of atmospheric CO2 and CO at Ahmedabad in western India. *Atmospheric Chemistry and Physics* 16(10), 6153–6173.

Chaucheyras-Durand, F., Walker, N. D. and Bach, A. (2008). Effects of active dry yeasts on the rumen microbial ecosystem: past, present and future. *Animal Feed Science and Technology* 145(1–4), 5–26.

Chung, Y. H., Walker, N. D., McGinn, S. M. and Beauchemin, K. A. (2011). Differing effects of 2 active dried yeast (Saccharomyces cerevisiae) strains on ruminal acidosis and methane production in nonlactating dairy cows. *Journal of Dairy Science* 94(5), 2431–2439.

Darabighane, B., Salem, A. Z. M., Mirzaei Aghjehgheshlagh, F., Mahdavi, A., Zarei, A., Elghandour, M. M. M. Y. and López, S. (2019). Environmental efficiency of Saccharomyces cerevisiae on methane production in dairy and beef cattle via a meta-analysis. *Environmental Science and Pollution Research International* 26(4), 3651–3658.

Deng, K. D., Xiao, Y., Ma, T., Tu, Y., Diao, Q. Y., Chen, Y. H. and Jiang, J. J. (2018). Ruminal fermentation, nutrient metabolism, and methane emissions of sheep in response to dietary supplementation with Bacillus licheniformis. *Animal Feed Science and Technology* 241, 38–44.

Deppenmeier, U. (2002). The unique biochemistry of methanogens. *Progress in Nucleic Acid Research and Molecular Biology* 71, 223–283.

Doyle, N., Mbandlwa, P., Kelly, W. J., Attwood, G., Li, Y., Ross, R. P., Stanton, C. and Leahy, S. (2019). Use of lactic acid Bacteria to reduce methane production in ruminants, a critical review general characteristics of lactic acid. *Frontiers in Microbiology* 10, 1–19.

Eckhardt, M., Knowles, B., Maker, E. and Stork, P. 2009. IHS U.S. Industry highlights. (IHS) Houston, TX, Feb–Mar 2009. Available at: http://www.gecionline.com/2009-prt-7-final-reviews.

Eickhout, B., Bouwman, A. F. and Zeijts, H. (2006). The role of nitrogen in world food production and environmental sustainability. *Agriculture, Ecosystems and Environment* 116(1–2), 4–14.

Environmental Protection Agency (2013). *Environmental Protection Agency Global Emissions Database. Global Emissions*. EPA, Johnstown Castle, Co. Wexford. Available at: www.epa.gov/climatechange/ghgemissions/global.

Environmental Protection Agency (2017). *Ireland's Provisional Greenhouse Gas Emissions 1990–2017*. EPA, Johnstown Castle, Co. Wexford. Available at: http://www.Epa.ie/pubs/reports/air/ airemissions/ghgemissions2017.

EPA (2014). Press releases, EPA survey shows dioxin levels in the Irish Environment remain well below EU limits, Fire at Oxigen Environmental Waste Facility, Ballymount, Dublin 22, Fire out at Oxigen, Merrywell, Dublin, Exploring Irish wetland habitats: EPA launches new schools' resources, Increased coal fired electricity generation increases emissions of key air pollutant, A Year- Highlights from 2013. Ireland's environment starts in your backyard, EPA calls for the elimination of raw sewage. Environmental Protection Agency (EPA), Johnstown Castle, Co. Wexford.

Ellis, J. L., Hindrichsen, I. K., Klop, G., Kinley, R. D., Milora, N., Bannink, A. and Dijkstra, J. (2016). Effects of lactic acid bacteria silage inoculation on methane emission and productivity of Holstein Friesian dairy cattle. *Journal of Dairy Science* 99(9), 7159–7174.

Forabosco, E., Chitchyan, Zh. and Mantovani, R. (2017). Methane, nitrous oxide and mitigation strategies for livestock in developing countries: a review. *South African Journal of Animal Science* 47, 268–280. (3)

Frumholk, P. P., Newbold, C. J. and Wallace, R. J. (1989). Influence of Aspergillus oryzae fermentation extract on the fermentation of a basal ration in the rumen simulation technique (Rusitec). *Journal of Agricultural Sciences* 113, 169.

Garsa, A. K., Choudhury, P. K., Puniya, A. K., Dhewa, T., Malik, R. K. and Tomar, S. K. (2019) Bovicins: the bacteriocins of streptococci and their potential in methane mitigation. *Probiotics and Antimicrobial Proteins* 11(4), 1403–1413.

Haisan, J., Sun, Y., Guan, L. L., Beauchemin, K. A., Iwaasa, A., Du-val, S., Barreda, D. R. and Oba, M. (2014). The effects of feeding 3-ni-trooxypropanol on methane emissions and productivity of Holstein cows in mid lactation. *Journal of Dairy Science* 97(5), 3110–3119.

Haque, M., Pervin, M., Sultana, S. and Huq, S. (2018). Towards establishing a National Mechanism to address losses anddamages: a case study from Bangladesh. In: *Loss and Damage from Climate Change: Concepts, Methods and Policy Options*. Springer Open.

Henderson, G., Cox, F., Ganesh, S., Jonker, A., Young, W., Janssen, P. H. and Zunino, P. (2015). Rumen microbial community composition varies with diet and host, but a core microbiome is found across a wide geographical range. *Scientific Reports* 5(April), 14567.

Henning, P. H., Horn, C. H., Leeuw, K.-J., Meissner, H. H. and Hagg, F. M. (2010). Effect of ruminal administration of the lactateutilizing strain Megasphaera elsdenii (Me) NCIMB 41125 on abrupt or gradual transition from forage to concentrate diets. *Animal Feed Science and Technology* 157(1–2), 20–29.

Hook, S. E., Wright, A.-D. and Mcbride, B. W. (2010). Methanogens: methane producers of the rumen and mitigation strategies. *Archaea* (Vancouver, BC) 2010, 945785.

Houghton, R. A. and Skole, D. L. (1990). Carbon. In:, Turner, B. L., Clark, W. C., Kates, R. W., Richards, J. F., Mathews, J. T. and Meyer, W. B. (Eds) *The Earth as Transformed by Human Action* (pp 393–408). Cambridge University Press, Cambridge, UK.

Hristov, A. N., Kebreab, E., Niu, M., Oh, J., Bannink, A., Bayat, A. R., Boland, T. M., Brito, A. F., Casper, D. P., Crompton, L. A., Dijkstra, J., Eugène, M., Garnsworthy, P. C., Haque, N., Hellwing, A. L. F., Huhtanen, P., Kreuzer, M., Kuhla, B., Lund, P., Madsen, J., Martin, C., Moate, P. J., Muetzel, S., Muñoz, C., Peiren, N., Powell, J. M., Reynolds, C. K., Schwarm, A., Shingfield, K. J., Storlien, T. M., Weisbjerg, M. R., Yáñez-Ruiz, D. R. and Yu, Z. (2018). Symposium review: Uncertainties in enteric methane inventories, measurement techniques, and prediction models. *Journal of Dairy Science*. 101(7), 6655–6674.

Hristov, A. N., Oh, J., Firkins, J. L., Dijkstra, J., Kebreab, E., Waghorn, G., Makkar, H. P. S., Adesogan, A. T., Yang, W., Lee, C., Gerber, P. J., Henderson, B. and Tricarico, J. M. (2013). Special topics--Mitigation of methane and nitrous oxide emissions from animal operations: a review of enteric methane mitigation options. *Journal of Animal Science* 91(11), 5045–5069.

Hristov, A. N., Oh, J., Giallongo, F., Frederick, T. W., Harper, M. T., Weeks, H. L., and Duval, S. (2015). An inhibitor persistently decreased enteric methane emission from dairy cows with no negative effect on milk production. *Proceedings of the National Academy of Sciences of the United States of America* 112(34), 10663–10668. Available at: http://www.pnas.org/content/112/34/10663.abstract.

Iguchi, H., Yurimoto, H. and Sakai, Y. (2015). Interactions of methylotrophs with plants and other heterotrophic bacteria. *Microorganisms* 3(2), 137–151.

IPCC (2013). *Climate Change 2013: The Physical Science Basis. Contribution of Working Group I to the Fifth Assessment Report of the Intergovernmental Panel on Climate Change* [Stocker, T. F., Qin, D., Plattner, G.-K., Tignor, M., Allen, S. K., Boschung, J., Nauels, A., Xia, Y., Bex, V. and Midgley, P. M. (Eds)]. Cambridge University Press, Cambridge, UK and New York, NY, 1535 pp.

IPCC (2014). Climate Change 2014: Mitigation of Climate Change. Summary for Policymakers and Technical Summary. Climate Change 2014 Mitigation of Climate Change. Part of the Working Group III Contribution to the Fifth Assessment Report of the Intergovernmental Panel on Climate Change. Available at: https://doi.org/10.1017/CBO9781107415416.005.

Iqbal, M. F., Cheng, Y. F., Zhu, W. Y. and Zeshan, B. (2008). Mitigation of ruminant methane production: current strategies, constraints and future options. *World Journal of Microbiology and Biotechnology* 24(12), 2747–2755.

Islam, M. and Lee, S. S. (2019). Advanced estimation and mitigation strategies: a cumulative approach to enteric methane abatement from ruminants. *Journal of Animal Science and Technology* 61(3), 122–137.

Jackson, R. B., Jobbágy, E. G., Avissar, R., Roy, S. B., Barrett, D. J., Cook, C. W., Farley, K. A., Le Maitre, D. C., McCarl, B. A. and Murray, B. C. (2005). Atmospheric science: trading water for carbon with biological carbon sequestration. *Science* 310(5756), 1944–1947.

Jeyanathan, J., Kirs, M., Ronimus, R. S., Hoskin, S. O. and Janssen, P. H. (2011). Methanogen community structure in the rumens of farmed sheep, cattle and red deer fed different diets. *FEMS Microbiology Ecology* Federation of European Microbiological Societies 76, 311–326.

Jeyanathan, J., Martin, C., Eugène, M., Ferlay, A., Popova, M. and Morgavi, D. P. (2019). Bacterial direct-fed microbials fail to reduce methane emissions in primiparous lactating dairy cows. *Journal of Animal Science and Biotechnology* 10, 41.

Jeyanathan, J., Martin, C. and Morgavi, D. P. (2014). The use of direct-fed microbials for mitigation of ruminant methane emissions: a review. *Animal* 8(2), 250–261.

Johnson, K. A. and Johnson, D. E. (1995). Methane emission from cattle. *Journal of Animal Science* 73(8), 2483–2492.

Kataria, R. P. (2015). Use of feed additives for reducing Greenhouse gas emissions from dairy farms. *Microbiology Research* 6(1), 19–25.

Khan, I. and Oh, D.-H. (2015). Integration of nisin into nanoparticles for application in foods. *Innovative Food Science and Emerging Technologies* 34, 376–384.

Klieve, A. V., Hennessy, D., Ouwerkerk, D., Forster, R. J., Mackie, R. I. and Attwood, G. T. (2003). Establishing populations of Megasphaera elsdenii YE 34 and Butyrivibrio fibrisolvens YE 44 in the rumen of cattle fed high grain diets. *Journal of Applied Microbiology* 95(3), 621–630.

König, H., Unden, G. and Fröhlich, J. (2009). *Biology of Microorganisms on Grapes, in Must and in Wine*. Springer, Berlin Heidelberg, Available at: https://books.google.ie/books?id=6CR1vQUMw-sC.

Krehbiel, C. R., Rust, S. R., Zhang, G. and Gilliland, S. E. (2003). Bacterial direct-fed microbials in ruminant diets: performance response and mode of action. *Journal of Animal Science* 81(E, Suppl. 2), E120–E132.

Kumar, S., Indugu, N., Vecchiarelli, B. and Pitta, D. W. (2015). Associative patterns among anaerobic fungi, methanogenic archaea, and bacterial communities in response to changes in diet and age in the rumen of dairy cows. *Frontiers in Microbiology* 6, 1-10.

Lassey, K. R., Lowe, D. C. and Smith, A. M. (2007). The atmospheric cycling of radiomethane and the "fossil fraction" of the methane source. *Atmospheric Chemistry and Physics* 7(8), 2141-2149

Lee, S. S., Hsu, J. T., Mantovani, H. C. and Russell, J. B. (2002). The effect of bovicin HC5, a bacteriocin from Streptococcus bovis HC5, on ruminal methane production in vitro. *FEMS Microbiology Letters* 217(1), 51-55.

Li, X., Norman, H. C., Kinley, R. D., Laurence, M., Wilmot, M., Bender, H., de Nys, R. and Tomkins, N. (2016). Asparagopsis taxiformis decreases enteric methane production from sheep. *Animal Production Science* 58, 681-688.

Lila, Z. A., Mohammed, N., Yasui, T., Kurokawa, Y., Kanda, S. and Itabashi, H. (2004). Effects of a twin strain of Saccharomyces cerevisiae live cells on mixed ruminal microorizanism fermentation in vitro. *Journal of Animal Science* 82(6), 1847-1854.

Martin, S. A. and Nisbet, D. J. (1990). Effects of Aspergillus oryzae fermentation extract on fermentation of amino acids, Bermudagrass and starch by mixed ruminal micro-organisms in vitro. *Journal of Animal Science* 68(7), 2142-2149.

Martinez-Fernandez, G., Denman, S. E., Yang, C., Cheung, J., Mitsumori, M. and McSweeney, C. S. (2016). Methane inhibition alters the microbial community, hydrogen flow, and fermentation response in the rumen of cattle. *Frontiers in Microbiology* 7(Jul), 1122.

Martinez-Fernandez, G., Duval, S., Kindermann, M., Schirra, H. J., Denman, S. E. and McSweeney, C. S. (2018). 3-NOP vs. halogenated compound: methane production, ruminal fermentation and microbial community response in forage fed cattle. *Frontiers in Microbiology* 9, 1582.

McAllister, T. A., Beauchemin, K. A., Alazzeh, A. Y., Baah, J., Teather, R. M. and Stanford, K. (2011). Review: the use of direct fed microbials to mitigate pathogens and enhance production in cattle. *Canadian Journal of Animal Science* 91(2), 193-211.

McAllister, T. A. and Newbold, C. J. (2008). Redirecting rumen fermentation to reduce methanogenesis. *Australian Journal of Experimental Agriculture* 48(1-2), 7-13.

McGinn, S. M., Beauchemin, K. A., Coates, T. and Colombatto, D. (2004). Methane emissions from beef cattle: effects of monensin, sunflower oil, enzymes, yeast, and fumaric acid. *Journal of Animal Science* 82(11), 3346-3356. The online version of this article , along with updated information and services , is located on the World Wide Web at: methane emissions from b, 4453(October 2015).

McMichael, A. J., Powles, J. W., Butler, C. D. and Uauy, R. (2007). Food, livestock production, energy, climate change, and health. *Lancet* 370(9594), 1253-1263.

Meieregger, A., Mayrhuber, E. and Lettner, H. (2010). Probiotics and health claims: the perspective of the feed industry. *Probiotics and Health Claims*, 223-248. doi: 10.1002/9781444329384.ch15.

Meller, R. A., Wenner, B. A., Ashworth, J., Gehman, A. M., Lakritz, J. and Firkins, J. L. (2019). Potential roles of nitrate and live yeast culture in suppressing methane emission and influencing ruminal fermentation, digestibility, and milk production in lactating Jersey cows. *Journal of Dairy Science* 102(7), 6144-6156.

Mirzaei-Aghsaghali, A. and Maheri-Sis, N. (2011). Factors affecting mitigation of methane emissions from Ruminants I : feeding strategies. *Asian Journal of Animal and Veterinary Advances* 6(9), 888–908.

Mosier, A., Kroeze, C., Nevison, C., Oenema, O., Seitzinger, S. and Cleemput, O. (1998). Closing the global N_2O budget: nitrous oxide emissions through the agricultural nitrogen cycle. *Nutrient Cycling in Agroecosystems* 52, 225–248.

Moss, A. R., Jouany, J. P. and Newbold, J. (2000). Methane production in ruminants: its contribution to global warming. *Annales de zootechnie* (INRA/EDP Sciences) 49(3), 231–253.

Mwenya, B., Santoso, B., Sar, C., Gamo, Y., Kobayashi, T., Arai, I. and Takahashi, J. (2004). Effects of including β 1 – 4 galacto-oligosaccharides, lactic acid bacteria or yeast culture on methanogenesis as well as energy and nitrogen metabolism in sheep. *Animal Feed Science and Technology* 115, 313–326.

Newbold, C. J. and Rode, L. M. (2006). Dietary additives to control methanogenesis in the rumen. *International Congress Series* 1293, 138–147.

Oh, J., Harper, M., Melgar, A., Compart, D. M. P. and Hristov, A. N. (2019). Effects of Saccharomyces cerevisiae-based direct-fed microbial and exogenous enzyme products on enteric methane emission and productivity in lactating dairy cows. *Journal of Dairy Science* 102(7), 6065–6075.

Patra, A. K. (2012). Enteric methane mitigation technologies for ruminant livestock: A synthesis of current research and future directions. *Environmental Monitoring and Assessment* 184(4), 1929–1952.

Paustian, K., Six, J., Elliott, E. and Hunt, H. (2000). Management options for reducing CO2 emissions from agricultural soils. *Biogeochemistry* 48, 147–163.

Phillipeau, C., Lettat, A., Martin, C., Silberberg, M., Morgavi, D. P., Ferlay, A., Berger, C. and Nozière, P. (2017). Effects of bacterial direct-fed microbials on ruminal characteristics, methane emission, and milk fatty acid composition in cows fed high- or low-starch diets. *Journal of Dairy Science* 100(4), 2637–2650.

Post, W. M., Peng, T.-H., Emanuel, W. R., King, A. W., Dale, V. H. and DeAngelis, D. L. (1990). The global carbon cycle. *American Scientist* 78, 310–326.

Reay, D. S., Davidson, E. A., Smith, K. A., Smith, P., Melillo, J. M., Dentener, F. and Crutzen, P. J. (2012). Global agriculture and nitrous oxide emissions. *Nature Climate Change* 2(6), 410–416.

Reynolds, C. K., Aikman, P. C., Lupoli, B., Humphries, D. J. and Beever, D. E. (2003). Splanchnic metabolism of dairy cows during the transition from late gestation through early lactation. *Journal of Dairy Science* 86(4), 1201–1217.

Rice, A. L., Butenhoff, C. L., Teama, D. G., Röger, F. H., Khalil, M. A. K. and Rasmussen, R. A. (2016). Atmospheric methane isotopic record favors fossil sources flat in 1980s and 1990s with recent increase. *Proceedings of the National Academy of Sciences of the United States of America* 113(39), 10791–10796.

Rooke, J. A., Miller, G., Flockhart, J., Mcdowell, M. and Macleod, M. (2016). Nutritional strategies to reduce enteric methane emissions. *Scottish Government, Climate Change and Agriculture* 1, 1–40.

Roque, B. M., Salwen, J. K., Kinley, R. and Kebreab, E. (2019). Inclusion of Asparagopsis armata in lactating dairy cows' diet reduces enteric methane emission by over 50 percent. *Journal of Cleaner Production* 234, 132–138.

Seo, J. K., Kim, S. W., Kim, M. H., Upadhaya, S. D., Kam, D. K. and Ha, J. K. (2010). Direct-fed microbials for ruminant animals. *Asian-Australasian Journal of Animal Sciences* 23(12), 1657–1667.

Singh, C. K., Kumar, A. and Roy, S. S. (2018). Quantitative analysis of the methane gas emissions from municipal solid waste in India. *Scientific Reports* 8(1), 2913.

Smith, P., Martino, D., Cai, Z., Gwary, D., Janzen, H., Kumar, P., McCarl, B., Ogle, S., O'Mara, F., Rice, C., Scholes, B., Sirotenko, O., Howden, S., Mcallister, T., Pan, G., Romanenkov, V., Schneider, U., Towprayoon, S., Wattenbach, M. and Smith, J. (2008). Greenhouse gas mitigation in agriculture. *Philosophical Transactions of the Royal Society of London. Series B, Biological Sciences* 363, 789–813.

Smith, P., Martino, D., Zucong, C., Gwary, D., Janzen, H., Kumar, P., McCarl, B., Ogle, S., O'Mara, F., Rice, C., Scholes, B., Sirotenko, O., Howden, M., McAllister, T., Pan, G., Romanenkov, V., Schneider, U., Towprayoon, S., Wattenbach, M., and Smith, J. (2007). Greenhouse gas mitigation in agriculture. *Philosophical Transactions of the Royal Society*. 363(1492):789–813.

Stein, D. R., Allen, D. T., Perry, E. B., Bruner, J. C., Gates, K. W., Rehberger, T. G., Mertz, K., Jones, D. and Spicer, L. J. (2006). Effects of feeding propionibacteria to dairy cows on milk yield, milk components, and reproduction. *Journal of Dairy Science* 89(1), 111–125.

Takahashi, J. (2011). Some prophylactic options to mitigate methane emission from animal agriculture in Japan. *Asian-Australasian Journal of Animal Sciences* 24(2), 285–294.

Tapio, I., Snelling, T. J., Strozzi, F. and Wallace, R. J. (2017). The ruminal microbiome associated with methane emissions from ruminant livestock. *Journal of Animal Science and Biotechnology* 8(1), 7.

Teagasc Silage Analysis, Sinead Deavany (2017). *Silage Analysis – Why It's Important and What It All Means*. Media Article. Available at: https://www.teagasc.ie/publications/2017/silage-analysis--why-its-important-and-what-it-all-means.php. Accessed: 23 November 2017.

Thota, P., Sarat, C. A., Mahender, M. and Ramana, D. B. V. (2017). Effect of probiotic supplementation on nutrient digestibilities, growth performance and enteric methane emissions in Deccani ram lambs. *Journal of Animal Research* 7(6), 1009–1017.

Wanapat, M., Cherdthong, A., Phesatcha, K. and Kang, S. (2015). Dietary sources and their effects on animal production and environmental sustainability. *Animal Nutrition* 1(3), 96–103.

Weiss, W. P., Wyatt, D. J. and McKelvey, T. R. (2008). Effect of feeding propionibacteria on milk production by early lactation dairy cows. *Journal of Dairy Science* 91(2), 646–652.

Wims, C. M., Deighton, M. H., Lewis, E., O'Loughlin, B., Delaby, L., Boland, T. M. and O'Donovan, M. (2010). Effect of pregrazing herbage mass on methane production, dry matter intake, and milk production of grazing dairy cows during the mid-season period. *Journal of Dairy Science* 93(10), 4976–4985.

Wolin, M. J. (1979). The rumen fermentation: a model for microbial interactions in anaerobic ecosystems. *Advances in Microbial Ecology* 3, 49–77.

Wood, S. and Cowie, A. (2004). *A Review of Greenhouse Gas Emission Factors for Fertiliser Production*.

World Resource Institute (WRI) (2014). *GHG Protocol Agricultural Guidance Interpreting the Corporate Accounting and Reporting Standard for the Agricultural Sector*, April 2014, p. 103.

Wright, L. A., Kemp, S. and Williams, I. (2011). 'Carbon footprinting': towards a universally accepted definition. *Carbon Management* 2(1), 61–72.

Yang, C., Rooke, J. A., Cabeza, I. and Wallace, R. J. (2016). Nitrate and inhibition of ruminal methanogenesis: microbial ecology, obstacles, and opportunities for lowering methane emissions from ruminant livestock. *Frontiers in Microbiology* 7(Feb), 132.

Yang, S. C., Lin, C. H., Sung, C. T. and Fang, J. Y. (2014). Antibacterial activities of bacteriocins: application in foods and pharmaceuticals. *Frontiers in Microbiology* 5, 241.

Chapter 11

Modifying the rumen environment to reduce greenhouse gas emissions

Yajing Ban, University of Alberta, Canada; André L. A. Neves, Embrapa Dairy Cattle, Brazilian Agricultural Research Corporation (Embrapa), Brazil; Le Luo Guan, University of Alberta, Canada; and Tim McAllister, Lethbridge Research and Development Centre, Agriculture and Agri-Food Canada, Canada

1 Introduction

The dramatic increase in the human population - estimated to reach 9.7 billion people by the year 2050 - will require an approximately 25% increase in gross agricultural output between 2020 and 2050 to meet the global food demand of humanity (FAO 2018; Nations 2019). However, increased animal production has placed an added strain on the environment as a result of the production of greenhouse gases (GHGs) (Huws et al. 2018) and nutrient accumulation in intensive livestock systems (Tullo et al. 2019). Ruminant production produces a number of GHGs including carbon dioxide (CO_2) mainly due to the use of fossil fuels, methane (CH_4) from enteric fermentation and manure, and nitrous oxide (N_2O) from manure and nitrogen fertilizer (Lynch and Pierrehumbert 2019). CH_4 is a particularly prominent GHG, as it has a global warming potential that is 28 times greater than that of CO_2 (Jackson et al. 2019). Globally, it has been shown that ruminants contribute about 11% of total anthropogenic GHG production, with approximately 6% arising from enteric CH_4 from ruminants (Rojas-Downing et al. 2017; Grossi et al. 2019; Beauchemin et al. 2020). In addition to the negative impacts on the environment, enteric CH_4 emissions can also represent a 2–12% loss in gross energy intake (Johnson and Johnson

http://dx.doi.org/10.19103/AS.2020.0077.16
Published by Burleigh Dodds Science Publishing Limited, 2021.

1995). Theoretically, if this energy loss could be reduced through mitigation technologies, the energy saved could be redirected toward meat and milk, improving production efficiency. Although the carbon footprint on an intensity basis from livestock husbandry has declined in the past 50 years owing to improvements in animal production efficiency, a trend that is expected to continue, the increasing demand for animal protein may limit the achievement of the Paris accord target of limiting temperature increases to 1.5°C above pre-industrial levels (Ripple et al. 2014; Beauchemin et al. 2020; Leahy et al. 2020). Therefore, the mitigation of enteric CH_4 emissions is essential to a reduction of agricultural GHG emissions.

This chapter contains three sections: the first section highlights the rumen microbiome and its contribution to CH_4 emissions; the second section summarizes the main factors (rumen- and feed-associated) that influence CH_4 production in ruminants; the final section discusses the most promising current strategies used to reduce CH_4 emissions in ruminants.

2 Greenhouse gas production and the role of the rumen microbiome

2.1 Greenhouse gas production in ruminants

The major GHGs associated with ruminant production are CO_2, CH_4, and nitrous oxide (N_2O), which are emitted either directly from enteric fermentation (CH_4) and manure (CH_4, N_2O) or indirectly from land-use change, fertilizer use and burning of fossil fuels (CH_4, N_2O, CO_2) during feed production (Schils et al. 2005; Grossi et al. 2019). CO_2 emitted by respiration is not considered as a net contributor to GHGs as its release to the atmosphere (via respiration) is offset by plants that capture CO_2 to form biomass through photosynthesis (Steinfeld 2006). However, CO_2 generated as a result of the use of fossil fuels for transport of ruminant feed or products (e.g. milk, beef), generation of electricity and other on-farm practices may exceed 41 million tonnes per year (Steinfeld 2006). Thus, the two major GHGs emitted directly by ruminants are enteric CH_4 and N_2O generated from manure (Gerber et al. 2013). Combined, these two GHGs account for approximately 80% of the total emissions arising from the livestock sector and thus are significant contributors to the global GHG budget (Opio et al. 2013). It is estimated that 44% of the total GHG emissions produced by ruminants are in the form of CH_4 (Fig. 1a), which arises as a result of the reduction of CO_2 to CH_4 by archaea in the rumen and in manure (Gerber et al. 2013). The main sources of N_2O are derived from chemical fertilizer and N deposition from manure, which are influenced by manure management and indirect emissions that arise from the volatilization of NH_3 from intensive livestock systems and its downwind deposition. Nitrous oxide comprises 31% of the 7.1 Gt of CO_2 equivalents that are produced annually by the livestock sector (Gerber et al.

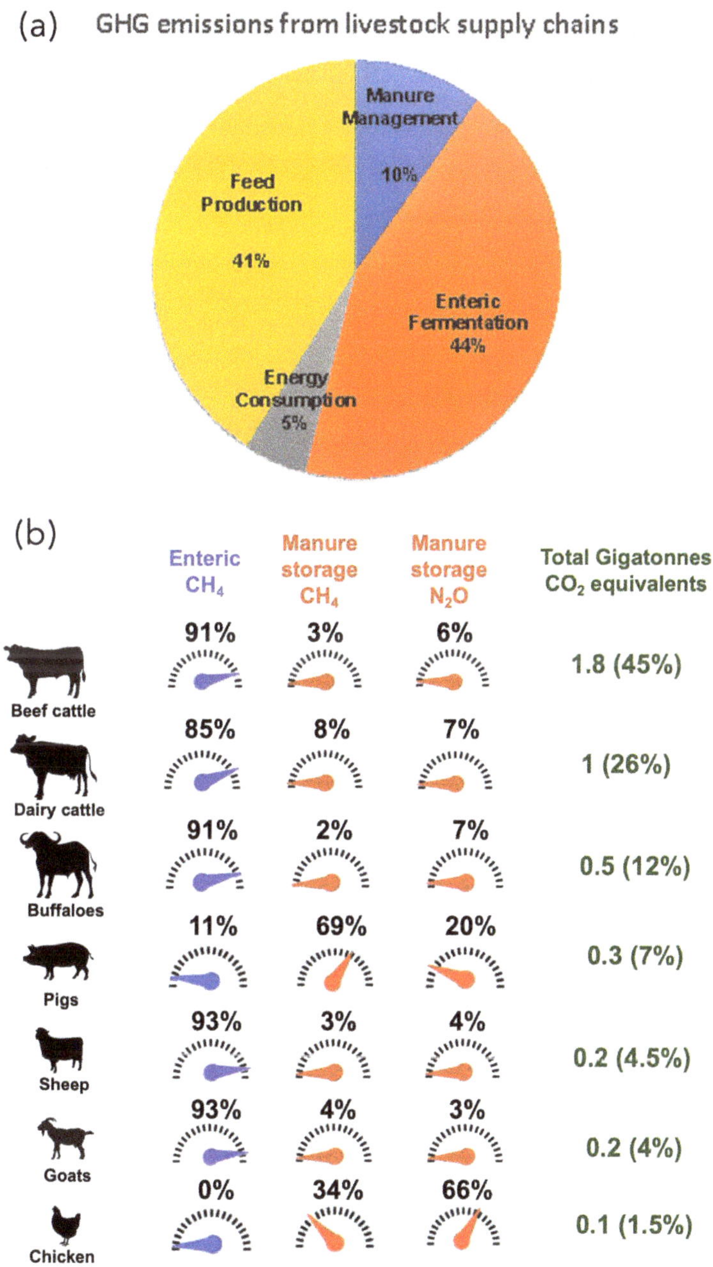

Figure 1 (a) Emissions in livestock supply chains from enteric fermentation, feed production, manure management and energy consumption (adapted from FAO 2017). (b) The farm-gate non-CO_2 emissions (%) from enteric fermentation and manure storage by livestock species in carbon dioxide equivalents (Grossi et al. 2019).

2013; Adler et al. 2015). On average, the enteric CH_4 in CO_2 equivalents accounts for over 90% of the farm-gate non-CO_2 emissions from ruminant production (Fig. 1b), with beef cattle producing the largest proportion of these emissions (Grossi et al. 2019). In all, these outcomes indicate that there is an urgent need for the development of strategies to mitigate GHG production by ruminants, while concurrently improving animal husbandry systems to meet the increasing demand for sustainable food production.

As enteric CH_4 is the largest contributor to GHG emissions from ruminants, research efforts have been primarily directed toward the development of strategies to reduce ruminal CH_4 production. CH_4 emissions are commonly expressed as CH_4 production (g/day), yield (g/kg dry matter intake) and intensity (g/kg average daily gain or energy corrected milk) (Niu et al. 2018; Ornelas et al. 2019). Rumen microbiota play a fundamental role in feed digestion and provision of energy for the host (Russell and Rychlik 2001), with CH_4 being a by-product of the rumen fermentation process. A variety of molecular, genomic, microbiological, nutritional and chemical approaches have been explored for their ability to reduce enteric CH_4 emissions, most of which have had limited success. The production and flow of reducing equivalents among the rumen microbiota is complex, and it is likely that any successful CH_4 mitigation strategy will be based on a fundamental understanding of the various roles of rumen microbiota in CH_4 production.

2.2 Rumen archaea and methanogenesis

In the rumen, archaea are one of the four main microbial groups. The domain archaea are divided into two different kingdoms: Euryarchaeota, consisting of methanogens and extreme halophytes, and Crenarchaeota, consisting of hyperthermophiles and nonthermophiles (Bayley et al. 1999). The assessment of the rumen archaeal community present in the ribosomal database project revealed 3516 archaeal sequences, representing an estimated 1500 species (Kim et al. 2011). Despite this diversity, studies have shown that 90% of rumen archaea are methanogens. Rumen methanogens are exclusively members of the Euryarchaeota, ranging from 10^6 to 10^8 cells per mL of rumen fluid, accounting for less than 4% of the rumen microbial population (Lin et al. 1997; Janssen and Kirs 2008). *Methanobrevibacter* (63.2% of the methanogenic population), *Methanomicrobium* (7.7%), *Methanosphaera* (9.8%), Rumen Cluster C (now referred as *Methanomassiliicoccaceae*, 7.4%), and *Methanobacterium* (1.2%) represent the majority of the methanogens in the rumen (Janssen and Kirs 2008; Patra et al. 2017).

Rumen methanogens produce CH_4 by transferring reducing equivalents formed by other microbiota (i.e. bacteria, fungi, and protozoa) to CO_2 (Deppenmeier 2002). Interspecies hydrogen transfer plays a key role in this process as it ensures a low partial pressure of H_2 (~162 Pa) in the rumen and avoids the inhibition of fermentation as a result of an accumulation of reduced co-factors (Janssen and Kirs 2008; Ungerfeld 2013; Rojas-Downing et al. 2017). During methanogenesis, reduced co-factors (e.g. NADH, NADPH, FADH and Fd) are oxidized and the reducing equivalents are transferred to methanogenic archaea through a series of biochemical steps to reduce CO_2 to CH_4 (Ungerfeld 2015). The hydrogenotrophic pathway is the predominant pathway for CH_4 production in the rumen, with *Methanobrevibacter* species accounting for the majority of this activity and typically accounting for over 90% of rumen archaeal 16S rRNA reads (Hristov et al. 2012). Most methanogens can also use formate as the electron donor for CO_2 reduction as a result of the activity of formate dehydrogenase. Formate is produced during fermentation by microorganisms that possess pyruvate-formate lyase or from the fermentation of plant metabolites like oxalic acid. Ruminal methanogens (e.g. *Methanosarcina, Methanosphaera, Methanimicrococcus*) can also utilize methanol, methylamines, and methylated sulphides as substrates for the synthesis of CH_4 (Ellis et al. 2008; Janssen and Kirs 2008; Poulsen et al. 2013; Tapio et al. 2017).

Recent studies have shown that the total number of methanogens in the rumen is not necessarily directly related to CH_4 yield, but rather it can be related to the abundance of particular archaeal species (e.g. *Methanobrevibacter gottschalkii*) (Zhou et al. 2011). However, it seems that the expression of genes involved in methanogenesis may be a more accurate predictor of rumen methanogenesis (Roehe et al. 2016) than the abundance of certain archaeal species (Zhou et al. 2011). The expression of genes within the methanogenesis pathway (e.g. *methyl-coenzyme M reductase alpha subunit - mcrA*) has been shown to be greater in high-CH_4-emitting cattle as compared to their low-emitting counterparts (Shi et al. 2014). It is worth noting that many factors influence rumen methanogenesis (e.g. diet, intake, rate of passage, rumen volume), and consequently the amount of CH_4 produced.

2.3 Other rumen microbiota and methane production

In the rumen, bacteria are the most abundant cellular microbiota (density of 10^{10}–10^{11} cells per mL rumen fluid), making up at least 50% of the microbial cell mass (Creevey et al. 2014). The rumen bacteria ferment feed to produce volatile fatty acids (VFA) and microbial protein, providing the main nutrients utilized by the host and other rumen microbiota (Ellis et al. 2008). Other bacterial fermentation end-products can also be used as substrates for methanogens (Zinder 1993).

Methanogens share a commensal relationship with rumen ciliates, and these protozoa-methanogen consortia account for approximately 30–40% of the enteric CH_4 produced in ruminants (Finlay et al. 1994). Protozoa are present at 10^4–10^6 cells per mL in rumen fluid and account for 20–50% of the rumen microbial biomass (McSweeney and Mackie 2012). They are involved in lipid hydrolysis and can produce large amounts of hydrogen via organelles known as hydrogenosomes (Tymensen et al. 2012), which provide substrates to methanogens via interspecies hydrogen transfer (Gijzen et al. 1988; Hobson and Stewart 1997; Kittelmann et al. 2015). Removal of protozoa from the rumen, a process known as defaunation, can suppress fiber digestion (Mosoni et al. 2011), but it is also consistently associated with a reduction in CH_4 production (Morgavi et al. 2012; Newbold et al. 2015). However, this reduction in CH_4 production is often transitory with emissions often returning to original levels within a month after defaunation as a result of an increase in hydrogen transfer between bacterial populations and methanogens (Guan et al. 2006).

Anaerobic rumen fungi are active players in the degradation of lignocellulosic feedstuffs and therefore produce reducing equivalents that are used by methanogens to reduce CO_2 to CH_4 (Gruninger et al. 2014). Anaerobic rumen fungi, such as *Neocallimastix frontalis*, have a synergistic relationship with methanogens through interspecies hydrogen transfer and also possess hydrogenosomes. This synergistic relationship increases the enzymatic activity of rumen fungi and their ability to degrade plant cell-wall carbohydrates and produce fermentation products (Hook et al. 2010). As a result of this relationship, greater fungal abundance has been associated with increased CH_4 emissions (Tapio et al. 2017). These findings suggest that rumen fungi are positively linked to methanogenesis through their provision of reducing equivalents for methanogens.

3 Factors influencing methane production in ruminants

As the largest forestomach in ruminants, the rumen is the primary site of CH_4 production within the digestive tract, with less than 5% of the emissions being emitted as flatulence (Lassey 2007; Hammond et al. 2016). Microbiological, physiological, nutritional and genetic factors all influence enteric CH_4 production in ruminants (Fig. 2).

3.1 Rumen-associated factors

3.1.1 Shifts in the rumen microbiome

It is clear that the rumen microbiome is associated with enteric CH_4 production and manipulating its structure and function through the diet is one of the simplest approaches in reducing CH_4 production in ruminants. For example, increasing the proportion of concentrate in the diet shifts the microbiota from

Published by Burleigh Dodds Science Publishing Limited, 2021.

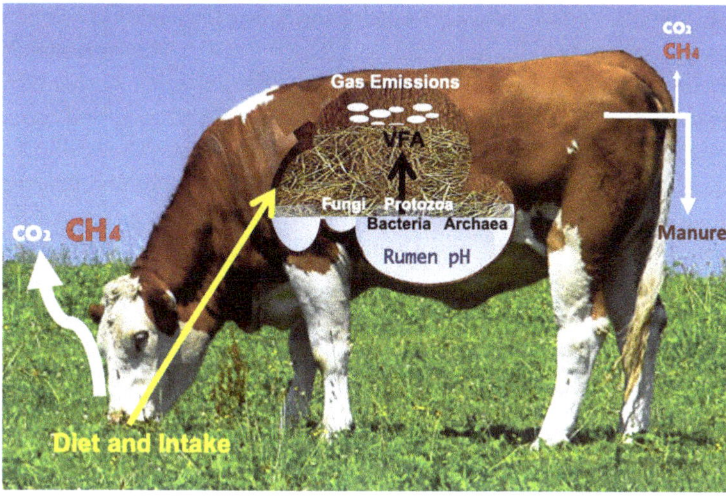

Figure 2 Rumen fermentation and factors associated with enteric CH_4 emissions. CH_4 is produced during rumen fermentation and influenced by rumen-associated factors (e.g. host genetics, rumen microbiota or physiological processes) as well as feed-associated factors (e.g. diet composition or feed intake). The weight of the white arrows represents the proportion of GHG emissions out of the mouth versus from manure and flatulence.

species that degrade plant cell-wall carbohydrates to those that degrade non-fiber carbohydrates (Zhang et al. 2017). This shift increases the production of propionate in the rumen and decreases the production of acetate. As propionate is a net sink of reducing equivalents, their availability to reduce CO_2 to CH_4 is decreased. A high-concentrate diet also has the capacity to change the rumen environment through shifts in ruminal pH, which in turn affects the diversity and composition of rumen microbiota. Unlike other members of the rumen microbiota, the diversity of archaeal communities appears to stay relatively stable during shifts from a high-forage to a high-concentrate diet (Kumar et al. 2015). This may reflect their role in the reduction of CO_2 to CH_4, a biochemical function that is common to both forage- and concentrate-based diets. It does appear that the composition of methanogenic archaea community does change with age of the host, with methylotrophic methanogens being predominant in the neonate, whereas hydrogenotrophic methanogens are responsible for the majority of CH_4 production in the mature rumen (Friedman et al. 2017).

3.1.2 Rumen pH

Rumen pH is one of the major factors influencing enteric CH_4 production. A rapid shift from a roughage-based to a grain-based diet can cause acute ruminal acidosis, shifting volatile fatty acid production toward lactate and dramatically reducing the diversity of rumen microbiota (Dijkstra et al. 2012; Petri et al. 2013). In this context, Slyter et al. (1966) studied the effect of ruminal pH on rumen fermentation and microbial population in vitro, finding a significant reduction in VFAs and CH_4 production at a pH below 6.0. Similarly, Van Kessel and Russell (1996) observed that CH_4 production was reduced in the rumen fluid from a cow fed a concentrate-based diet when the pH declined to 5.45, but if the pH was restored to 7.0 with NaOH, CH_4 production resumed to a level similar to that of the rumen fluid which was collected when the cow was fed forage. The increase in propionate production with concentrate-based diets could account for a portion of the decline in CH_4 production, as propionate-producing bacteria which are abundant in these diets also utilize reducing equivalents, while the growth of methanogens is inhibited (Janssen and Kirs 2008). The optimal pH for the growth of methanogens in vitro ranges from 6.0 to 7.2 (Paynter and Hungate 1968; Jarvis et al. 2000; Kumar et al. 2009; Hook et al. 2011). However, responses of methanogens to pH in vitro may differ from those observed in vivo. Studies with heifers fed a high concentrate diet found that CH_4 production was the highest during the period after feeding, when the ruminal pH was the lowest (Fig. 3; Hünerberg et al. 2015). Methanogens exist within biofilms on feed, protozoa and the rumen wall and metabolic conditions within these microenvironments may differ from that in free rumen fluid or in in vitro cultures.

Consequently, although high-concentrate diets can reduce the intensity of CH_4 production (i.e. CH_4 per kg of meat or milk) they may not result in a reduction in absolute emissions. Furthermore, care must be taken to ensure that ruminal pH does not decline to levels that promote subclinical or clinical acidosis.

3.1.3 VFA production and absorption capacity of the rumen wall

Carbohydrates are broken down into VFA during ruminal fermentation to supply the host animal with the energy required for growth and maintenance. In this process, feed particles are fermented by rumen microbes to simple sugars that are then converted to VFAs, CH_4 and CO_2. The formation of acetic acid and butyric acid releases reducing equivalents, while the formation of propionic acid consumes reducing equivalents. Therefore, the types of VFAs produced in the rumen are closely linked to methanogenesis, such that CH_4 production can theoretically be predicted by the following equation (Moss et al. 2000):

$$CH_4(\text{mol}) = 0.45\ \text{acetic acid}(\text{mol}) + 0.40\ \text{butyric acid}(\text{mol}) - 0.275\ \text{propionic acid}(\text{mol}).$$

This equation shows that the production of acetate and butyrate enhances CH_4 production while propionate formation inhibits methanogenesis by

Published by Burleigh Dodds Science Publishing Limited, 2021.

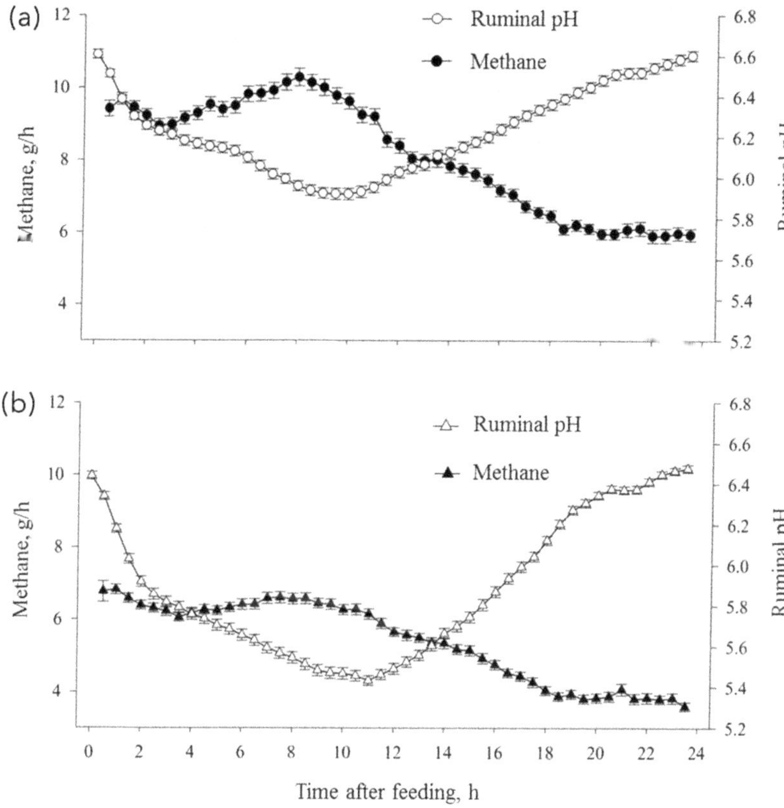

Figure 3 Diurnal CH_4 emission rate (g/h) and ruminal pH of beef heifers fed high-forage (a) or high-concentrate (b) diets. CH_4 was measured using 4 open-circuit respiratory chambers, with each chamber housing 2 heifers. Continuous rumen pH of individual heifers was measured using indwelling pH loggers. The ruminal pH and CH_4 emission rates were averaged between heifers housed in the same chamber (n = 8; means ± SEM) (Hünerberg et al. 2015).

competing for reducing equivalents in the rumen. The production of VFAs, however, is not necessarily directly related to their concentrations in rumen fluid due to variations in their absorption rate across the rumen wall and their rate of passage from the rumen. Indeed, Dijkstra et al. (1993) showed that concentrations of VFA in rumen fluid may not reflect VFA production, particularly at low rumen pH where rates of VFA absorption are increased. In this context, Penner (2014) reported that VFA absorption through the ruminal wall is linked to ruminal pH and is influenced by the passive diffusion of protons or by the secretion of bicarbonate through anion exchange mechanisms. In general, ruminal VFA concentrations have been shown to be positively correlated with VFA production (Leng 1970) and as a result some studies have attempted to predict ruminal CH_4 production based on VFA concentrations. Williams et al. (2019) developed a model that considered the production and removal of VFAs via absorption through the ruminal wall, and concluded that CH_4 yield (MY, g/kg DMI) in dairy and beef cattle could be predicted using the following equations:

$$MY = 4.08 \times (\text{acetate/propionate}) + 7.05;$$

$$MY = 3.28 \times (\text{acetate} + \text{butyrate})/\text{propionate} + 7.6;$$

$$MY = 316 / \text{propionate} + 4.4;$$

where the unit of VFAs is mol/100 mol total VFA. However, VFA concentrations may not be suitable predictors of daily CH_4 production in all feeding situations, as it has been shown to be the case for grazing sheep (Robinson et al. 2010). The lack of information on molar proportions of VFAs, rumen pH and fractional rates of VFA absorption in pastured ruminants may account for this shortcoming (Bannink et al. 2006).

3.2 Feed-associated factors

3.2.1 Feed intake

It is well known that total daily CH_4 production is primarily affected by feed intake and differs widely among ruminant species (Kirchgessner et al. 1991; Shibata et al. 1993; Benchaar et al. 2001). Past studies showed that CH_4 production can be estimated from dry matter intake (DMI), particularly when feed intake is less than 1.5 times maintenance energy requirements (Moe and Tyrrell 1979). The positive correlation between CH_4 production and DMI at lower intake levels (< 1.5 times the maintenance) was used by Shibata et al. (1992) to develop an equation to estimate CH_4 production in ruminants:

$$CH_4\,(\text{L/day}) = 0.0305 \times \text{DMI}(\text{g/day}) - 4.441\,(r = 0.992).$$

However, CH_4 production per unit of feed intake is negatively correlated with an increase in feed consumption for cattle and sheep, although the total output of CH_4 may increase at higher intakes (Hammond et al. 2013; Warner et al. 2017; Goopy et al. 2020). Higher levels of intake are usually achieved by increasing the level of concentrate in the diet, which can increase the production of propionate production and decrease the production of CH_4 (Coppock et al. 1964). A reduction in retention time of feed in the rumen as a result of increased

DMI may also account for the decline in CH_4 production per unit of feed intake (Lechner-Doll et al. 1991). Therefore, maximizing feed intake can be an effective strategy to reduce CH_4 yield per kg DM consumed.

CH_4 production differs among ruminant species, and Shibata et al. (1993) were one of the first authors to use a simple quadratic equation to accurately describe the relationship of CH_4 production and DMI at different feeding levels in dairy cattle, beef cattle, sheep and goats:

$$CH_4 \text{ production} \left(\text{L/day}\right) = -17.766 + 42.793X - 0.8749X^2 \left(r = 0.966\right);$$

where X is DMI (kg/day).

Charmley et al. (2016) reported the equation to predict CH_4 production of forage-fed Australian cattle:

$$CH_4 \text{ production} \left(\text{g/day}\right) = 20.7 \left(\pm 0.28\right) \times \text{DMI} \left(\text{kg/day}\right).$$

Several equations to estimate CH_4 production have been developed based on diet composition across a number of ruminant species (Mills et al. 2003; Beauchemin and McGinn 2006; Ellis et al. 2007; Patra et al. 2016; Bell et al. 2016; Niu et al. 2018). These equations consider not only DMI, but also other nutritional traits such as digestible energy, non-fiber carbohydrate, and neutral detergent fiber. Consequently, the chemical composition of the feed (i.e. forage to concentrate ratio, feed type and quality) can also have a direct influence on CH_4 emissions in ruminants.

3.2.2 Diet composition

Moe and Tyrrell (1979) found that the production of CH_4 per unit of structural carbohydrates (i.e. cellulose and hemicellulose) was higher than that for soluble carbohydrates (i.e. starch and sugars). Later, Johnson and Johnson (1995) suggested to further separate non-cell wall carbohydrates into soluble sugar and starch to improve the estimation of CH_4 production per unit of feed. This differentiation is necessary as, according to Johnson and Johnson (1995), soluble sugars tend to produce more CH_4 than starch. For example, high grain, starch-rich diets can exhibit increased passage rates, propionate production and decreased ruminal pH. As a result, CH_4 production is usually lower with high concentrate than high forage diets. Chemical composition of the feed can also influence enteric CH_4 production. For example, corn grain results in lower CH_4 production than barley grain (Beauchemin and McGinn 2005), and likewise cattle fed corn silage produce less CH_4 than those fed barley silage (Benchaar et al. 2014). These differences likely reflect the high fiber and lower starch content of barley grain and barley silage as compared to corn grain and corn silage. Ruminal CH_4 yields have also been reported to be lower from legume forages (alfalfa, clover, etc.) than grass forages (McCaughey et al. 1999; Benchaar et al. 2001). This may also reflect the lower structural carbohydrate levels of legume forages versus grass forages and possibly their higher rate of passage through the rumen. Finally, plant maturity is also a critical factor in determining the amount of CH_4 produced by ruminants, as the fiber content of forages increases with maturity (Russell et al. 2007). Consequently, DMI decreases and CH_4 production per unit of forage typically increases with advancing maturity (Pinares-Patiño et al. 2007; Molano and Clark 2008). Additional diet components which may impact CH_4 yield (g/kg DMI) in ruminants include the level of digestible organic matter, lipid, neutral detergent fiber, acid detergent fiber and lignin content (Ellis et al. 2007; Bell et al. 2016). As a result, feed quality can be a major determinant of CH_4 emissions in ruminant production systems.

It is well known that CH_4 production is usually lower with high concentrate than high forage diets as has been shown in lactating cows (Kurihara et al. 1997; Lovett et al. 2005; Aguerre et al. 2011), beef cattle (Lovett et al. 2003; Doreau et al. 2011) and goats (Kurihara et al. 1997). Beauchemin and McGinn (2005) reported that CH_4 emissions as a percentage of gross energy intake dropped from 7.42% with a high-forage diet to about 3% with a high-grain diet fed to Angus cattle. However, this practice may result in subacute or acute rumen acidosis (Owens et al. 1998; Plaizier et al. 2008), as well as higher CH_4 emissions from manure (Hindrichsen et al. 2006). Moreover, grain supplementation did not influence daily CH_4 production from grazing beef steers (Boadi et al. 2002), and similar observations were also reported as there was no difference in enteric CH_4 emissions when goats were fed diets at maintenance that differed in forage to concentrate (Lima et al. 2013). This may reflect differences in the quality and quantity of the forage selected during grazing.

Since feed quality affects enteric CH_4 production, corn grain-based diets can be used to replace barley grain diets to lower CH_4 emissions (Beauchemin and McGinn 2005; Fellner et al. 2008; Yurtseven and Ozturk 2009). Similarly, the consumption of legume forages needs to be encouraged in regions where legume forages are suitable for cultivation as ruminant feed, as CH_4 emissions are generally lower for legume forages (McCaughey et al. 1999; Benchaar et al. 2001). Secondary compounds of plants such as condensed tannins may also contribute to the lowering of CH_4 emissions associated with legume forages (Woodward et al. 2002; Beauchemin et al. 2009). Other plants or plant extracts with high levels of bioactive compounds (e.g. saponins, flavonoids, and essential oils) can reduce ruminal CH_4 emissions by affecting metabolic processes and/or the activity of rumen microorganisms, but the effectiveness varies depending on the source, type and concentration of the bioactive (Patra and Saxena 2010; Bodas et al. 2012; Kim et al. 2015). In addition, feed intake and the maturity of the plants need to be considered in this strategy. Feeding

Published by Burleigh Dodds Science Publishing Limited, 2021.

alfalfa hay harvested at the mid-bloom stage is recommended compared to using vegetative-stage hay due to the 15% decrease in CH_4 production relative to gross energy intake (Benchaar et al. 2001).

In addition, forage processing methods may reduce enteric CH_4 production from ruminants if feed intake is not restricted. Grinding or pelleting forages leads to a faster rate of passage and lower fiber digestibility, reducing CH_4 production by 20–40% (Blaxter 1989; Johnson and Johnson 1995; Boadi et al. 2004). However, the extent of physically effective fiber needs to be considered while using forage processing methods to ensure the minimal quantity of fiber required to maintain rumen health and function. Increased ruminal passage rates with pelleted forages can also result in reduced ruminal fiber digestion.

4 Modifying the rumen environment to reduce methane emissions

A number of strategies have been investigated for their ability to reduce enteric CH_4 emissions, either by directly targeting methanogens or limiting the availability of reducing equivalents to convert CO_2 to CH_4. A number of approaches have been undertaken including diet manipulation, a variety of feed additives, immunization, defaunation, genomics technologies, and genetic selection of the host (Table 1).

4.1 Rumen manipulation

4.1.1 Specialty feed and additives

4.1.1.1 Fat supplementation

Addition of fat increases the energy density of the diet and has been shown to effectively mitigate CH_4 emissions in ruminants. Supplementing fat for fermentable carbohydrates can reduce rumen fermentation and decrease the growth of ruminal methanogens and protozoa (Prins et al. 1972; Czerkawski et al. 1975; Zhou et al. 2013). For example, CH_4 emissions (g/day) were reduced by 22% when sunflower oil (400 g/d, 5% of DMI) was added to the diets of Holstein steers (McGinn et al. 2004). In addition to sunflower oil, coconut oil, crushed whole oilseeds (rapeseed, sunflower and linseed) and rumen-protected crystalline fat have been assessed for their ability to lower CH_4 emissions. Machmüller et al. (2000) found that coconut oil and oilseeds decreased CH_4 production (ml/kg live weight) by 10–27% in lambs, a response that was associated with lower numbers of protozoa and a decline in the acetate to propionate ratio. Beauchemin and McGinn (2006) reported that adding canola oil at up to 4.6% of DMI reduced CH_4 emissions (g/d) by 32%, but care needs to be taken as an inclusion of total dietary fat at levels above 6% of diet DM can suppress feed intake and fiber digestibility.

The extent to which fat suppresses CH_4 production depends on the source (i.e. oilseed, tallow, nuts), the form (i.e. whole seed, extruded, extracted oil, protected fat), fatty acid profile, supplementation level, and the composition of the diet (Machmüller et al. 2000, 2003; Machmüller 2006; Jordan et al. 2006a; Petrie et al. 2009). Unsaturated fatty acids can serve as a minor hydrogen sink via biohydrogenation of the double bonds, decreasing the availability of reducing equivalents for methanogens (Johnson and Johnson 1995). Fat supplements that contain high levels of medium-chain fatty acids (e.g. coconut oil, palm kernel oil) are particularly effective at reducing CH_4 emissions as they appear to be toxic to ruminal methanogens (Machmüller and Kreuzer 1999; Machmüller et al. 2003). Long-chain fatty acids may also reduce CH_4 production in an undesirable fashion by lowering DMI and fiber digestion (McGinn et al. 2004; Jordan et al. 2006b). Beauchemin et al. (2007) reported that adding 3% of either saturated or unsaturated long-chain fatty acids in the form of tallow, sunflower oil or whole sunflower seeds to a forage diet reduced CH_4 emissions by 14%, 14% and 33%, respectively. Of these sources, sunflower oil inhibited fiber digestion the least. Fat supplementation has also been shown to reduce CH_4 emissions in grazing cattle (Pinares-Patiño et al. 2016; Beck et al. 2019). However, ensuring uniform intake of fat supplements among individuals in a herd is difficult and care must be taken to ensure that fiber digestion is not inhibited. Furthermore, oilseeds can be expensive and often do not fit into diets that are formulated on a least-cost basis.

4.1.1.2 Ionophores

Ionophores (i.e. monensin, lasalocid, salinomycin, narasin, maduramicin, laidlomycin and semduramicin) are antibiotics that are frequently added to ruminant diets to cause favorable changes in rumen fermentation (Novilla et al. 2017). Monensin, produced by *Streptomyces cinnamonensis,* is the most widely used ionophore and has been extensively used to improve ruminal fermentation and feed efficiency in beef and dairy cattle (Russell and Strobel 1989; Sauer et al. 1998). Monensin reduces CH_4 production by decreasing the ratio of acetate to propionate and inhibiting the growth of ruminal hydrogen-producing bacteria and disrupting the association between methanogens and protozoa (Russell and Houlihan 2003; Guan et al. 2006; Beauchemin et al. 2008). Some studies have shown that monensin can lower CH_4 production by up to 25% (Van Nevel and Demeyer 1995; Tedeschi et al. 2003), but reductions are usually less than this and often do not persist. Guan et al. (2006) tested the effect of monensin on steers fed either a low- or a high-concentrate diet and observed that its inhibitory effects on CH_4 emissions declined over time. Moreover, the effect of monensin on CH_4 production is dose-dependent. Beauchemin et al. (2008) conducted a meta-analysis (Sauer et al. 1998; McGinn et al. 2004; Van Vugt 2005; Guan et al. 2006; Odongo et al. 2007; Waghorn 2007) and concluded that a dose less than 15 ppm of monensin had no effect on CH_4 production (g/kg DMI).

4.1.1.3 Nitrate

Nitrate (NO_3^-) acts as an electron acceptor which utilizes H_2 as it is reduced to ammonia (NH_3) via nitrite (NO_2^-), a reaction that is thermodynamically more favorable than the reduction of CO_2 to CH_4 by methanogens (Ungerfeld and Kohn 2006). Nitrate can also be directly toxic to methanogens and thus stands out as a potential feed additive to suppress ruminal CH_4 production. However, the reduction of NO_3^- to NO_2^- occurs more rapidly than the reduction of NO_2^- to NH_3 and as a result NO_2^- in the rumen can enter the bloodstream and reach toxic levels (Latham et al. 2016). Nitrite can convert Fe^{2+} to Fe^{3+} within hemoglobin, forming methaemoglobin, which is no longer capable of transporting oxygen (Lee and Beauchemin 2014). The clinical symptoms of acute NO_3^- toxicity include reduced feed intake and gain, reproductive failure or even death (Bruning-Fann and Kaneene 1993).

Despite the risk of toxicity, several studies have evaluated the effect of NO_3^- on CH_4 emissions in ruminants. In general, NO_3^- has been shown to reduce CH_4 production in sheep (Sar et al. 2004; Van Zijderveld et al. 2010; Nolan et al. 2010), dairy cows (Van Zijderveld et al. 2011; Lund et al. 2014), and beef cattle (Hulshof et al. 2012; Velazco et al. 2014; Duthie et al. 2018). Van Zijderveld et al. (2011) reported that NO_3^- (21 g/kg DM) consistently decreased CH_4 emission by 16% in lactating dairy cows. Lund et al. (2014) also tested diets with NO_3^- (20 g/kg DM) and found a significant reduction in CH_4 production both in vivo (31%) with dairy cows and in vitro (10–16%). However, if NO_3^- was removed from the diet, CH_4 emissions returned to pretreatment levels. If the concentration of NO_3^- in the diet is gradually increased, rumen microbes can adapt and accelerate the conversion of NO_2^- to NH_3, reducing the risk of toxicity (Lee and Beauchemin 2014). However, even with this consideration, the risks of NO_2^- toxicity may be too high for NO_3 to be used to lower CH_4 emissions in ruminants under practical production conditions.

4.1.1.4 3-Nitrooxypropanol (3-NOP)

3-Nitrooxypropanol (3-NOP) is a small molecule that binds to methyl-coenzyme M reductase (MCR) and inhibits its active site nickel (I), thereby inhibiting the final step in methanogenesis (Duin et al. 2016). The main advantage of using 3-NOP is that it specifically targets *MCR* within methanogenic archaea without

Table 1 Summary of strategies for reducing enteric GHG emissions in ruminants

Strategy	Mechanism	Advantages/ Limitations	Feasibility of application
Diet Manipulation			
Increased concentrate	Alters VFA production toward propionate; decreases rumen pH and inhibits rumen microbes associated with methanogenesis	Can reduce CH_4 production; Improves production Decreases enteric CH_4 production but may not reduce total GHG emissions; Animals may have acidosis risk	Feasible for housing or grazing systems
Feed types and Forage quality/ maturity	Influences intake and digestibility; some feedstuffs contain secondary compounds that inhibit rumen microbes	Can reduce CH_4 production per unit of product; improves digestibility and animal production May increase feed costs; variable influence on CH_4 production	Feasible for multiple systems but highly dependent on feedstuff availability and environmental factors (e.g. rainfall, temperature)
Forage processing	Influences intake and digestibility	Can reduce CH_4 production immediately Animals may have acidosis risk; increases costs and energy to process forages	May not be feasible for grazing system
Feed Additives			
Fat supplementation	Inhibits microorganisms associated with methanogenesis	Can reduce CH_4 production; can increase diet energy density and animal performance Potential negative impacts on intake, fiber digestibility; needs more information on animal productivity	May not feasible for grazing system
Ionophores	Inhibits microorganisms associated with methanogenesis	Improves feed efficiency; mild to moderate reduction in CH_4 production Lack of long-term inhibitory effect; not approved in some countries	Feasible for housing or grazing system
Nitrate	Reduces H_2 availability, directly toxic to methanogens	Can reduce CH_4 production; can be used to replace urea as a protein source Risk of toxicity; may increase N excretion	Feasible for housing or grazing system

Published by Burleigh Dodds Science Publishing Limited, 2021.

3-NOP	Inhibits methanogens	Reduces CH_4 effectively Needs more information on animal performance, dose suggestions and food safety; not approved in the market	May not be feasible for grazing system
Algae	Inhibits methanogens	Reduces CH_4 effectively Lack of more in vivo studies to show animal performance and long-term effect; may require energy to process the algae	Experimental, but may not feasible for grazing system
Immunization	Inhibits methanogens	Applicable for grazing ruminants, easy to operate in farms Lack of effect information on CH_4 production, animal health and production	Feasible for different systems
Defaunation	Eliminates ruminal protozoa	Reduces CH_4 effectively Adequate level of defaunation depends on diet compositions; some chemical agents are toxic; Needs more research on animal health and performance	Difficult to implement for grazing system
Genomic technology	Inhibits microorganisms associated with methanogenesis	Important for developing vaccines or inhibitors to target microorganisms associated with methanogenesis; may improve feed efficiency and animal productivity by manipulating microbes and their pathways Research is in early stages	Experimental, may be feasible for different systems
Genetic selection	Animal breeding for low CH_4 production per unit of product by manipulating rumen microbiota	Increases the efficiency of production in multiple ruminant systems; may have long-term persistency on reducing CH_4 efficiently Needs to determine genotype and diet interactions; lack of ability to measure CH_4 production on a large population of animals; lack of information on long-term productivity and animal performance	Feasible for different systems

impacting the activity of other microbes involved in the fermentation of feed. To support licensing and application in ruminants, 3-NOP has been experimentally evaluated in dairy cows (Reynolds et al. 2014; Haisan et al. 2014; Hristov et al. 2015), beef cattle (Romero-Perez et al. 2014; Vyas et al. 2018; Martinez-Fernandez et al. 2018) and sheep (Martínez-Fernández et al. 2014) at various doses and with different diets. Early experiments studied the effect of 3-NOP (40 μL/L and 80 μL/L) in vitro using sheep rumen fluid and reported a reduction in CH_4 production of 86–96%, with no effect on total VFA concentration (Martínez-Fernández et al. 2014). Then, 3-NOP was tested in sheep at a dose of 100 mg/d per animal, resulting in a 26% decrease in CH_4 production (g/kg DMI). Haisan et al. (2014) also observed that CH_4 production was decreased by 60% in lactating dairy cows fed 3-NOP at 2.5 g/d, with no negative effect on DMI or milk production. A 30% reduction in CH_4 emissions was also reported in lactating dairy cows fed mixed rations with 40, 60 or 80 mg 3-NOP per kg of DM over 12 weeks (Hristov et al. 2015). In this study, body weight gain increased, indicating the potential repartitioning of energy that would have been lost as CH_4 toward tissue deposition (Hristov et al. 2015). 3-nitrooxypropanol was also shown to improve the feed efficiency of beef cattle fed high-forage diets and high-grain diets, with a 42% and a 37% reduction in CH_4 yield (g/kg DMI), respectively (Vyas et al. 2018). Based on these studies, the optimal dose of 3-NOP was proposed to range between 1 g/day and 2 g/day. Incorporation of 3-NOP into a pelleted supplement versus mixing it into a complete diet reduced CH_4 emissions in lactating dairy cows by 23% and 28%, respectively (Van Wesemael et al. 2019). As no significant difference was observed between the two supplementation methods, it is feasible to administer 3-NOP either in the complete diet or in a pelleted supplement.

When compared to other *MCR* inhibitors (e.g. bromoethane sulphonate – BES) (Grawert et al. 2014), 3-NOP did not reduce feed intake, digestibility or animal performance. Moreover, 3-NOP- (200 mg/kg DM for backgrounding or 125 mg/kg for finishing phase) mediated reductions in enteric CH_4 in beef cattle fed high-grain diets for at least 105 days (Vyas et al. 2018) and in dairy cattle for over 12 weeks (Hristov et al. 2015; Vyas et al. 2018). It has also been reported that 3-NOP decreases total VFA concentration, particular acetate, whereas it increases butyrate levels with no adverse effects on growth of the host (Melgar et al. 2020; Kim et al. 2020). While applications for registration of 3-NOP (Bovaer®) have been submitted to regulatory authorities in Europe and other regions of the world, further studies are required to define the optimum concentrations of 3-NOP that optimize reductions in enteric CH_4 emissions across a range of diets and ruminant production conditions. The development of systems to administer 3-NOP to grazing ruminants will be key, if meaningful reductions in CH_4 are to be achieved throughout the production system. Fortunately, recent studies have shown that administration of 3-NOP is unlikely to result in the release of residues into meat or milk (Thiel et al. 2019a) or cause toxicity (Thiel et al. 2019b).

4.1.1.5 Algae

Algae can be classified according to their size (i.e. microalgae, macroalgae), or based on their pigmentation (i.e. green, red, brown) and the habitat where they grow (i.e. freshwater, marine). All algae produce secondary metabolites that exhibit antimicrobial activity, such as halogenated methane analogues (HMAs) (Plaza et al. 2008; Blunt et al. 2011; Machado et al. 2018). The HMAs (e.g. bromochloromethane, bromoform or chloroform) exert antimethanogenic activity by binding to reduced vitamin B_{12}, blocking the cobamide-dependent methyltransferase reaction needed for the formation of methyl-coenzyme M, and thus consequently mitigating CH_4 emissions (Wood et al. 1968; Machado et al. 2018). Because of these antimicrobial properties, some algae (e.g. *Ascophyllum nodosum* and *Asparagopsis taxiformis*) have been tested in vitro and in vivo for their ability to reduce enteric CH_4 emissions (Wang et al. 2008; Machado et al. 2014; Li et al. 2018a). Machado et al. (2014) evaluated 20 species of tropical macroalgae through in vitro methods and concluded that marine algae were more effective than freshwater algae at inhibiting methanogenesis. Among the examined species, *Asparagopsis* spp. was the most effective at reducing enteric CH_4 production, with reductions as high as 98.9% in rumen fluid after 72 h-incubation (Machado et al. 2014). This result was attributed to secondary metabolites in *Asparagopsis* that inhibited both ruminal bacteria and protozoa (Paul et al. 2006; Genovese et al. 2009). Machado et al. (2018) reported that *Asparagopsis taxoformis* biomass containing 1.3 μM bromoform in vitro was similar to the addition of 5 μM bromoform in reducing CH_4 production, suggesting that bromoform was not the only CH_4 inhibitor in *A. taxiformis*. Li et al. (2018a) showed that addition of *A. taxiformis* at up to 3% of dietary DM lowered CH_4 emissions in sheep by as much as 80%, without impacting weight gain and by as much as 98% in steers with a 42% improvement in weight gain (Kinley et al. 2020). Supplementing *Asparagopsis armata* at 1% of dietary OM reduced CH_4 intensity (g/kg milk yield) in lactating dairy cows by 60% in the short term (Roque et al. 2019). Belanche et al. (2016) proposed that *Ascophyllum nodosum* possessed anti-protozoal activity and showed that it decreased CH_4 emissions from rumen fluid in vitro. Another study also reported that rumen bacteria and methanogens were linearly reduced, while protozoa were increased, when increasing levels (1%, 3% and 5 %) of sun-dried *Ascophyllum nodosum* were fed to sheep (Zhou et al. 2018). Although CH_4 emissions were not measured in this study, the 30–75% decline in total methanogens may have decreased enteric CH_4 emissions.

As can be observed, the effectiveness of algae in reducing rumen CH_4 emissions has been tested in several experiments, but investigations of their efficacy in large-scale beef and dairy production systems are needed. Harvesting of these aquatic plants from natural ecosystems may not be sustainable and scale-up of commercial production systems to a level that could satisfy demand could be both challenging and expensive. Drying and transportation of microalgae and macroalgae can also contribute to the GHG footprint of these additives.

4.1.2 Immunization

Vaccination against rumen methanogens has been studied for its ability to mitigate CH_4 emissions, an approach that is particularly appealing for use in grazing ruminants. Vaccines induce the immune system of the host animal to produce antibodies, which could potentially enter the rumen via saliva and bind to antigenic proteins located on the surface of methanogens, inhibiting their growth and activity (Williams et al. 2009; Wedlock et al. 2010, 2013; Subharat et al. 2016). Antibodies have been shown to be stable in the rumen for at least 8 h (Williams et al. 2008), but sufficient diversity and quantities of antibodies must enter the rumen so as to impact the diversity and the majority of the methanogens within the rumen.

Whole methanogen cells from different methanogen species have been investigated for their immunogenicity in grazing sheep in Australia and New Zealand, but results have been inconsistent (Wright et al. 2004; Leslie et al. 2008; Williams et al. 2009). Zhang et al. (2015) vaccinated goats with a new recombinant protein (*EhaF*, energy-converting hydrogenase A subunit F), which is a *Methanobrevibacter ruminantium* M1 adhesin-like protein responsible for hydrogenotrophic methanogenesis. The vaccine had no apparent impact on enteric CH_4 emissions or ruminal methanogens, although it did alter the composition of the bacterial community. Also, whole fixed protozoal cells have been evaluated as antigens in vaccines to control protozoa in the rumen of sheep, but this approach also was not shown to be effective (Williams et al. 2008). Improving the specificity of antibody targets against methanogens and enhancing immune responses so as to increase the concentration of antibodies that enter the rumen could improve the efficacy of anti-methanogen vaccines.

4.1.3 Defaunation

Defaunation is a process whereby ruminal protozoa are reduced or eliminated from the rumen by dietary, chemical or biological agents. Several chemicals (e.g. copper sulfate, dioctyl sodium sulfosuccinate and detergents) have been used to remove protozoa from the rumen (Becker and Everett 1930; Burggraaf and Leng 1980; Rowe et al. 1985). However, the most promising approach for removing protozoa from the rumen is through dietary manipulation as chemical agents may be harmful to host health (Williams and Coleman 1992; Ushida et al. 1997). Removal of protozoa from the rumen has been proposed as a strategy to reduce enteric CH_4, as 9–37% of total ruminal CH_4 emissions are attributed to interspecies hydrogen transfer between protozoa and methanogens (Newbold et al. 2015). Prolonged feeding of high grain and the resulting lower ruminal pH can inhibit the growth and reduce the diversity of protozoal populations (Whitelaw et al. 1984; Kreuzer and Kirchgessner 1987). This outcome could be one of the reasons why CH_4 emissions tend to be lower in ruminants fed high concentrate versus high forage diets. Addition of fat to the diet can also lower protozoa numbers. Coconut oil, in combination with high concentrate diets, has been recommended as an alternative feeding strategy that could reduce CH_4 emissions by up to 57% (Machmüller et al. 1998; Nguyen and Hegarty 2017). The effectiveness of defaunation at lowering CH_4 emissions depends on the composition of the diet, as suppression of protozoa in ruminants fed high-forage diets can significantly inhibit fiber digestion (Li et al. 2018b). Furthermore, most defaunation procedures do not remove all protozoa from the rumen, and if they do it is challenging to ensure that ruminants remain defaunated under practical production conditions. Some studies have suggested that programming microbial colonization in the neonate could result in life-long reductions in enteric CH_4 emissions (Hegarty et al. 2008; Yáñez-Ruiz et al. 2015). Although these studies have shown that these approaches can alter microbial community structure and/or activity post-weaning, they have yet to demonstrate a life-long reduction in enteric CH_4 emissions.

4.1.4 Genomic approaches in characterizing and reducing methanogenesis

It is challenging to find a universal approach for effectively mitigating enteric CH_4 production in different farming systems (housing or grazing) without influencing nutrient digestibility and animal productivity. Following the advancement of omics technologies, more in-depth information of the rumen microbiome has revealed remarkable host-microbe interactions and how they contribute to CH_4 production. Whole genome sequencing, as the basis of these technologies, plays a critical role in targeting methanogens and other ruminal microbes that produce substrates for methanogenesis. Culturomics is thought to be a useful technology for characterizing, as of yet, uncharacterized rumen microorganisms. A 'Hungate1000' project was launched in 2012 aiming to sequence 1000 cultured rumen microbial genomes to enhance our knowledge about the diversity and function of these microbes (Seshadri et al. 2018). Currently, 501 genome sequences of rumen microorganisms have been generated, of which 480 are bacterial genomes and 21 are archaea (Seshadri et al. 2018; Newbold and Ramos-Morales 2019). This includes members of 9 phyla, 48 families and 82 genera, with many species represented by only one or a few isolates. Although, multiple isolates of some polysaccharide-degrading genera such as *Butyrivibrio*, *Prevotella* and *Ruminococcus* were sequenced. Next-generation sequencing helps researchers predict rumen microbial function from metataxonomic data. In this context, the available rumen microbial genomics of a single species can be applied to identify and explain the process of methanogenesis based on microbial genomes (i.e. methanogens and other microbes-contributing substrates to methanogens) and the specific proteins that are expressed during methanogenesis (Huws et al. 2018). For example, Gilmore et al. (2017) reported that assembly and analysis of the methanogen genomes enabled the metabolism of four methanogens (*Methanobacterium bryantii*, *Methanosarcina spelaei*, *Methanosphaera cuniculi* and *Methanocorpusculum parvum*) to be predicted. They found all sequenced methanogens were enriched for genes in energy production and conversion, coenzyme transport and metabolism, but were lower in genes assigned to carbohydrate and lipid transport and metabolism. With a greater understanding of methanogen metabolism, CH_4 production may be modified by optimal designing of biochemical processes. Furthermore, with the revealing of the presence of a prophage (*peiR*) in the rumen methanogen (*M. ruminantium* M1) genome, another potential approach using the phage-encoded lytic enzyme (endoisopeptidase PeiR) to specifically target and lyse the pseudomurein cell walls of rumen methanogens could provide another viable strategy to reducing enteric CH_4 emissions (Leahy et al. 2013; Altermann et al. 2018; Gilbert et al. 2020). Microbial genomes can also enhance our understanding of microbial interactions that drive methanogenesis, such as the commensal relationship between methanogens and ciliates. However, to date, unlike rumen bacteria and archaea, our knowledge of the genomes of protozoa and fungi is limited as a result of their G-C rich genomes which are difficult to assemble.

Tapio et al. (2017) revealed that methanogenesis may be associated with the composition of rumen archaeal community but not archaea abundance, especially with the *Methanobrevibacterium gottschalkii* clade. Therefore, metagenomics would be a useful tool to understand methanogenesis mechanisms in the rumen by predicting microbial metabolic functions. By characterizing methanogens in the buffalo rumen through metagenomics, genes encoding key steps in the process of methanogenesis were identified (Singh et al. 2015). In a beef cattle study, the abundance of archaeal genes in rumen digesta differed between high and low CH_4-emitting steers, with eight genes being differentially abundant between these two populations (Wallace et al. 2015). *Succinovibrionaceae* were more abundant in the low CH_4 emitters, whereas *Methanobrevibacter* were more abundant in high CH_4 emitters.

Researchers have characterized the rumen microbiota of high CH_4-emitting and low CH_4-emitting sheep, finding that lactic acid-producing *Sharpea azabuensis* were more abundant in low CH_4 emitters (Kamke et al. 2016). These studies support the possibility of there being value in selecting ruminants for specific rumen microbiota that are associated with lower enteric CH_4 emissions.

McAllister et al. (2015) outlined several metagenomic and metatranscriptomic approaches that could be used to mitigate enteric CH_4 emissions. The chemogenomics approach is based on finding chemical compounds that inhibit specific proteins expressed by genes involved in methanogenesis. This methodology resulted in the development of 3-NOP (Duval and Kindermann 2012) as described above. Using genomics to identify extracellular proteins with improved antigenicity in the host could improve the efficacy of vaccines against methanogens.

4.2 Improved genetic selection

It has been reported that CH_4 yield (g/kg DMI) varies among individual ruminants fed the same diet. Pinares-Patiño et al. (2003) found that some pastured sheep were persistent-high or low CH_4 emitters in four separate measurement periods. This might be due to genetic differences in CH_4 production among individuals and variation in rumen methanogenic ecology (Zhou et al. 2009). This strategy aims to link genetic traits (e.g. weaning weight, dag score, muscle depth) with the rumen microbiota in order to predict CH_4 yield phenotypes across ruminant species (Kamke et al. 2016). Through the identification of rumen microbiome of these grazing sheep, they discovered that species closely related to the lactic acid-producing, *Sharpea azabuensis*, were more abundant in low CH_4-emitting sheep. In contrast, H_2 producing *Ruminoccocaceae*, *Lachnospiraceae* and *Verricomicrobia* were more prevalent in the rumen of high CH_4-yielding sheep (Kittelmann et al. 2014; Kamke et al. 2016). These findings suggest that the selection of ruminants that yield less CH_4 per unit of DMI may be a viable strategy to lower herd emissions. Residual feed intake (RFI), as an indicator of cattle feed efficiency, is the difference between actual feed intake and the expected feed intake estimated from body weight, growth rate and production over a specific period (Jones et al. 2011). Researchers have demonstrated that cattle with low RFI emit less CH_4 than animals with high RFI (Hegarty et al. 2007; Zhou et al. 2009). This finding promoted studies focusing on the genetic selection for low RFI animals (Weber et al. 2016). However, some studies have contradicted these findings, suggesting that CH_4 yields may actually be higher in low-RFI cattle (McDonnell et al. 2016; Velazco et al. 2017; Flay et al. 2019). It has been suggested that this strategy of selecting animals with low RFI for the suppression of CH_4 emissions could be more advantageous for systems that rely on the use of low-quality feeds in ruminant production (Patra 2012).

Nucleic acids have been used as genetic markers for rumen microbial processes (Huws et al. 2018) and those genetic markers related to ruminant CH_4 production have been studied in recent years. Golder et al. (2018) studied the genetic markers associated with rumen microbiome and metabolome of dairy heifers fed high grain diets. They observed that some metabolites (e.g. acetate to propionate ratio, lactate) and rumen microbial phyla (e.g. Actinobacteria, Euryarchaeota, Fibrobacteres) had associated markers and quantitative trait loci, which may enable selection for acidosis-resistant cattle with lower CH_4 emissions. A high genetic correlation between DMI and predicted CH_4 emissions was revealed by Pickering et al. (2015) using a data set that included 1726 dairy cows, where one single nucleotide polymorphism (SNP) on chromosome 7 was reported to be associated with CH_4 emission traits. Singh et al. (2015) identified the functional genes involved in rumen methanogenesis and acetogenesis pathways in Indian buffalo, which were key enzymes in CH_4 production. Their results also contributed new knowledge about acetogenic bacteria that compete with methanogens and thus may be useful to reduce CH_4 emissions. Roehe et al. (2016) suggested using archaeal abundance in ruminal digesta to genetically select animals for lower enteric CH_4 emission, avoiding the need for direct measurement of enteric CH_4 emissions. They also identified 20 genes (e.g. *mcrA* and *fmdB*) associated with CH_4 production through metagenomic analysis of rumen contents. These results contradicted with those of Carberry et al. (2014) where they found that diversity rather than abundance of methanogens determined the extent of enteric CH_4 emissions. Auffret et al. (2018) identified genes associated with the hydrogenotrophic methanogenesis pathway in beef cattle of different breeds fed different diets and suggested that they could be used as biomarkers of CH_4 emissions. The inconsistencies across these studies may reflect the challenges of comparing divergent data sets from different animals fed various diets as well as differences in sampling method, sample preparation and data analysis (Huws et al. 2018).

Nevertheless, as some rumen microbial features have been reported to be heritable and influenced by host genetics in beef cattle (Li et al. 2019), it has been suggested that it should be possible to genetically select for ruminants with low CH_4 emission traits. Wallace et al. (2019) collected samples from over 1000 dairy cows to investigate the relationship of host genomics and rumen microbiome as well as their interactions on CH_4 emissions. Difford et al. (2018) studied the individual variation in CH_4 production influenced by both host genetics and rumen microbiome in dairy cows. They reported that host genotypes can explain 21% of the individual variance while the cumulative effect of rumen bacteria and archaea was 13%, and these two were independent. The core rumen microbiome was found to be correlated with host genetics, and the possibly heritable rumen microbes raise the possibilities of selecting for low CH_4-emitting cattle. However, the complexity of measuring enteric CH_4 emissions may make it difficult to employ this approach in commercial ruminant production systems.

5 Conclusion

Research in the genetics, nutrition, microbiology, and physiology of ruminants has identified several strategies that are effective at reducing CH_4 emissions per unit of animal product (i.e. milk, meat or wool). Altering the composition of the diet and some feed additives are capable of reducing enteric CH_4 emissions, particularly in ruminants fed total mixed diets where additives can be easily administered. Some of these technologies only suppress CH_4 emission in the short term and almost all must be administered daily to be effective. Administration of these additives is notoriously difficult in more extensive grazing-based ruminant productions systems, which incidentally are often responsible for the largest portion of CH_4 emissions in ruminant production systems. Immunization and genetic selection are CH_4 mitigation approaches that could be applicable to grazing ruminants. However, evidence that vaccination can reduce enteric CH_4 emissions is still lacking and the complexities of measuring enteric CH_4 production has hampered the adoption of this selection criterion in commercial breeding programs.

Overall, a combined approach may be the best strategy to achieve a consistent decrease in CH_4 emissions in ruminants. For example, dietary management in a feed-efficient herd will help to reduce the intensity of enteric CH_4 emissions. Furthermore, novel strategies to reduce CH_4 production per unit of milk or meat in grazing systems need to be emphasized. Besides, consumers' preference regarding animal welfare, food quality and safety need to be taken into consideration while developing and selecting strategies that reduce enteric CH_4 emissions.

6 Where to look for further information

The Global Research Alliance harbors a livestock research group that is focused on reducing the greenhouse gas emissions intensity of livestock production systems and increasing the quantity of carbon stored in soils. It is home to a number of relevant networks including the Animal Health and Greenhouse Gas Emissions Intensity Network; Animal Selection and Genetic and Genomics Network; Feed and Nutrition Network; Manure Management Network and Rumen Microbial Genomics Network. The Global Rumen Census, Low Emission Livestock Development and Measuring, Reporting and Verifying Greenhouse Gases activities were also developed within this organization.

Published by Burleigh Dodds Science Publishing Limited, 2021.

- https://globalresearchalliance.org/research/livestock/.

The Livestock Environmental Assessment and Performance Partnership has also produced a number of documents related to sustainability in livestock production systems. Particularly relevant documents would include guidelines on: Environmental performance of feed additives in livestock supply chains; Measuring and modelling soil carbon stocks and stock changes in livestock production systems; Environmental performance of animal feeds supply chains; Greenhouse gas emissions and fossil energy demand from small ruminant supply chains; Environmental performance of large ruminant supply chains.

- http://www.fao.org/partnerships/leap/resources/guidelines/en/.

The Intergovernmental Panel on Climate Change can also be a useful source on overall anthropogenic impacts on climate change. Some sections of the report are related to agricultural production and the role of livestock in greenhouse gas emissions. Searching the site using the term 'Agriculture' will bring up most relevant documents.

- https://www.ipcc.ch/.

7 References

Adler, A. A., Doole, G. J., Romera, A. J. and Beukes, P. C. (2015). Managing greenhouse gas emissions in two major dairy regions of New Zealand: a system-level evaluation, *Agricultural Systems* 135, 1-9. DOI: 10.1016/j.agsy.2014.11.007.

Aguerre, M. J., Wattiaux, M. A., Powell, J. M., Broderick, G. A. and Arndt, C. (2011). Effect of forage-to-concentrate ratio in dairy cow diets on emission of methane, carbon dioxide, and ammonia, lactation performance, and manure excretion, *Journal of Dairy Science* 94(6), 3081-3093. DOI: 10.3168/jds.2010-4011.

Altermann, E., Schofield, L. R., Ronimus, R. S., Beattie, A. K. and Reilly, K. (2018). Inhibition of rumen methanogens by a novel archaeal lytic enzyme displayed on tailored bionanoparticles, *Frontiers in Microbiology* 9, 2378. DOI: 10.3389/fmicb.2018.02378.

Auffret, M. D., Stewart, R., Dewhurst, R. J., Duthie, C. A., Rooke, J. A., Wallace, R. J., Freeman, T. C., Snelling, T. J., Watson, M. and Roehe, R. (2018). Identification, comparison, and validation of robust rumen microbial biomarkers for methane emissions using diverse Bos taurus breeds and basal diets, *Frontiers in Microbiology* 8, 2642. DOI: 10.3389/fmicb.2017.02642.

Bannink, A., Kogut, J., Dijkstra, J., France, J., Kebreab, E., Van Vuuren, A. M. and Tamminga, S. (2006). Estimation of the stoichiometry of volatile fatty acid production in the rumen of lactating cows, *Journal of Theoretical Biology* 238(1), 36-51. DOI: 10.1016/j.jtbi.2005.05.026.

Bayley, D. P., Correia, J. D., Jarrell, K. F. and Thomas, N. A. (1999). Recent Excitement about the Archaea, *BioScience* 49(7), 530-541. DOI: 10.2307/1313474.

Beauchemin, K. A., Kreuzer, M., O'Mara, F. and McAllister, T. A. (2008). Nutritional management for enteric methane abatement: a review, *Australian Journal of Experimental Agriculture* 48(2), 21-27. DOI: 10.1071/EA07199.

Beauchemin, K. A., McAllister, T. A. and McGinn, S. M. (2009). Dietary mitigation of enteric methane from cattle, *CAB Reviews: Perspectives in Agriculture, Veterinary Science, Nutrition and Natural Resources* 4(35), 1-18. DOI: 10.1079/PAVSNNR20094035.

Beauchemin, K. A. and McGinn, S. M. (2005). Methane emissions from feedlot cattle fed barley or corn diets, *Journal of Animal Science* 83(3), 653-661. DOI: 10.2527/2005.833653x.

Beauchemin, K. A. and McGinn, S. M. (2006). Enteric methane emissions from growing beef cattle as affected by diet and level of intake, *Canadian Journal of Animal Science* 86(3), 401-408. DOI: 10.4141/A06-021.

Beauchemin, K. A., McGinn, S. M. and Petit, H. V. (2007). Methane abatement strategies for cattle: lipid supplementation of diets, *Canadian Journal of Animal Science* 87(3), 431-440. DOI: 10.4141/CJAS07011.

Beauchemin, K. A., Ungerfeld, E. M., Eckard, R. J. and Wang, M. (2020). Review: Fifty years of research on rumen methanogenesis: lessons learned and future challenges for mitigation, *Animal* 14(S1), s2-s16. DOI: 10.1017/S1751731119003100.

Beck, M. R., Thompson, L. R., Williams, G. D., Place, S. E., Gunter, S. A. and Reuter, R. R. (2019). Fat supplements differing in physical form improve performance but divergently influence methane emissions of grazing beef cattle, *Animal Feed Science and Technology* 254, 114210. DOI: 10.1016/j.anifeedsci.2019.114210.

Becker, E. R. and Everett, R. C. (1930). Comparative growths of normal and infusoria-free lambs, *American Journal of Epidemiology* 11(2), 362-370. DOI: 10.1093/oxfordjournals.aje.a113838.

Belanche, A., Ramos-Morales, E. and Newbold, C. J. (2016). In vitro screening of natural feed additives from crustaceans, diatoms, seaweeds and plant extracts to manipulate rumen fermentation, *Journal of the Science of Food and Agriculture* 96(9), 3069-3078. DOI: 10.1002/jsfa.7481.

Bell, M., Eckard, R., Moate, P. J. and Yan, T. (2016). Modelling the effect of diet composition on enteric methane emissions across sheep, beef cattle and dairy cows, *Animals* 6(9), 54. DOI: 10.3390/ani6090054.

Benchaar, C., Hassanat, F., Gervais, R., Chouinard, P. Y., Petit, H. V. and Massé, D. I. (2014). Methane production, digestion, ruminal fermentation, nitrogen balance, and milk production of cows fed corn silage-or barley silage-based diets, *Journal of Dairy Science* 97(2), 961-974. DOI: 10.3168/jds.2013-7122.

Benchaar, C., Pomar, C. and Chiquette, J. (2001). Evaluation of dietary strategies to reduce methane production in ruminants: a modelling approach, *Canadian Journal of Animal Science* 81(4), 563-574. DOI: 10.4141/A00-119.

Blaxter, K. (1989). *Energy Metabolism in Animals and Man*, CUP Archive. Cambridge University Press, Cambridge.

Blunt, J. W., Copp, B. R., Munro, M. H., Northcote, P. T. and Prinsep, M. R. (2011). Marine natural products, *Natural Product Reports* 28(2), 196-268. DOI: 10.1039/C005001F.

Boadi, D., Benchaar, C., Chiquette, J. and Massé, D. (2004). Mitigation strategies to reduce enteric methane emissions from dairy cows: update review, *Canadian Journal of Animal Science* 84(3), 319-335. DOI: 10.4141/A03-109.

Boadi, D. A., Wittenberg, K. M. and McCaughey, W. (2002). Effects of grain supplementation on methane production of grazing steers using the sulphur (SF6) tracer gas technique, *Canadian Journal of Animal Science* 82(2), 151-157. DOI: 10.4141/A01-038.

Bodas, R., Prieto, N., García-González, R., Andrés, S., Giráldez, F. J. and López, S. (2012). Manipulation of rumen fermentation and methane production with plant secondary metabolites, *Animal Feed Science and Technology* 176(1-4), 78-93. DOI: 10.1016/j.anifeedsci.2012.07.010.

Bruning-Fann, C. S. and Kaneene, J. B. (1993). The effects of nitrate, nitrite, and N-nitroso compounds on animal health, *Veterinary and Human Toxicology* 35(3), 237-253.

Burggraaf, W. and Leng, R. A. (1980). Antiprotozoal effects of surfactant detergents in the rumen of sheep, *New Zealand Journal of Agricultural Research* 23(3), 287-291. DOI: 10.1080/00288233.1980.10425358.

Carberry, C. A., Kenny, D. A., Kelly, A. K. and Waters, S. M. (2014). Quantitative analysis of ruminal methanogenic microbial populations in beef cattle divergent in phenotypic residual feed intake (RFI) offered contrasting diets, *Journal of Animal Science and Biotechnology* 5(1), 41. DOI: 10.1186/2049-1891-5-41.

Charmley, E. S. R. O., Williams, S. R. O., Moate, P. J., Hegarty, R. S., Herd, R. M., Oddy, V. H., Reyenga, P., Staunton, K. M., Anderson, A. and Hannah, M. C. (2016). A universal equation to predict methane production of forage-fed cattle in Australia, *Animal Production Science* 56(3), 169-180. DOI: 10.1071/AN15365.

Coppock, C. E., Flatt, W. P., Moore, L. A. and Stewart, W. E. (1964). Relationships between end products of rumen fermentation and utilization of metabolizable energy for milk Production1, *Journal of Dairy Science* 47(12), 1359-1364. DOI: 10.3168/jds.S0022-0302(64)88918-9.

Creevey, C. J., Kelly, W. J., Henderson, G. and Leahy, S. C. (2014). Determining the culturability of the rumen bacterial microbiome, *Microbial Biotechnology* 7(5), 467-479. DOI: 10.1111/1751-7915.12141.

Czerkawski, J. W., Christie, W. W., Breckenridge, G. and Hunter, M. L. (1975). Changes in the rumen metabolism of sheep given increasing amounts of linseed oil in their diet, *British Journal of Nutrition* 34(1), 25-44. DOI: 10.1017/S0007114575000074.

Deppenmeier, U. (2002). The unique biochemistry of methanogenesis, *Progress in Nucleic Acid Research and Molecular Biology* 71, 223-283. DOI: 10.1016/S0079-6603(02)71045-3.

Difford, G. F., Plichta, D. R., Løvendahl, P., Lassen, J., Noel, S. J., Højberg, O., Wright, A.-D. G., Zhu, Z., Kristensen, L., Nielsen, H. B., Guldbrandtsen, B. and Sahana, G. (2018). Host genetics and the rumen microbiome jointly associate with methane emissions in dairy cows, *PLoS Genetics* 14(10), e1007580. DOI: 10.1371/journal.pgen.1007580.

Dijkstra, J., Boer, H., Van Bruchem, J., Bruining, M. and Tamminga, S. (1993). Absorption of volatile fatty acids from the rumen of lactating dairy cows as influenced by volatile fatty acid concentration, pH and rumen liquid volume, *British Journal of Nutrition* 69(2), 385-396. DOI: 10.1079/BJN19930041.

Dijkstra, J., Ellis, J. L., Kebreab, E., Strathe, A. B., López, S., France, J. and Bannink, A. (2012). Ruminal pH regulation and nutritional consequences of low pH, *Animal Feed Science and Technology* 172(1-2), 22-33. DOI: 10.1016/j.anifeedsci.2011.12.005.

Doreau, M., Van Der Werf, H. M., Micol, D., Dubroeucq, H., Agabriel, J., Rochette, Y. and Martin, C. (2011). Enteric methane production and greenhouse gases balance of diets differing in concentrate in the fattening phase of a beef production system, *Journal of Animal Science* 89(8), 2518-2528. DOI: 10.2527/jas.2010-3140.

Duin, E. C., Wagner, T., Shima, S., Prakash, D., Cronin, B., Yáñez-Ruiz, D. R., Duval, S., Rümbeli, R., Stemmler, R. T., Thauer, R. K. and Kindermann, M. (2016). Mode of action uncovered for the specific reduction of methane emissions from ruminants by the small molecule 3-nitrooxypropanol, *Proceedings of the National Academy of Sciences of the United States of America* 113(22), 6172-6177. DOI: 10.1073/pnas.1600298113.

Duthie, C. A., Troy, S. M., Hyslop, J. J., Ross, D. W., Roehe, R. and Rooke, J. A. (2018). The effect of dietary addition of nitrate or increase in lipid concentrations, alone or in combination, on performance and methane emissions of beef cattle, *Animal: An International Journal of Animal Bioscience* 12(2), 280-287. DOI: 10.1017/S175173111700146X.

Duval, S. and Kindermann, M. (2012). Use of nitrooxy organic molecules in feed for reducing enteric methane emissions in ruminants, and/or to improve ruminant performance, International patent application WO, 84629, A1.

Ellis, J., Dijkstra, J., Kebreab, E., Bannink, A., Odongo, N., McBride, B. and France, J. (2008). Aspects of rumen microbiology central to mechanistic modelling of methane production in cattle, *The Journal of Agricultural Science* 146(2), 213-233. DOI: 10.1017/S0021859608007752.

Ellis, J. L., Kebreab, E., Odongo, N. E., McBride, B. W., Okine, E. K. and France, J. (2007). Prediction of methane production from dairy and beef cattle, *Journal of Dairy Science* 90(7), 3456-3466. DOI: 10.3168/jds.2006-675.

FAO (2017). Global livestock environmental assessment model (GLEAM). Rome: Food and Agriculture Organization of the United Nations (FAO). Available from www.fao.org/gleam/en/ (accessed 3 September 2018).

FAO (2018). *The Future of Food and Agriculture - Alternative Pathways to 2050*. Rome. Licence: CC BY-NC-SA 3.0 IGO.

Fellner, V., Burns, J. C. and Marshall, D. S. (2008). Effect of feeding corn, hull-less or hulled barley on fermentation by mixed cultures of ruminal microorganisms, *Journal of Dairy Science* 91(5), 1936-1941. DOI: 10.3168/jds.2007-0078.

Finlay, B. J., Esteban, G., Clarke, K. J., Williams, A. G., Embley, T. M. and Hirt, R. P. (1994). Some rumen ciliates have endosymbiotic methanogens, *FEMS Microbiology Letters* 117(2), 157-161. DOI: 10.1111/j.1574-6968.1994.tb06758.x.

Flay, H. E., Kuhn-Sherlock, B., Macdonald, K. A., Camara, M., Lopez-Villalobos, N., Donaghy, D. J. and Roche, J. R. (2019). Hot topic: selecting cattle for low residual feed intake did not affect daily methane production but increased methane yield, *Journal of Dairy Science* 102(3), 2708-2713. DOI: 10.3168/jds.2018-15234.

Friedman, N., Jami, E. and Mizrahi, I. (2017). Compositional and functional dynamics of the bovine rumen methanogenic community across different developmental stages, *Environmental Microbiology* 19(8), 3365-3373. DOI: 10.1111/1462-2920.13846.

Genovese, G., Tedone, L., Hamann, M. T. and Morabito, M. (2009). The Mediterranean red alga Asparagopsis: a source of compounds against Leishmania, *Marine Drugs* 7(3), 361-366. DOI: 10.3390/md7030361.

Gerber, P. J., Steinfeld, H., Henderson, B., Mottet, A., Opio, C., Dijkman, J., Falcucci, A. and Tempio, G. (2013). *Tackling Climate Change through Livestock: a Global Assessment of Emissions and Mitigation Opportunities*, Rome: Food and Agriculture Organization of the United Nations.

Gijzen, H. J., Lubberding, H. J., Gerhardus, M. J. T. and Vogels, G. D. (1988). Contribution of rumen protozoa to fibre degradation and cellulase activity in vitro, *FEMS Microbiology Letters* 53(1), 35–43. DOI: 10.1111/j.1574-6968.1988.tb02645.x.

Gilbert, R. A., Townsend, E. M., Crew, K. S., Hitch, T. C. A., Friedersdorff, J. C. A., Creevey, C. J., Pope, P. B., Ouwerkerk, D. and Jameson, E. (2020). Rumen virus populations: technological advances enhancing current understanding, *Frontiers in Microbiology* 11, 450. DOI: 10.3389/fmicb.2020.00450.

Gilmore, S. P., Henske, J. K., Sexton, J. A., Solomon, K. V., Seppälä, S., Yoo, J. I., Huyett, L. M., Pressman, A., Cogan, J. Z., Kivenson, V., Peng, X., Tan, Y., Valentine, D. L. and O'Malley, M. A. (2017). Genomic analysis of methanogenic archaea reveals a shift towards energy conservation, *BMC Genomics* 18(1), 639. DOI: 10.1186/s12864-017-4036-4.

Golder, H. M., Thomson, J. M., Denman, S. E., McSweeney, C. S. and Lean, I. J. (2018). Genetic markers are associated with the ruminal microbiome and metabolome in grain and sugar challenged dairy heifers, *Frontiers in Genetics* 9, 62. DOI: 10.3389/fgene.2018.00062.

Goopy, J. P., Korir, D., Pelster, D., Ali, A. I. M., Wassie, S. E., Schlecht, E., Dickhoefer, U., Merbold, L. and Butterbach-Bahl, K. (2020). Severe below-maintenance feed intake increases methane yield from enteric fermentation in cattle, *British Journal of Nutrition* 123(11), 1239–1246. DOI: 10.1017/S0007114519003350.

Grawert, T., Hohmann, H. P., Kindermann, M., Duval, S., Bacher, A. and Fischer, M. (2014). Inhibition of methyl-CoM reductase from Methanobrevibacter ruminantium by 2-bromoethanesulfonate, *Journal of Agricultural and Food Chemistry* 62(52), 12487–12490. DOI: 10.1021/jf505056g.

Grossi, G., Goglio, P., Vitali, A. and Williams, A. G. (2019). Livestock and climate change: impact of livestock on climate and mitigation strategies, *Animal Frontiers: The Review Magazine of Animal Agriculture* 9(1), 69–76. DOI: 10.1093/af/vfy034.

Gruninger, R. J., Puniya, A. K., Callaghan, T. M., Edwards, J. E., Youssef, N., Dagar, S. S., Fliegerova, K., Griffith, G. W., Forster, R., Tsang, A., McAllister, T. and Elshahed, M. S. (2014). Anaerobic fungi (phylum Neocallimastigomycota): advances in understanding their taxonomy, life cycle, ecology, role and biotechnological potential, *FEMS Microbiology Ecology* 90(1), 1–17. DOI: 10.1111/1574-6941.12383.

Guan, H., Wittenberg, K. M., Ominski, K. H. and Krause, D. O. (2006). Efficacy of ionophores in cattle diets for mitigation of enteric methane, *Journal of Animal Science* 84(7), 1896–1906. DOI: 10.2527/jas.2005-652.

Haisan, J., Sun, Y., Guan, L. L., Beauchemin, K. A., Iwaasa, A., Duval, S., Barreda, D. R. and Oba, M. (2014). The effects of feeding 3-nitrooxypropanol on methane emissions and productivity of Holstein cows in mid lactation, *Journal of Dairy Science* 97(5), 3110–3119. DOI: 10.3168/jds.2013-7834.

Hammond, K. J., Burke, J. L., Koolaard, J. P., Muetzel, S., Pinares-Patiño, C. S. and Waghorn, G. C. (2013). Effects of feed intake on enteric methane emissions from sheep fed fresh white clover (Trifolium repens) and perennial ryegrass (Lolium perenne) forages, *Animal Feed Science and Technology* 179(1–4), 121–132. DOI: 10.1016/j.anifeedsci.2012.11.004.

Hammond, K. J., Crompton, L. A., Bannink, A., Dijkstra, J., Yáñez-Ruiz, D. R., O'Kiely, P., Kebreab, E., Eugène, M. A., Yu, Z., Shingfield, K. J., Schwarm, A., Hristov, A. N. and Reynolds, C. K. (2016). Review of current in vivo measurement techniques for quantifying enteric methane emission from ruminants, *Animal Feed Science and Technology* 219, 13–30. DOI: 10.1016/j.anifeedsci.2016.05.018.

Hegarty, R. S., Bird, S. H., Vanselow, B. A. and Woodgate, R. (2008). Effects of the absence of protozoa from birth or from weaning on the growth and methane production of lambs, *British Journal of Nutrition* 100(6), 1220–1227. DOI: 10.1017/S0007114508981435.

Hegarty, R. S., Goopy, J. P., Herd, R. M. and McCorkell, B. (2007). Cattle selected for lower residual feed intake have reduced daily methane production, *Journal of Animal Science* 85(6), 1479–1486. DOI: 10.2527/jas.2006-236.

Hindrichsen, I. K., Wettstein, H. -R., Machmüller, A. and Kreuzer, M. (2006). Methane emission, nutrient degradation and nitrogen turnover in dairy cows and their slurry at different milk production scenarios with and without concentrate supplementation, *Agriculture, Ecosystems and Environment* 113(1–4), 150–161. DOI: 10.1016/j.agee.2005.09.004.

Hobson, P. N. and Stewart, C. S. (1997). *The Rumen Microbial Ecosystem*, London: Blackie Academic & Professional.

Hook, S. E., Steele, M. A., Northwood, K. S., Wright, A. D. and McBride, B. W. (2011). Impact of high-concentrate feeding and low ruminal pH on methanogens and protozoa in the rumen of dairy cows, *Microbial Ecology* 62(1), 94–105. DOI: 10.1007/s00248-011-9881-0.

Hook, S. E., Wright, A. D. and McBride, B. W. (2010). Methanogens: methane producers of the rumen and mitigation strategies, *Archaea* 2010, 945785. DOI: 10.1155/2010/945785.

Hristov, A. N., Callaway, T. R., Lee, C. and Dowd, S. E. (2012). Rumen bacterial, archaeal, and fungal diversity of dairy cows in response to ingestion of lauric or myristic acid, *Journal of Animal Science* 90(12), 4449–4457. DOI: 10.2527/jas.2011-4624.

Hristov, A. N., Oh, J., Giallongo, F., Frederick, T. W., Harper, M. T., Weeks, H. L., Branco, A. F., Moate, P. J., Deighton, M. H., Williams, S. R. O., Kindermann, M. and Duval, S. (2015). An inhibitor persistently decreased enteric methane emission from dairy cows with no negative effect on milk production, *Proceedings of the National Academy of Sciences of the United States of America* 112(34), 10663–10668. DOI: 10.1073/pnas.1504124112.

Hulshof, R. B., Berndt, A., Gerrits, W. J., Dijkstra, J., Van Zijderveld, S. M., Newbold, J. R. and Perdok, H. B. (2012). Dietary nitrate supplementation reduces methane emission in beef cattle fed sugarcane-based diets, *Journal of Animal Science* 90(7), 2317–2323. DOI: 10.2527/jas.2011-4209.

Hünerberg, M., McGinn, S. M., Beauchemin, K. A., Entz, T., Okine, E. K., Harstad, O. M. and McAllister, T. A. (2015). Impact of ruminal pH on enteric methane emissions, *Journal of Animal Science* 93(4), 1760–1766. DOI: 10.2527/jas.2014-8469.

Huws, S. A., Creevey, C. J., Oyama, L. B., Mizrahi, I., Denman, S. E., Popova, M., Muñoz-Tamayo, R., Forano, E., Waters, S. M., Hess, M., Tapio, I., Smidt, H., Krizsan, S. J., Yáñez-Ruiz, D. R., Belanche, A., Guan, L., Gruninger, R. J., McAllister, T. A., Newbold, C. J., Roehe, R., Dewhurst, R. J., Snelling, T. J., Watson, M., Suen, G., Hart, E. H., Kingston-Smith, A. H., Scollan, N. D., do Prado, R. M., Pilau, E. J., Mantovani, H. C., Attwood, G. T., Edwards, J. E., McEwan, N. R., Morrisson, S., Mayorga, O. L., Elliott, C. and Morgavi, D. P. (2018). Addressing global ruminant agricultural challenges Through understanding the rumen microbiome: past, present, and future, *Frontiers in Microbiology* 9, 2161. DOI: 10.3389/fmicb.2018.02161.

Jackson, R. B., Solomon, E. I., Canadell, J. G., Cargnello, M. and Field, C. B. (2019). Methane removal and atmospheric restoration, *Nature Sustainability* 2(6), 436–438. DOI: 10.1038/s41893-019-0299-x.

Janssen, P. H. and Kirs, M. (2008). Structure of the archaeal community of the rumen, *Applied and Environmental Microbiology* 74(12), 3619-3625. DOI: 10.1128/AEM.02812-07.

Jarvis, G. N., Strömpl, C., Burgess, D. M., Skillman, L. C., Moore, E. R. and Joblin, K. N. (2000). Isolation and identification of ruminal methanogens from grazing cattle, *Current Microbiology* 40(5), 327–332. DOI: 10.1007/s002849910065.

Johnson, K. A. and Johnson, D. E. (1995). Methane emissions from cattle, *Journal of Animal Science* 73(8), 2483–2492. DOI: 10.2527/1995.7382483x.

Jones, F. M., Phillips, F. A., Naylor, T. and Mercer, N. B. (2011). Methane emissions from grazing Angus beef cows selected for divergent residual feed intake, *Animal Feed Science and Technology* 166-167, 302–307. DOI: 10.1016/j.anifeedsci.2011.04.020.

Jordan, E., Kenny, D., Hawkins, M., Malone, R., Lovett, D. K. and O'Mara, F. P. (2006a). Effect of refined soy oil or whole soybeans on intake, methane output, and performance of young bulls, *Journal of Animal Science* 84(9), 2418–2425. DOI: 10.2527/jas.2005-354.

Jordan, E., Lovett, D. K., Monahan, F. J., Callan, J., Flynn, B. and O'Mara, F. P. (2006b). Effect of refined coconut oil or copra meal on methane output and on intake and performance of beef heifers, *Journal of Animal Science* 84(1), 162–170. DOI: 10.2527/2006.841162x.

Kamke, J., Kittelmann, S., Soni, P., Li, Y., Tavendale, M., Ganesh, S., Janssen, P. H., Shi, W., Froula, J., Rubin, E. M. and Attwood, G. T. (2016). Rumen metagenome and metatranscriptome analyses of low methane yield sheep reveals a Sharpea-enriched microbiome characterised by lactic acid formation and utilisation, *Microbiome* 4(1), 56. DOI: 10.1186/s40168-016-0201-2.

Kim, E. T., Guan, le L., Lee, S. J., Lee, S. M., Lee, S. S., Lee, I. D., Lee, S. K. and Lee, S. S. (2015). Effects of flavonoid-rich plant extracts on in vitro ruminal methanogenesis, microbial populations and fermentation characteristics, *Asian-Australasian Journal of Animal Sciences* 28(4), 530–537. DOI: 10.5713/ajas.14.0692.

Kim, H., Lee, H. G., Baek, Y. C., Lee, S. and Seo, J. (2020). The effects of dietary supplementation with 3-nitrooxypropanol on enteric methane emissions, rumen fermentation, and production performance in ruminants: a meta-analysis, *Journal of Animal Science and Technology* 62(1), 31–42. DOI: 10.5187/jast.2020.62.1.31.

Kim, M., Morrison, M. and Yu, Z. (2011). Status of the phylogenetic diversity census of ruminal microbiomes, *FEMS Microbiology Ecology* 76(1), 49–63. DOI: 10.1111/j.1574-6941.2010.01029.x.

Kinley, R. D., Martinez-Fernandez, G., Matthews, M. K., de Nys, R., Magnusson, M. and Tomkins, N. W. (2020). Mitigating the carbon footprint and improving productivity of ruminant livestock agriculture using red seaweed, *Journal of Cleaner Production* 259, DOI: 10.1016/j.jclepro.2020.120836.

Kirchgessner, M., Windisch, W., Müller, H. and Kreuzer, M. (1991). Release of methane and of carbon dioxide by dairy cattle, *Agricultural and Biological Research* 44(2), 91-102.

Kittelmann, S., Devente, S. R., Kirk, M. R., Seedorf, H., Dehority, B. A. and Janssen, P. H. (2015). Phylogeny of intestinal ciliates, including Charonina ventriculi, and comparison of microscopy and 18S rRNA gene pyrosequencing for rumen ciliate community structure analysis, *Applied and Environmental Microbiology* 81(7), 2433-2444. DOI: 10.1128/AEM.03697-14.

Kittelmann, S., Pinares-Patino, C. S., Seedorf, H., Kirk, M. R., Ganesh, S., McEwan, J. C. and Janssen, P. H. (2014). Two different bacterial community types are linked with the low-methane emission trait in sheep, *PLoS ONE* 9(7), e103171. DOI: 10.1371/journal.pone.0103171.

Kreuzer, M. and Kirchgessner, M. (1987). The nutritive defaunation of the rumen in ruminants, *Archiv fur Tierernahrung* 37(6), 489–503. DOI: 10.1080/17450398709421064.

Kumar, S., Indugu, N., Vecchiarelli, B. and Pitta, D. W. (2015). Associative patterns among anaerobic fungi, methanogenic archaea, and bacterial communities in response to changes in diet and age in the rumen of dairy cows, *Frontiers in Microbiology* 6, 781. DOI: 10.3389/fmicb.2015.00781.

Kumar, S., Puniya, A. K., Puniya, M., Dagar, S. S., Sirohi, S. K., Singh, K. and Griffith, G. W. (2009). Factors affecting rumen methanogens and methane mitigation strategies, *World Journal of Microbiology and Biotechnology* 25(9), 1557–1566. DOI: 10.1007/s11274-009-0041-3.

Kurihara, M., Shibata, M., Nishida, T., Purnomoadi, A. and Terada, F. (1997). Methane production and its dietary manipulation in ruminants. In: Oncera, R., Itabashi, H., Ushida, K., Tano, H. and Sasaki, Y. (Eds), *Rumen Microbes and Digestive Physiology in Ruminants*. Satellite Symposium of the 8th Animal Science Congress, Kyoto, Japan.

Lassey, K. R. (2007). Livestock methane emission: from the individual grazing animal through national inventories to the global methane cycle, *Agricultural and Forest Meteorology* 142(2–4), 120–132. DOI: 10.1016/j.agrformet.2006.03.028.

Latham, E. A., Anderson, R. C., Pinchak, W. E. and Nisbet, D. J. (2016). Insights on alterations to the rumen ecosystem by nitrate and nitrocompounds, *Frontiers in Microbiology* 7, 228. DOI: 10.3389/fmicb.2016.00228.

Leahy, S. C., Clark, H. and Reisinger, A. (2020). Challenges and prospects for agricultural greenhouse gas mitigation pathways consistent with the Paris Agreement, *Frontiers in Sustainable Food Systems* 4, 69. DOI: 10.3389/fsufs.2020.00069.

Leahy, S. C., Kelly, W. J., Ronimus, R. S., Wedlock, N., Altermann, E. and Attwood, G. T. (2013). Genome sequencing of rumen bacteria and archaea and its application to methane mitigation strategies, *Animal* 7(s2), 235–243. DOI: 10.1017/S1751731113000700.

Lechner-Doll, M., Kaske, M. and Engelhardt, W. (1991). Factors affecting the mean retention time of particles in the forestomach of ruminants and camelids. In: *Physiological Aspects of Digestion and Metabolism in Ruminants*, 455–482. New York: Academic Press. DOI: 10.1016/B978-0-12-702290-1.50027-8.

Lee, C. and Beauchemin, K. A. (2014). A review of feeding supplementary nitrate to ruminant animals: nitrate toxicity, methane emissions, and production performance, *Canadian Journal of Animal Science* 94(4), 557–570. DOI: 10.4141/cjas-2014-069.

Leng, R. (1970). Formation and production of volatile fatty acids in the rumen. In: Phillipson, A. T., Annison, E. F., Armstrong, D. G., Balch, C. C., Comline, R. S., Hardy, R. N., Hobson, P. N. and Keynes, R. D. (Eds), *Physiology of Digestion and Metabolism in the Ruminant*, 406–421. PA: Oriel Press, Pennsylvania State University.

Leslie, M., Aspin, M. and Clark, H. (2008). Greenhouse gas emissions from New Zealand agriculture: issues, perspectives and industry response, *Australian Journal of Experimental Agriculture* 48(2), 1–5. DOI: 10.1071/EA07306.

Li, F., Li, C., Chen, Y., Liu, J., Zhang, C., Irving, B., Fitzsimmons, C., Plastow, G. and Guan, L. L. (2019). Host genetics influence the rumen microbiota and heritable rumen microbial features associate with feed efficiency in cattle, *Microbiome* 7(1), 92. DOI: 10.1186/s40168-019-0699-1.

Li, X., Norman, H. C., Kinley, R. D., Laurence, M., Wilmot, M., Bender, H., de Nys, R. and Tomkins, N. (2018a). Asparagopsis taxiformis decreases enteric methane production from sheep, *Animal Production Science* 58(4), 681-688. DOI: 10.1071/AN 15883.

Li, Z., Deng, Q., Liu, Y., Yan, T., Li, F., Cao, Y. and Yao, J. (2018b). Dynamics of methanogenesis, ruminal fermentation and fiber digestibility in ruminants following elimination of protozoa: a meta-analysis, *Journal of Animal Science and Biotechnology* 9(1), 89. DOI: 10.1186/s40104-018-0305-6.

Lima, A., Resende, K. T., Teixeira, I. A., Fernandes, M., Frighetto, R., Bonfá, H., Junior, C., Dias, E. and Silva, N. (2013). Methane production of goats fed different forage: concentrate ratio. In: *Annual Meeting Brazilian Society of Animal Science: The Integration of Knowledge in Animal Production*. Campinas: SBZ. 1 CD ROM. Ref. P2-A2. 87.

Lin, C., Raskin, L. and Stahl, D. A. (1997). Microbial community structure in gastrointestinal tracts of domestic animals: comparative analyses using rRNA-targeted oligonucleotide probes, *FEMS Microbiology Ecology* 22(4), 281-294. DOI: 10.1111/j.1574-6941 .1997.tb00380.x.

Lovett, D., Lovell, S., Stack, L., Callan, J., Finlay, M., Conolly, J. and O'Mara, F. P. (2003). Effect of forage/concentrate ratio and dietary coconut oil level on methane output and performance of finishing beef heifers, *Livestock Production Science* 84(2), 135-146. DOI: 10.1016/j.livprodsci.2003.09.010.

Lovett, D. K., Stack, L. J., Lovell, S., Callan, J., Flynn, B., Hawkins, M. and O'Mara, F. P. (2005). Manipulating enteric methane emissions and animal performance of late-lactation dairy cows through concentrate supplementation at pasture, *Journal of Dairy Science* 88(8), 2836-2842. DOI: 10.3168/jds.S0022-0302(05)72964-7.

Lund, P., Dahl, R., Yang, H. J., Hellwing, A. L. F., Cao, B. B. and Weisbjerg, M. R. (2014). The acute effect of addition of nitrate on in vitro and in vivo methane emission in dairy cows, *Animal Production Science* 54(9), 1432. DOI: 10.1071/AN14339.

Lynch, J. and Pierrehumbert, R. (2019). Climate impacts of cultured meat and beef cattle, *Frontiers in Sustainable Food Systems* 3(5). DOI: 10.3389/fsufs.2019.00005.

Machado, L., Magnusson, M., Paul, N. A., de Nys, R. and Tomkins, N. (2014). Effects of marine and freshwater macroalgae on in vitro total gas and methane production, *PLoS ONE* 9(1), e85289. DOI: 10.1371/journal.pone.0085289.

Machado, L., Tomkins, N., Magnusson, M., Midgley, D. J., de Nys, R. and Rosewarne, C. P. (2018). In vitro response of rumen microbiota to the antimethanogenic red macroalga Asparagopsis taxiformis, *Microbial Ecology* 75(3), 811-818. DOI: 10.1007/s00248-017-1086-8.

Machmüller, A. (2006). Medium-chain fatty acids and their potential to reduce methanogenesis in domestic ruminants, *Agriculture, Ecosystems and Environment* 112(2-3), 107-114. DOI: 10.1016/j.agee.2005.08.010.

Machmüller, A. and Kreuzer, M. (1999). Methane suppression by coconut oil and associated effects on nutrient and energy balance in sheep, *Canadian Journal of Animal Science* 79(1), 65-72. DOI: 10.4141/A98-079.

Machmüller, A., Machmüller, A., Soliva, C. R. and Kreuzer, M. (2003). Methane-suppressing effect of myristic acid in sheep as affected by dietary calcium and forage proportion, *British Journal of Nutrition* 90(3), 529-540. DOI: 10.1079/BJN2003932.

Machmüller, A., Ossowski, D. A. and Kreuzer, M. (2000). Comparative evaluation of the effects of coconut oil, oilseeds and crystalline fat on methane release, digestion and energy balance in lambs, *Animal Feed Science and Technology* 85(1-2), 41-60. DOI: 10.1016/S0377-8401(00)00126-7.

Machmüller, A., Ossowski, D. A., Wanner, M. and Kreuzer, M. (1998). Potential of various fatty feeds to reduce methane release from rumen fermentation in vitro (Rusitec), *Animal Feed Science and Technology* 71(1-2), 117-130. DOI: 10.1016/S0377-8401(97)00126-0.

Martínez-Fernández, G., Abecia, L., Arco, A., Cantalapiedra-Hijar, G., Martín-García, A. I., Molina-Alcaide, E., Kindermann, M., Duval, S. and Yáñez-Ruiz, D. R. (2014). Effects of ethyl-3-nitrooxy propionate and 3-nitrooxypropanol on ruminal fermentation, microbial abundance, and methane emissions in sheep, *Journal of Dairy Science* 97(6), 3790-3799. DOI: 10.3168/jds.2013-7398.

Martinez-Fernandez, G., Duval, S., Kindermann, M., Schirra, H. J., Denman, S. E. and McSweeney, C. S. (2018). 3-NOP vs. halogenated compound: methane production, ruminal fermentation and microbial community response in forage fed cattle, *Frontiers in Microbiology* 9, 1582. DOI: 10.3389/fmicb.2018.01582.

McAllister, T. A., Meale, S. J., Valle, E., Guan, L. L., Zhou, M., Kelly, W. J., Henderson, G., Attwood, G. T. and Janssen, P. H. (2015). Ruminant nutrition symposium: use of genomics and transcriptomics to identify strategies to lower ruminal methanogenesis, *Journal of Animal Science* 93(4), 1431-1449. DOI: 10.2527/jas.2014-8329.

McCaughey, W. P., Wittenberg, K. and Corrigan, D. (1999). Impact of pasture type on methane production by lactating beef cows, *Canadian Journal of Animal Science* 79(2), 221-226. DOI: 10.4141/A98-107.

McDonnell, R. P., Hart, K. J., Boland, T. M., Kelly, A. K., McGee, M. and Kenny, D. A. (2016). Effect of divergence in phenotypic residual feed intake on methane emissions, ruminal fermentation, and apparent whole-tract digestibility of beef heifers across three contrasting diets, *Journal of Animal Science* 94(3), 1179-1193. DOI: 10.2527/jas.2015-0080.

McGinn, S. M., Beauchemin, K. A., Coates, T. and Colombatto, D. (2004). Methane emissions from beef cattle: effects of monensin, sunflower oil, enzymes, yeast, and fumaric acid, *Journal of Animal Science* 82(11), 3346-3356. DOI: 10.2527/2004.82113346x.

McSweeney, C. and Mackie, R. I. (2012). Commission on genetic resources for food and agriculture. Micro-organisms and ruminant digestion: state of knowledge, trends and future prospects, *Background Study Paper (FAO)* 61, 1-62.

Melgar, A., Harper, M. T., Oh, J., Giallongo, F., Young, M. E., Ott, T. L., Duval, S. and Hristov, A. N. (2020). Effects of 3-nitrooxypropanol on rumen fermentation, lactational performance, and resumption of ovarian cyclicity in dairy cows, *Journal of Dairy Science* 103(1), 410-432. DOI: 10.3168/jds.2019-17085.

Mills, J. A., Kebreab, E., Yates, C. M., Crompton, L. A., Cammell, S. B., Dhanoa, M. S., Agnew, R. E. and France, J. (2003). Alternative approaches to predicting methane emissions from dairy cows, *Journal of Animal Science* 81(12), 3141-3150. DOI: 10.2527/2003.81123141x.

Moe, P. W. and Tyrrell, H. F. (1979). Methane production in dairy cows, *Journal of Dairy Science* 62(10), 1583-1586. DOI: 10.3168/jds.S0022-0302(79)83465-7.

Molano, G. and Clark, H. (2008). The effect of level of intake and forage quality on methane production by sheep, *Australian Journal of Experimental Agriculture* 48(2), 219-222. DOI: 10.1071/EA07253.

Morgavi, D. P., Martin, C., Jouany, J. P. and Ranilla, M. J. (2012). Rumen protozoa and methanogenesis: not a simple cause-effect relationship, *British Journal of Nutrition* 107(3), 388–397. DOI: 10.1017/S0007114511002935.

Mosoni, P., Martin, C., Forano, E. and Morgavi, D. P. (2011). Long-term defaunation increases the abundance of cellulolytic ruminococci and methanogens but does not affect the bacterial and methanogen diversity in the rumen of sheep, *Journal of Animal Science* 89(3), 783–791. DOI: 10.2527/jas.2010-2947.

Moss, A. R., Jouany, J.-P. and Newbold, J. (2000). Methane production by ruminants: its contribution to global warming, *INRA/EDP Sciences, Annales de Zootechnie* 49(3), 231–253. DOI: 10.1051/animres:2000119.

Newbold, C. J., De La Fuente, G., Belanche, A., Ramos-Morales, E. and McEwan, N. R. (2015). The role of ciliate protozoa in the rumen, *Frontiers in Microbiology* 6, 1313. DOI: 10.3389/fmicb.2015.01313.

Newbold, C. J. and Ramos-Morales, E. (2019). Microbiome of the rumen. Available at: https://pure.sruc.ac.uk/ws/portalfiles/portal/16673373/045424636_08_Newbold_Ramos_Morales_Microbiome_of_the_rumen_BY_ND_.pdf.

Nguyen, S. H. and Hegarty, R. S. (2017). Effects of defaunation and dietary coconut oil distillate on fermentation, digesta kinetics and methane production of Brahman heifers, *Journal of Animal Physiology and Animal Nutrition* 101(5), 984–993. DOI: 10.1111/jpn.12534.

Niu, M., Kebreab, E., Hristov, A. N., Oh, J., Arndt, C., Bannink, A., Bayat, A. R., Brito, A. F., Boland, T., Casper, D., Crompton, L. A., Dijkstra, J., Eugène, M. A., Garnsworthy, P. C., Haque, M. N., Hellwing, A. L. F., Huhtanen, P., Kreuzer, M., Kuhla, B., Lund, P., Madsen, J., Martin, C., McClelland, S. C., McGee, M., Moate, P. J., Muetzel, S., Muñoz, C., O'Kiely, P., Peiren, N., Reynolds, C. K., Schwarm, A., Shingfield, K. J., Storlien, T. M., Weisbjerg, M. R., Yáñez-Ruiz, D. R. and Yu, Z. (2018). Prediction of enteric methane production, yield, and intensity in dairy cattle using an intercontinental database, *Global Change Biology* 24(8), 3368–3389. DOI: 10.1111/gcb.14094.

Nolan, J. V., Hegarty, R. S., Hegarty, J., Godwin, I. R. and Woodgate, R. (2010). Effects of dietary nitrate on fermentation, methane production and digesta kinetics in sheep, *Animal Production Science* 50(8), 801–806. DOI: 10.1071/AN09211.

Novilla, M. N., McClary, D. and Laudert, S. B. (2017). Ionophores. In: Gupta, R. C. (Ed.), *Reproductive and Developmental Toxicology*, 503–518. New York: Academic Press. DOI: 10.1016/B978-0-12-804239-7.00029-9.

Odongo, N. E., Bagg, R., Vessie, G., Dick, P., Or-Rashid, M. M., Hook, S. E., Gray, J. T., Kebreab, E., France, J. and McBride, B. W. (2007). Long-term effects of feeding monensin on methane production in lactating dairy cows, *Journal of Dairy Science* 90(4), 1781–1788. DOI: 10.3168/jds.2006-708.

Opio, C., Gerber, P., Mottet, A., Falcucci, A., Tempio, G., MacLeod, M., Vellinga, T., Henderson, B. and Steinfeld, H. (2013). *Greenhouse Gas Emissions from Ruminant Supply Chains–A Global Life Cycle Assessment*. Rome: Food and Agriculture Organization of the United Nations. Available at: www.fao.org/3/i3461e/i3461e.pdf.

Ornelas, L. T. C., Silva, D. C., Tomich, T. R., Campos, M. M., Machado, F. S., Ferreira, A. L., Maurício, R. M. and Pereira, L. G. R. (2019). Differences in methane production, yield and intensity and its effects on metabolism of dairy heifers, *Science of the Total Environment* 689, 1133–1140. DOI: 10.1016/j.scitotenv.2019.06.489.

Owens, F. N., Secrist, D. S., Hill, W. J. and Gill, D. R. (1998). Acidosis in cattle: a review, *Journal of Animal Science* 76(1), 275–286. DOI: 10.2527/1998.761275x.

Patra, A. K. (2012). Enteric methane mitigation technologies for ruminant livestock: a synthesis of current research and future directions, *Environmental Monitoring and Assessment* 184(4), 1929–1952. DOI: 10.1007/s10661-011-2090-y.

Patra, A. K., Lalhriatpuii, M. and Debnath, B. C. (2016). Predicting enteric methane emission in sheep using linear and non-linear statistical models from dietary variables, *Animal Production Science* 56(3), 574–584. DOI: 10.1071/AN15505.

Patra, A. K., Park, T., Kim, M. and Yu, Z. (2017). Rumen methanogens and mitigation of methane emission by anti-methanogenic compounds and substances, *Journal of Animal Science and Biotechnology* 8(1), 13. DOI: 10.1186/s40104-017-0145-9.

Patra, A. K. and Saxena, J. (2010). A new perspective on the use of plant secondary metabolites to inhibit methanogenesis in the rumen, *Phytochemistry* 71(11–12), 1198–1222. DOI: 10.1016/j.phytochem.2010.05.010.

Paul, N. A., de Nys, R. and Steinberg, P. (2006). Chemical defence against bacteria in the red alga Asparagopsis armata: linking structure with function, *Marine Ecology Progress Series* 306, 87–101. DOI: 10.3354/meps306087.

Paynter, M. J. and Hungate, R. E. (1968). Characterization of Methanobacterium mobilis, sp. n., isolated from the bovine rumen, *Journal of Bacteriology* 95(5), 1943–1951. DOI: 10.1128/JB.95.5.1943-1951.1968.

Penner, G. B. (2014). Mechanisms of volatile fatty acid absorption and metabolism and maintenance of a stable rumen environment. In: *25th Florida Ruminant Nutrition Symposium*, Gainesville, FL vol. 4, 92–104.

Petri, R. M., Schwaiger, T., Penner, G. B., Beauchemin, K. A., Forster, R. J., McKinnon, J. J. and McAllister, T. A. (2013). Characterization of the core rumen microbiome in cattle during transition from forage to concentrate as well as during and after an acidotic challenge, *PLoS ONE* 8(12), e83424. DOI: 10.1371/journal.pone.0083424.

Petrie, K., Hart, K., Callan, J., Boland, T. and Kenny, D. (2009). The effect of level of dietary fish oil inclusion on intake and methane emissions of beef steers. In: *Proceedings of the British Society of Animal Science*, Cambridge University Press, vol. 2009, 25. DOI: 10.1017/S1752756200028647.

Pickering, N. K., Chagunda, M. G., Banos, G., Mrode, R., McEwan, J. C. and Wall, E. (2015). Genetic parameters for predicted methane production and laser methane detector measurements, *Journal of Animal Science* 93(1), 11–20. DOI: 10.2527/jas.2014-8302.

Pinares-Patiño, C. S., D'hour, P., Jouany, J.-P. and Martin, C. (2007). Effects of stocking rate on methane and carbon dioxide emissions from grazing cattle, *Agriculture, Ecosystems and Environment* 121(1–2), 30–46. DOI: 10.1016/j.agee.2006.03.024.

Pinares-Patiño, C. S., Franco, F. E., Molano, G., Kjestrup, H., Sandoval, E., MacLean, S., Battistotti, M., Koolaard, J. and Laubach, J. (2016). Feed intake and methane emissions from cattle grazing pasture sprayed with canola oil, *Livestock Science* 184, 7–12. DOI: 10.1016/j.livsci.2015.11.020.

Pinares-Patiño, C. S., Ulyatt, M. J., Lassey, K. R., Barry, T. N. and Holmes, C. W. (2003). Persistence of differences between sheep in methane emission under generous grazing conditions, *The Journal of Agricultural Science* 140(2), 227–233. DOI: 10.1017/S0021859603003071.

Plaizier, J. C., Krause, D. O., Gozho, G. N. and McBride, B. W. (2008). Subacute ruminal acidosis in dairy cows: the physiological causes, incidence and consequences, *The Veterinary Journal* 176(1), 21-31. DOI: 10.1016/j.tvjl.2007.12.016.

Plaza, M., Cifuentes, A. and Ibáñez, E. (2008). In the search of new functional food ingredients from algae, *Trends in Food Science and Technology* 19(1), 31-39. DOI: 10.1016/j.tifs.2007.07.012.

Poulsen, M., Schwab, C., Borg Jensen, B. B., Engberg, R. M., Spang, A., Canibe, N., Højberg, O., Milinovich, G., Fragner, L., Schleper, C., Weckwerth, W., Lund, P., Schramm, A. and Urich, T. (2013). Methylotrophic methanogenic Thermoplasmata implicated in reduced methane emissions from bovine rumen, *Nature Communications* 4, 1428. DOI: 10.1038/ncomms2432.

Prins, R. A., Van Nevel, C. J. and Demeyer, D. I. (1972). Pure culture studies of inhibitors for methanogenic bacteria, *Antonie van Leeuwenhoek* 38(3), 281-287. DOI: 10.1007/BF02328099.

Reynolds, C. K., Humphries, D. J., Kirton, P., Kindermann, M., Duval, S. and Steinberg, W. (2014). Effects of 3-nitrooxypropanol on methane emission, digestion, and energy and nitrogen balance of lactating dairy cows, *Journal of Dairy Science* 97(6), 3777-3789. DOI: 10.3168/jds.2013-7397.

Ripple, W. J., Smith, P., Haberl, H., Montzka, S. A., McAlpine, C. and Boucher, D. H. (2014). Ruminants, climate change and climate policy, *Nature Climate Change* 4(1), 2-5. DOI: 10.1038/nclimate2081.

Robinson, D. L., Goopy, J. and Hegarty, R. S. (2010). Can rumen methane production be predicted from volatile fatty acid concentrations?, *Animal Production Science* 50(6), 630-636. DOI: 10.1071/AN09214.

Roehe, R., Dewhurst, R. J., Duthie, C. A., Rooke, J. A., McKain, N., Ross, D. W., Hyslop, J. J., Waterhouse, A., Freeman, T. C., Watson, M. and Wallace, R. J. (2016). Bovine host genetic variation influences rumen microbial methane production with best selection criterion for low methane emitting and efficiently feed converting hosts based on metagenomic gene abundance, *PLoS Genetics* 12(2), e1005846. DOI: 10.1371/journal.pgen.1005846.

Rojas-Downing, M. M., Nejadhashemi, A. P., Harrigan, T. and Woznicki, S. A. (2017). Climate change and livestock: impacts, adaptation, and mitigation, *Climate Risk Management* 16, 145-163. DOI: 10.1016/j.crm.2017.02.001.

Romero-Perez, A., Okine, E. K., McGinn, S. M., Guan, L. L., Oba, M., Duval, S. M., Kindermann, M. and Beauchemin, K. A. (2014). The potential of 3-nitrooxypropanol to lower enteric methane emissions from beef cattle, *Journal of Animal Science* 92(10), 4682-4693. DOI: 10.2527/jas.2014-7573.

Roque, B. M., Salwen, J. K., Kinley, R. and Kebreab, E. (2019). Inclusion of Asparagopsis armata in lactating dairy cows' diet reduces enteric methane emission by over 50 percent, *Journal of Cleaner Production* 234, 132-138. DOI: 10.1016/j.jclepro.2019.06.193.

Rowe, J. B., Davies, A. and Broome, A. W. (1985). Quantitative effects of defaunation on rumen fermentation and digestion in sheep, *British Journal of Nutrition* 54(1), 105-119. DOI: 10.1079/BJN19850097.

Russell, J., Barnett, J., Desilets, E. and Bertrand, S. (2007). 'Mitigation strategies to reduce GHG emissions from the dairy industry, *Fil-Idf Bulletin: Federation Internationale de Laiterie International Dairy Federation* 422, 30.

Russell, J. B. and Houlihan, A. J. (2003). Ionophore resistance of ruminal bacteria and its potential impact on human health, *FEMS Microbiology Reviews* 27(1), 65–74. DOI: 10.1016/S0168-6445(03)00019-6.

Russell, J. B. and Rychlik, J. L. (2001). Factors that alter rumen microbial ecology, *Science* 292(5519), 1119–1122. DOI: 10.1126/science.1058830.

Russell, J. B. and Strobel, H. J. (1989). Effect of ionophores on ruminal fermentation, *Applied and Environmental Microbiology* 55(1), 1–6. DOI: 10.1128/AEM.55.1.1-6.1989.

Sar, C., Santoso, B., Mwenya, B., Gamo, Y., Kobayashi, T., Morikawa, R., Kimura, K., Mizukoshi, H. and Takahashi, J. (2004). Manipulation of rumen methanogenesis by the combination of nitrate with β1-4 galacto-oligosaccharides or nisin in sheep, *Animal Feed Science and Technology* 115(1–2), 129–142. DOI: 10.1016/j.anifeedsci.2004.01.006.

Sauer, F. D., Fellner, V., Kinsman, R., Kramer, J. K. G., Jackson, H. A., Lee, A. J. and Chen, S. (1998). Methane output and lactation response in Holstein cattle with monensin or unsaturated fat added to the diet, *Journal of Animal Science* 76(3), 906–914. DOI: 10.2527/1998.763906x.

Schils, R. L. M., Verhagen, A., Aarts, H. F. M. and Šebek, L. B. J. (2005). A farm level approach to define successful mitigation strategies for GHG emissions from ruminant livestock systems, *Nutrient Cycling in Agroecosystems* 71(2), 163–175. DOI: 10.1007/s10705-004-2212-9.

Seshadri, R., Leahy, S. C., Attwood, G. T., Teh, K. H., Lambie, S. C., Cookson, A. L., Eloe-Fadrosh, E. A., Pavlopoulos, G. A., Hadjithomas, M., Varghese, N. J., Paez-Espino, D., Hungate1000 project collaborators, Perry, R., Henderson, G., Creevey, C. J., Terrapon, N., Lapebie, P., Drula, E., Lombard, V., Rubin, E., Kyrpides, N. C., Henrissat, B., Woyke, T., Ivanova, N. N. and Kelly, W. J. (2018). Cultivation and sequencing of rumen microbiome members from the Hungate1000 Collection, *Nature Biotechnology* 36(4), 359–367. DOI: 10.1038/nbt.4110.

Shi, W., Moon, C. D., Leahy, S. C., Kang, D., Froula, J., Kittelmann, S., Fan, C., Deutsch, S., Gagic, D., Seedorf, H., Kelly, W. J., Atua, R., Sang, C., Soni, P., Li, D., Pinares-Patiño, C. S., McEwan, J. C., Janssen, P. H., Chen, F., Visel, A., Wang, Z., Attwood, G. T. and Rubin, E. M. (2014). Methane yield phenotypes linked to differential gene expression in the sheep rumen microbiome, *Genome Research* 24(9), 1517–1525. DOI: 10.1101/gr.168245.113.

Shibata, M., Terada, F., Iwasaki, K., Kurihara, M. and Nishida, T. (1992). Methane production in heifers, sheep and goats consuming diets of various hay-concentrate ratios, *Nihon Chikusan Gakkaiho* 63(12), 1221–1227. DOI: 10.2508/chikusan.63.1221.

Shibata, M., Terada, F., Kurihara, M., Nishida, T. and Iwasaki, K. (1993). Estimation of methane production in ruminants, *Nihon Chikusan Gakkaiho* 64(8), 790–796. DOI: 10.2508/chikusan.64.790.

Singh, K. M., Patel, A. K., Shah, R. K., Reddy, B. and Joshi, C. G. (2015). Potential functional gene diversity involved in methanogenesis and methanogenic community structure in Indian buffalo (Bubalus bubalis) rumen, *Journal of Applied Genetics* 56(3), 411–426. DOI: 10.1007/s13353-015-0270-0.

Slyter, L. L., Bryant, M. P. and Wolin, M. J. (1966). Effect of pH on population and fermentation in a continuously cultured rumen ecosystem, *Applied Microbiology* 14(4), 573–578. DOI: 10.1128/AEM.14.4.573-578.1966.

Steinfeld, H. (2006). *Livestock's Long Shadow: Environmental Issues and Options*. Rome: Food and Agriculture Organization of the United Nations. Available at: http://www.fao.org/3/a-a0701e.pdf.

Subharat, S., Shu, D., Zheng, T., Buddle, B. M., Kaneko, K., Hook, S., Janssen, P. H. and Wedlock, D. N. (2016). Vaccination of sheep with a methanogen protein provides insight into levels of antibody in saliva needed to target ruminal methanogens, *PLoS ONE* 11(7), e0159861. DOI: 10.1371/journal.pone.0159861.

Tapio, I., Snelling, T. J., Strozzi, F. and Wallace, R. J. (2017). The ruminal microbiome associated with methane emissions from ruminant livestock, *Journal of Animal Science and Biotechnology* 8(1), 7. DOI: 10.1186/s40104-017-0141-0.

Tedeschi, L. O., Fox, D. G. and Tylutki, T. P. (2003). Potential Environmental Benefits of Ionophores in Ruminant Diets, *Journal of Environmental Quality* 32(5), 1591–1602. DOI: 10.2134/jeq2003.1591.

Thiel, A., Rumbeli, R., Mair, P., Yeman, H.. and Beilstein, P. (2019a). 3-NOP: ADME studies in rats and ruminanting animals, *Food and Chemical Toxicology* 125, 528–539. DOI: 10.1016/j.fct.2019.02.002.

Thiel, A., Schoenmakers, A. C. M., Verbaan, I. A. J., Chenal, E., Etheve, S. and Beilstein, P. (2019b). 3-NOP: mutagenicity and genotoxcity assessment, *Food and Chemical Toxicology* 123, 566–573. DOI: 10.1016/j.fct.2018.11.010.

Tullo, E., Finzi, A. and Guarino, M. (2019). Review: Environmental impact of livestock farming and Precision Livestock Farming as a mitigation strategy, *Science of the Total Environment* 650(2), 2751–2760. DOI: 10.1016/j.scitotenv.2018.10.018.

Tymensen, L. D., Beauchemin, K. A. and McAllister, T. A. (2012). Structures of free-living and protozoa-associated methanogen communities in the bovine rumen differ according to comparative analysis of 16S rRNA and mcrA genes, *Microbiology* 158(7), 1808–1817. DOI: 10.1099/mic.0.057984-0.

Ungerfeld, E. (2013). A theoretical comparison between two ruminal electron sinks, *Frontiers in Microbiology* 4, 319. DOI: 10.3389/fmicb.2013.00319.

Ungerfeld, E. and Kohn, R. (2006). The role of thermodynamics in the control of ruminal fermentation. In: Sejrsen, K., Hvelplund, T. and Nielsen, M. O. (Eds), *Ruminant Physiology: Digestion, Metabolism and Impact of Nutrition on Gene Expression, Immunology and Stress*, 55–85. Wageningen, The Netherlands: Wageningen Academic Press.

Ungerfeld, E. M. (2015). Shifts in metabolic hydrogen sinks in the methanogenesis-inhibited ruminal fermentation: a meta-analysis, *Frontiers in Microbiology* 6, 37. DOI: 10.3389/fmicb.2015.00037.

United Nations, U. (2019). World population prospects 2019. New York: United Nations Department for Economic and Social Affairs. Available at: https://population.un.org/wpp/.

Ushida, K., Newbold, C. J. and Jouany, J. P. (1997). Interspecies hydrogen transfer between the rumen ciliate Polyplastron multivesiculatum and Methanosarcina barkeri, *The Journal of General and Applied Microbiology* 43(2), 129–131. DOI: 10.2323/jgam.43.129.

Van Kessel, J. A. S. and Russell, J. B. (1996). The effect of pH on ruminal methanogenesis, *FEMS Microbiology Ecology* 20(4), 205–210. DOI: 10.1016/0168-6496(96)00030-X.

Van Nevel, C. and Demeyer, D. I. (1995). Lipolysis and biohydrogenation of soybean oil in the rumen in vitro: inhibition by antimicrobials, *Journal of Dairy Science* 78(12), 2797–2806. DOI: 10.3168/jds.S0022-0302(95)76910-7.

Van Vugt, C. (2005). Impact of monensin on methane production and performance of cows fed forage diets. In: Proceedings-New Zealand Society of Animal Production, New Zealand Society of Animal Production, 1999, vol. 65, 362.

Van Wesemael, D., Vandaele, L., Ampe, B., Cattrysse, H., Duval, S., Kindermann, M., Fievez, V., De Campeneere, S. and Peiren, N. (2019). Reducing enteric methane emissions from dairy cattle: two ways to supplement 3-nitrooxypropanol, *Journal of Dairy Science* 102(2), 1780–1787. DOI: 10.3168/jds.2018-14534.

Van Zijderveld, S. M., Gerrits, W. J., Apajalahti, J. A., Newbold, J. R., Dijkstra, J., Leng, R. A. and Perdok, H. B. (2010). Nitrate and sulfate: effective alternative hydrogen sinks for mitigation of ruminal methane production in sheep, *Journal of Dairy Science* 93(12), 5856–5866. DOI: 10.3168/jds.2010-3281.

Van Zijderveld, S. M., Gerrits, W. J., Dijkstra, J., Newbold, J. R., Hulshof, R. B. and Perdok, H. B. (2011). Persistency of methane mitigation by dietary nitrate supplementation in dairy cows, *Journal of Dairy Science* 94(8), 4028–4038. DOI: 10.3168/jds.2011-4236.

Velazco, J. I., Cottle, D. J. and Hegarty, R. S. (2014). Methane emissions and feeding behaviour of feedlot cattle supplemented with nitrate or urea, *Animal Production Science* 54(10), 1737–1740. DOI: 10.1071/AN14345.

Velazco, J. I., Herd, R. M., Cottle, D. J. and Hegarty, R. S. (2017). Daily methane emissions and emission intensity of grazing beef cattle genetically divergent for residual feed intake, *Animal Production Science* 57(4), 627–635. DOI: 10.1071/AN15111.

Vyas, D., Alemu, A. W., McGinn, S. M., Duval, S. M., Kindermann, M. and Beauchemin, K. A. (2018). The combined effects of supplementing monensin and 3-nitrooxypropanol on methane emissions, growth rate, and feed conversion efficiency in beef cattle fed high-forage and high-grain diets, *Journal of Animal Science* 96(7), 2923–2938. DOI: 10.1093/jas/sky174.

Waghorn, G. C. (2007). Monensin controlled release capsules for improved production and mitigating methane in dairy cows fed pasture. In: Proceedings-New Zealand Society of Animal Production, New Zealand Society of Animal Production, 1999, vol. 67, 266.

Wallace, R. J., Rooke, J. A., McKain, N., Duthie, C. A., Hyslop, J. J., Ross, D. W., Waterhouse, A., Watson, M. and Roehe, R. (2015). The rumen microbial metagenome associated with high methane production in cattle, *BMC Genomics* 16(1), 839. DOI: 10.1186/s12864-015-2032-0.

Wallace, R. J., Sasson, G., Garnsworthy, P. C., Tapio, I., Gregson, E., Bani, P., Huhtanen, P., Bayat, A. R., Strozzi, F., Biscarini, F., Snelling, T. J., Saunders, N., Potterton, S. L., Craigon, J., Minuti, A., Trevisi, E., Callegari, M. L., Cappelli, F. P., Cabezas-Garcia, E. H., Vilkki, J., Pinares-Patino, C., Fliegerová, K. O., Mrázek, J., Sechovcová, H., Kopečný, J., Bonin, A., Boyer, F., Taberlet, P., Kokou, F., Halperin, E., Williams, J. L., Shingfield, K. J. and Mizrahi, I. (2019). A heritable subset of the core rumen microbiome dictates dairy cow productivity and emissions, *Science Advances* 5(7), eaav8391. DOI: 10.1126/sciadv.aav8391.

Wang, Y., Xu, Z., Bach, S. J. and McAllister, T. A. (2008). Effects of phlorotannins from Ascophyllum nodosum (brown seaweed) on in vitro ruminal digestion of mixed forage or barley grain, *Animal Feed Science and Technology* 145(1–4), 375–395. DOI: 10.1016/j.anifeedsci.2007.03.013.

Warner, D., Bannink, A., Hatew, B., Van Laar, H. and Dijkstra, J. (2017). Effects of grass silage quality and level of feed intake on enteric methane production in lactating dairy cows, *Journal of Animal Science* 95(8), 3687–3700. DOI: 10.2527/jas.2017.1459.

Weber, K. L., Welly, B. T., Van Eenennaam, A. L., Young, A. E., Porto-Neto, L. R., Reverter, A. and Rincon, G. (2016). Identification of gene networks for residual feed intake in Angus cattle using genomic prediction and RNA-seq, *PLoS ONE* 11(3), e0152274. DOI: 10.1371/journal.pone.0152274.

Wedlock, D. N., Janssen, P. H., Leahy, S. C., Shu, D. and Buddle, B. M. (2013). Progress in the development of vaccines against rumen methanogens, *Animal* 7(s2), 244–252. DOI: 10.1017/S1751731113000682.

Wedlock, D. N., Pedersen, G., Denis, M., Dey, D., Janssen, P. H. and Buddle, B. M. (2010). Development of a vaccine to mitigate greenhouse gas emissions in agriculture: vaccination of sheep with methanogen fractions induces antibodies that block methane production in vitro, *New Zealand Veterinary Journal* 58(1), 29–36. DOI: 10.1080/00480169.2010.65058.

Whitelaw, F. G., Eadie, J. M., Bruce, L. A. and Shand, W. J. (1984). Methane formation in faunated and ciliate-free cattle and its relationship with rumen volatile fatty acid proportions, *British Journal of Nutrition* 52(2), 261–275. DOI: 10.1079/BJN19840094.

Williams, A. G. and Coleman, G. S. (1992). Role of protozoa in the rumen. In:, *Brock/Springer Series in Contemporary Bioscience*, Springer, 317–347. DOI: 10.1007/978-1-4612-2776-2_10.

Williams, S. R. O., Hannah, M. C., Jacobs, J. L., Wales, W. J. and Moate, P. J. (2019). Volatile fatty acids in ruminal fluid can be used to predict methane yield of dairy cows, *Animals* 9(12), 1006. DOI: 10.3390/ani9121006.

Williams, Y. J., Popovski, S., Rea, S. M., Skillman, L. C., Toovey, A. F., Northwood, K. S. and Wright, A. D. (2009). A vaccine against rumen methanogens can alter the composition of archaeal populations, *Applied and Environmental Microbiology* 75(7), 1860–1866. DOI: 10.1128/AEM.02453-08.

Williams, Y. J., Rea, S. M., Popovski, S., Pimm, C. L., Williams, A. J., Toovey, A. F., Skillman, L. C. and Wright, A. D. (2008). Reponses of sheep to a vaccination of entodinial or mixed rumen protozoal antigens to reduce rumen protozoal numbers, *British Journal of Nutrition* 99(1), 100–109. DOI: 10.1017/S0007114507801553.

Wood, J. M., Kennedy, F. S. and Wolfe, R. S. (1968). Reaction of multihalogenated hydrocarbons with free and bound reduced vitamin B12, *Biochemistry* 7(5), 1707–1713. DOI: 10.1021/bi00845a013.

Woodward, S., Waghorn, G., Lassey, K. and Laboyrie, P. (2002). Does feeding sulla (Hedysarum coronarium) reduce methane emissions from dairy cows?, *New Zealand Society of Animal Production* 62, 227–230.

Wright, A. D., Kennedy, P., O'Neill, C. J., Toovey, A. F., Popovski, S., Rea, S. M., Pimm, C. L. and Klein, L. (2004). Reducing methane emissions in sheep by immunization against rumen methanogens, *Vaccine* 22(29–30), 3976–3985. DOI: 10.1016/j.vaccine.2004.03.053.

Yáñez-Ruiz, D. R., Abecia, L. and Newbold, C. J. (2015). Manipulating rumen microbiome and fermentation through interventions during early life: a review, *Frontiers in Microbiology* 6, 1133. DOI: 10.3389/fmicb.2015.01133.

Yurtseven, S. and Ozturk, I. (2009). Influence of two sources of cereals (corn or barley), in free choice feeding on diet selection, milk production indices and gaseous products (CH4 and CO2) in lactating sheep, *Asian Journal of Animal and Veterinary Advances* 4(2), 76–85. DOI: 10.3923/ajava.2009.76.85.

Zhang, J., Shi, H., Wang, Y., Li, S., Cao, Z., Ji, S., He, Y. and Zhang, H. (2017). Effect of dietary forage to concentrate ratios on dynamic profile changes and interactions of ruminal microbiota and metabolites in Holstein heifers, *Frontiers in Microbiology* 8, 2206. DOI: 10.3389/fmicb.2017.02206.

Zhang, L., Huang, X., Xue, B., Peng, Q., Wang, Z., Yan, T. and Wang, L. (2015). Immunization against rumen methanogenesis by vaccination with a new recombinant protein, *PLoS ONE* 10(10), e0140086. DOI: 10.1371/journal.pone.0140086.

Zhou, M., Chung, Y. H., Beauchemin, K. A., Holtshausen, L., Oba, M., McAllister, T. A. and Guan, L. L. (2011). Relationship between rumen methanogens and methane production in dairy cows fed diets supplemented with a feed enzyme additive, *Journal of Applied Microbiology* 111(5), 1148–1158. DOI: 10.1111/j.1365-2672.2011.05126.x.

Zhou, M., Hernandez-Sanabria, E. and Guan, L. L. (2009). Assessment of the microbial ecology of ruminal methanogens in cattle with different feed efficiencies, *Applied and Environmental Microbiology* 75(20), 6524–6533. DOI: 10.1128/AEM.02815-08.

Zhou, M., Hünerberg, M., Chen, Y., Reuter, T., McAllister, T. A., Evans, F., Critchley, A. T. and Guan, L. L. (2018). Air-dried brown seaweed, Ascophyllum nodosum, alters the rumen microbiome in a manner that changes rumen fermentation profiles and lowers the prevalence of foodborne pathogens, *mSphere* 3(1), e00017-18. DOI: 10.1128/mSphere.00017-18.

Zhou, X., Meile, L., Kreuzer, M. and Zeitz, J. (2013). The effect of lauric acid on methane production and cell viability of Methanobrevibacter ruminantium. In: *Proceedings of the 5th Greenhouse Gases and Animal Agriculture Conference (GGAA 2013): Advances in Animal Biosciences Animal*, Cambridge University Press, vol. 4(2), 458–458. DOI: 10.1155/2013/106916.

Zinder, S. H. (1993). Physiological ecology of methanogens. In: Ferry, J. G. (Ed.) *Methanogenesis*. Boston, MA: Springer, 128–206. DOI: 10.1007/978-1-4615-2391-8_4.

Index

www.ingramcontent.com/pod-product-compliance
Ingram Content Group UK Ltd.
Pitfield, Milton Keynes, MK11 3LW, UK
UKHW021826100926
8556IPUK00003B/177

9 781786 764393